An Introduction to the Mechanical Properties of Solid Polymers

Second Edition

An Introduction to the
Mechanical Properties of
Solid Polymers

Second Edition

I. M. Ward
IRC in Polymer Science and Technology, School of Physics and Astronomy, University of Leeds, UK

and

J. Sweeney
IRC in Polymer Science and Technology, School of Engineering, Design and Technology, University of Bradford, UK

John Wiley & Sons, Ltd

Other Wiley Editorial Offices

John Wiley & Sons Inc., 111 River Street, Hoboken, NJ 07030, USA

Jossey-Bass, 989 Market Street, San Francisco, CA 94103-1741, USA

Wiley-VCH Verlag GmbH, Boschstr. 12, D-69469 Weinheim, Germany

John Wiley & Sons Australia Ltd, 33 Park Road, Milton, Queensland 4064, Australia

John Wiley & Sons (Asia) Pte Ltd, 2 Clementi Loop #02-01, Jin Xing Distripark, Singapore 129809

John Wiley & Sons Canada Ltd, 22 Worcester Road, Etobicoke, Ontario, Canada M9W 1L1

Wiley also publishes its books in a variety of electronic formats. Some content that appears in print may not
be available in electronic books.

Library of Congress Cataloging-in-Publication Data
Ward, I. M. (Ian Macmillan), 1928-
    An Introduction to the mechanical properties of solid polymers / I. M. Ward and J. Sweeney. – 2nd ed.
       p. cm.
    Includes bibliographical references and index.
    ISBN 0-471-49625-1 (cloth : alk. paper) – ISBN 0-471-49626-X (pbk. : alk. paper)
       1. Polymers–Mechanical properties. I. Sweeney, John, 1952-II.
    Title. TA455,P58W36 2004
    620. 1'9204292–dc22

                                                                        2004003363

British Library Cataloguing in Publication Data

A catalogue record for this book is available from the British Library

ISBN 10: 0471 49625 1 (HB)   ISBN 13: 978 0471 49625 0 (HB)
        10: 0471 49626 X (PB)   13: 978 0471 49626 7 (PB)

Typeset in $10\frac{1}{2}/12\frac{1}{2}$ pt Times by Keytec Typesetting, Bridport
Printed and bound in Great Britain by TJ International Ltd., Padstow, Cornwall
This book is printed on acid-free paper responsibly manufactured from sustainable forestry
in which at least two trees are planted for each one used for paper production.

Contents

An Introduction to the Mechanical Properties of Solid Polymers I. M. Ward and J. Sweeney
© 2004 John Wiley & Sons, Ltd   ISBN: 0471 49625 1 (HB); 0471 49626 X (PB)

Preface

This book is the second edition of *An Introduction to the Mechanical Properties of Solid Polymers*. Its aim is to provide an introduction to the mechanical behaviour of solid polymers at a fairly elementary level for research workers in polymer science and for postgraduate students with first degrees in physics, chemistry, engineering or materials science. It follows the approach of the first edition in developing the mechanics of behaviour first and then discussing molecular and structural interpretations. The individual chapters are self-contained so that they can be read as reviews of progress in different areas.

Since the publication of the first edition, in 1993, the subject has advanced and this has been dealt with in some instances by adding additional sections with the latest developments. In other cases, although the overall original format has been retained, there has been substantial rewriting and many additions to the text.

We are very grateful to Margaret Ward for undertaking the majority of the initial typing of the new text. We also wish to thank Colin Morath and Jagan Mohanraj for their assistance with the preparation of new diagrams.

I. M. Ward
J. Sweeney

An Introduction to the Mechanical Properties of Solid Polymers I. M. Ward and J. Sweeney
© 2004 John Wiley & Sons, Ltd   ISBN: 0471 49625 1 (HB); 0471 49626 X (PB)

1

Structure of Polymers

The mechanical properties that form the subject of this book are a consequence of the chemical composition of the polymer and also of its structure at the molecular and supermolecular levels. We shall therefore introduce a few elementary ideas concerning these aspects.

1.1 Chemical composition

1.1.1 Polymerization

Linear polymers consist of long molecular chains of covalently bonded atoms, each chain being a repetition of much smaller chemical units. One of the simplest polymers is polyethylene, which is an *addition* polymer made by polymerizing the monomer ethylene, $CH_2{=}CH_2$, to form the polymer

$$\left[-CH_2-CH_2-\right]_n$$

Note that the double bond is removed during the polymerization (Figure 1.1). The well-known *vinyl* polymers are made by polymerizing compounds of the form

$$\overset{\displaystyle X}{\underset{\displaystyle |}{CH_2{=}CH}}$$

where X represents a chemical group; examples are as follows:

$$\text{polypropylene} \quad \left[-CH_2-\overset{\displaystyle CH_3}{\underset{\displaystyle |}{CH}}-\right]_n$$

An Introduction to the Mechanical Properties of Solid Polymers I. M. Ward and J. Sweeney
© 2004 John Wiley & Sons, Ltd   ISBN: 0471 49625 1 (HB); 0471 49626 X (PB)

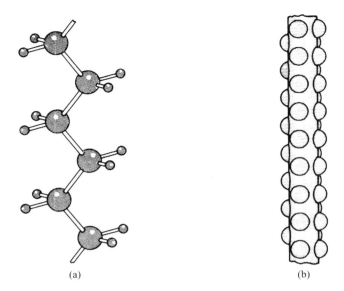

(a)   (b)

Figure 1.1   (a) The polyethylene chain $(CH_2)_n$ in schematic form (larger spheres, carbon; smaller spheres, hydrogen) and (b) sketch of a molecular model of a polyethylene chain

$$\text{polystyrene} \quad \left[-CH_2-\overset{\overset{\displaystyle C_6H_5}{|}}{CH}- \right]_n$$

and

$$\text{polyvinyl chloride} \quad \left[-CH_2-\overset{\overset{\displaystyle Cl}{|}}{CH}- \right]_n$$

Natural rubber, polyisoprene, is a diene, and its repeat unit

$$\left[-CH_2-CH=\overset{\overset{\displaystyle CH_3}{|}}{C}-CH_2- \right]_n$$

contains a double bond that gives it added flexibility.

Condensation (or step-growth) polymers are formed by reacting difunctional molecules, usually with the elimination of water. One example is the formation of polyethylene terephthalate (the polyester used for Terylene and Dacron fibres and transparent films and bottles) from ethylene glycol and terephthalic acid:

$$n(HO\!-\!CH_2\!-\!CH_2\!-\!OH) + n(HOOC\langle\bigcirc\rangle COOH)$$

$$\Rightarrow H\left[-O\!-\!CH_2\!-\!CH_2\!-\!O\!-\!\underset{O}{\overset{\displaystyle O}{\underset{\|}{C}}}\langle\bigcirc\rangle\overset{\displaystyle O}{\overset{\|}{C}}-\right]_n OH + nH_2O$$

Another common condensation polymer is nylon 6:6

$$\left[-NH\!-\!(CH_2)_6\!-\!NH\!-\!\underset{O}{\overset{\displaystyle O}{\underset{\|}{C}}}\!-\!(CH_2)_4\!-\!\overset{\displaystyle O}{\overset{\|}{C}}-\right]_n$$

1.1.2   Cross-linking and chain-branching

Linear polymers can be joined by other chains at points along their length to make a cross-linked structure (Figure 1.2). Chemical cross-linking produces a thermosetting polymer, so called because the cross-linking agent is normally activated by heating, after which the material does not soften and melt when heated further; examples are bakelite and epoxy resins. A small amount of cross-linking through sulphur bonds is needed to give natural rubber its characteristic feature of rapid recovery from a large extension.

Very long molecules in linear polymers can entangle to form temporary physical cross-links, and we shall show later that a number of the characteristic

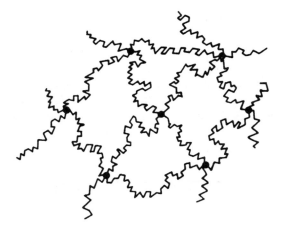

Figure 1.2   Schematic diagram of a cross-linked polymer

properties of solid polymers are explicable in terms of the behaviour of a deformed network.

A less extreme complication is chain branching, where a secondary chain initiates from a point on the main chain, as is illustrated for polyethylene in Figure 1.3. Low-density polyethylene, as distinct from the high-density linear polyethylene shown in Figure 1.1, possesses on average one long branch per molecule and a larger number of small branches, mainly ethyl (—CH₂—CH₃) or butyl (—(CH₂)₃—CH₃) side groups. The presence of these branch points leads to considerable differences in mechanical behaviour compared with linear polyethylene.

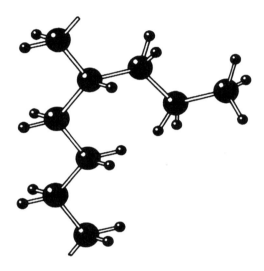

Figure 1.3   A chain branch in polyethylene

1.1.3   Average molecular mass and molecular mass distribution

Each sample of a polymer contains molecular chains of varying lengths, i.e. of varying molecular mass (Figure 1.4). The mass (length) distribution is of importance in determining the properties of the polymer, but until the advent of gel permeation chromatography [1, 2] it could be determined only by tedious fractionation procedures. Most investigations therefore quoted different types of average molecular mass, the commonest being the number average $\overline{M}_n$ and the weight average $\overline{M}_w$, defined as

$$\overline{M}_n = \frac{\sum N_i M_i}{\sum N_i} \qquad \overline{M}_w = \frac{\sum (N_i M_i) M_i}{\sum N_i M_i}$$

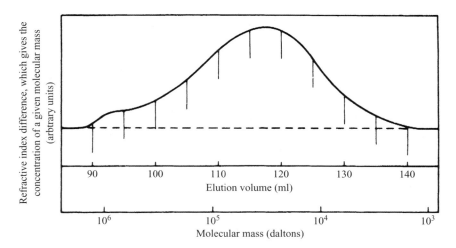

Figure 1.4 The gel permeation chromatograph trace gives a direct indication of the molecular distribution. (Results obtained in Marlex 6009 by Dr T. Williams)

where N_i is the number of molecules of molecular mass M_i, and $\sum$ denotes summation over all i molecular masses.

The weight average molecular mass is always higher than the number average, as the former is strongly influenced by the relatively small number of very long (massive) molecules. The ratio of the two averages gives a general idea of the width of the molecular mass distribution.

Fundamental measurements of average molecular mass must be performed on solutions so dilute that intermolecular interactions can be ignored or compensated for. The commonest techniques are osmotic pressure for the number average and light scattering for the weight average. Both methods are rather lengthy, so in practice an average molecular mass was often deduced from viscosity measurements of either a dilute solution of the polymer (which relates to $\overline{M}_n$) or a polymer melt (which relates to $\overline{M}_w$). Each method yielded a different average value, which made it difficult to correlate specimens characterized by different groups of workers.

The molecular mass distribution is important in determining flow properties, and so may affect the mechanical properties of a solid polymer indirectly by influencing the final physical state. Direct correlations of molecular mass to viscoelastic behaviour and brittle strength have also been obtained.

1.1.4 Chemical and steric isomerism and stereoregularity

A further complication of the chemical structure of polymers lies in the possibility of different chemical isomeric forms within a repeat unit or between a series of repeat units. Natural rubber and gutta percha are chemically both polyisoprene, but the former is the *cis* form and the latter is the *trans* form (see Figure 1.5). The characteristic properties of rubber are a consequence of the loose packing of molecules (i.e. large *free volume*) that arises from its structure.

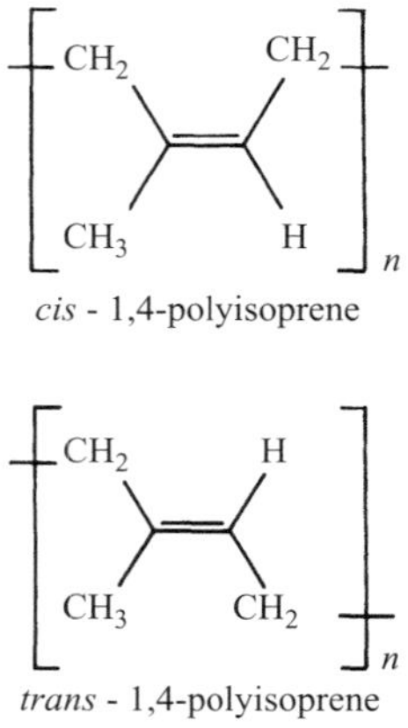

cis - 1,4-polyisoprene

trans - 1,4-polyisoprene

Figure 1.5   *cis*- 1,4-Polyisoprene and *trans*- 1,4-polyisoprene

Vinyl monomer units

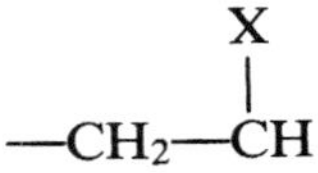

can be added to a growing chain either head-to-tail:

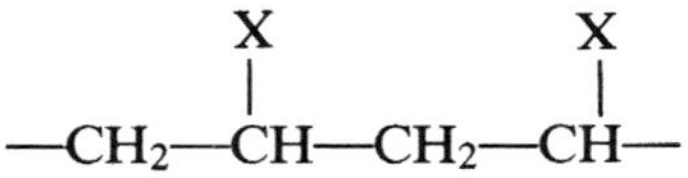

or head-to-head:

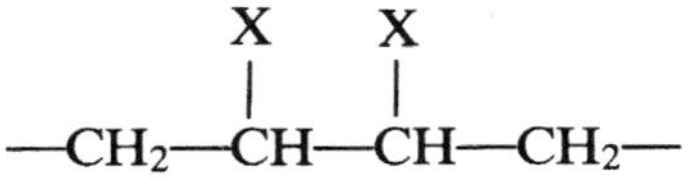

Head-to-tail substitution is usual, and only a small proportion of head-to-head linkages can produce a reduction in the tensile strength because of the loss of regularity.

Stereoregularity provides a more complex situation, which we will examine in terms of the simplest type of vinyl polymer (Figure 1.6) and for which we shall

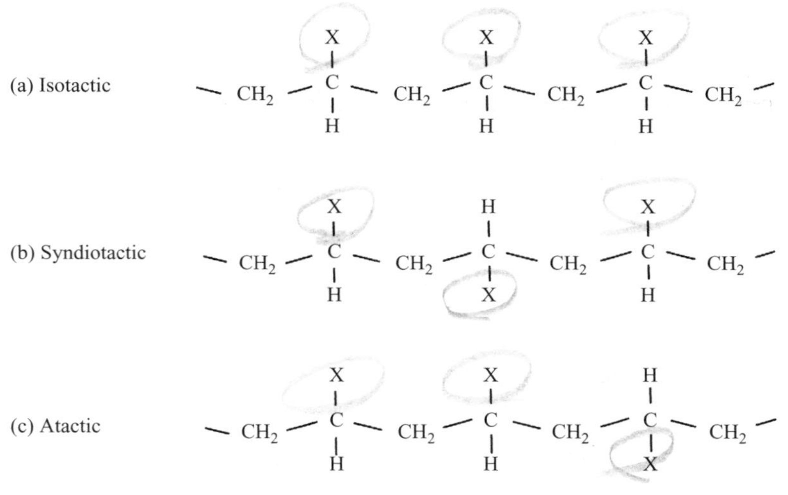

(a) Isotactic

$$-CH_2 - \overset{\overset{\displaystyle X}{|}}{\underset{\underset{\displaystyle H}{|}}{C}} - CH_2 - \overset{\overset{\displaystyle X}{|}}{\underset{\underset{\displaystyle H}{|}}{C}} - CH_2 - \overset{\overset{\displaystyle X}{|}}{\underset{\underset{\displaystyle H}{|}}{C}} - CH_2 -$$

(b) Syndiotactic

$$-CH_2 - \overset{\overset{\displaystyle X}{|}}{\underset{\underset{\displaystyle H}{|}}{C}} - CH_2 - \overset{\overset{\displaystyle H}{|}}{\underset{\underset{\displaystyle X}{|}}{C}} - CH_2 - \overset{\overset{\displaystyle X}{|}}{\underset{\underset{\displaystyle H}{|}}{C}} - CH_2 -$$

(c) Atactic

$$-CH_2 - \overset{\overset{\displaystyle X}{|}}{\underset{\underset{\displaystyle H}{|}}{C}} - CH_2 - \overset{\overset{\displaystyle X}{|}}{\underset{\underset{\displaystyle H}{|}}{C}} - CH_2 - \overset{\overset{\displaystyle H}{|}}{\underset{\underset{\displaystyle X}{|}}{C}} - CH_2 -$$

Figure 1.6 A substituted α-olefin can take three stereosubstituted forms

suppose that the polymer chain is a planar zigzag. Two very simple regular polymers can be constructed. In the first (Figure 1.6(a)) the substituent groups are all added in an identical manner to give an *isotactic* polymer. In the second regular polymer (Figure 1.6(b)) there is an inversion of the manner of substitution between consecutive units, giving a *syndiotactic* polymer for which the substituent groups alternate regularly on opposite sides of the chain. The regular sequence of units is called *stereoregularity*, and stereoregular polymers are crystalline and can possess high melting points. The working range of a polymer is thereby extended compared with the amorphous form, whose range is limited by the lower softening point. The final alternative structure is formed when the orientation of successive substituents takes place randomly (Figure 1.5(c)) to give an irregular *atactic* polymer that is incapable of crystallizing. Polypropylene ($-CH_2CHCH_3-$)$_n$ was for many years obtainable only as an atactic polymer, and its widespread use began only when stereospecific catalysts were developed to produce the isotactic form. Even so, some faulty substitution occurs and atactic chains can be separated from the rest of the polymer by solvent extraction.

1.1.5 Liquid crystalline polymers

Liquid crystals (or plastic crystals as they are sometimes called) are low molecular mass materials that show molecular alignment in one direction but not three-dimensional crystalline order. During the last 20 years, liquid crystalline polymers have been developed where the polymer chains are so straight and rigid that small regions of almost uniform orientation (domains) separated by distinct boundaries are produced. In the case where these domains occur in solution, polymers are termed *lyotropic*. Where the domains occur in the melt, the polymers are termed *thermotropic*.

An important class of lyotropic liquid crystal polymers are the aramid polymers such as polyparabenzamide

and polyparaphenylene terephthalamide

better known as Kevlar, which is a commercially produced high stiffness and high strength fibre. It is important to emphasize that although Kevlar fibres are prepared by spinning a lyotropic liquid crystalline phase, the final fibre shows clear evidence of three-dimensional order.

Important examples of thermotropic liquid crystalline polymers are copolyesters produced by condensation of hydroxybenzoic acid (HBA)

and 2-6 hydroxynaphthoic acid (HNA)

most usually in the proportions HBA:HNA = 73:27.

In addition to these main-chain liquid crystalline polymers, there are also side-chain liquid crystalline polymers, where the liquid crystalline nature arises from the presence of rigid straight side-chain units (called the mesogens) chemically linked to an existing polymer backbone either directly or via flexible spacer units.

The review by Noël and Navard [3] gives further information on liquid crystalline polymers, including methods of preparation.

1.1.6 Blends, grafts and copolymers

A *blend* is a physical mixture of two or more polymers. A *graft* is formed when long side chains of a second polymer are chemically attached to the base polymer. A *copolymer* is formed when chemical combination exists in the main chain between two or more polymers, $[A]_n$, $[B]_n$, etc. The two principal forms are block copolymers ($[AAAA...]$ $[BBB...]$) and *random* copolymers, the latter having no long sequences of A or B units.

All these processes are commonly used to enhance the ductility and toughness of brittle homopolymers or increase the stiffness of rubbery polymers. An example of a blend is acrylonitrile–butadiene–styrene copolymer (ABS), where the separate rubber phase gives much improved impact resistance.

The basic properties of polymers may be enhanced by physical as well as chemical means. An important example is the use of finely divided carbon black as a filler in rubber compounds. Polymers may be combined with stiffer filaments, such as glass or carbon fibres, to form a composite. We shall show later that some semicrystalline polymers may be treated as composites at a molecular level.

It must not be forgotten that all useful polymers contain small quantities of additives to aid processing and increase the resistance to degradation. The physical properties of the base polymer may be modified by the presence of such additives.

1.2 Physical structure

The physical properties of a polymer of a given chemical composition are dependent on two distinct aspects of the arrangement of the molecular chains in space.

1. The arrangement of a single chain without regard to its neighbour: rotational isomerism.

2. The arrangment of chains with respect to each other: orientation and crystallinity.

1.2.1 Rotational isomerism

Rotational isomerism arises because of the alternative conformations of a molecule that can result from the possibility of hindered rotation about the many single bonds in the structure. Spectroscopic techniques [4] developed in small molecules have been extended to polymers, and as an example we illustrate (Figure 1.7) the alternative *trans* and *gauche* conformations in the glycol residue of polyethylene

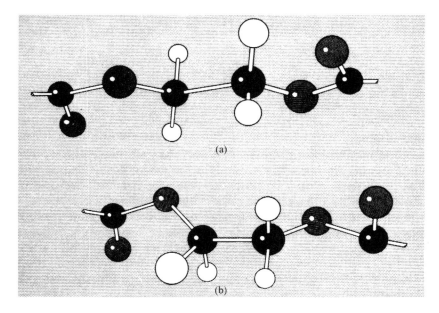

Figure 1.7   Polyethylene terephthalate in the crystalline *trans* conformation (a) and in the *gauche* conformation present in 'amorphous' regions (b). (After Grime and Ward [5])

terephthalate [5]: the former is a crystalline conformation, but the latter is present in amorphous regions.

To pass from one rotational isomeric form to another requires that an energy barrier be surmounted (Figure 1.8), so that the possibility of the chain molecules

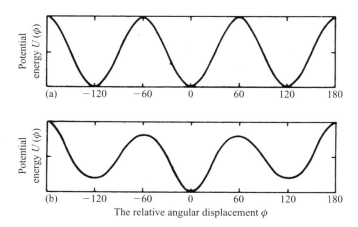

Figure 1.8   Potential energy for rotation (a) around the C—C bond in ethane and (b) around the central C—C bond in *n*-butane. (Reproduced with permission from McCrum, Read and Williams, *Anelastic and Dielectric Effects in Polymeric Solids*, Wiley, London, 1967)

changing their conformations depends on the relative magnitude of the energy barrier compared with thermal energies and the perturbing effects of applied stress. Hence arises the possibility of linking molecular flexibility to deformation mechanisms, a theme to which we will return on several occasions.

1.2.2 Orientation and crystallinity

When we consider the arrangement of molecular chains with respect to each other there are again two largely separate aspects, those of molecular orientation and crystallinity. In semicrystalline polymers this distinction may at times be an artificial one.

When cooled from the melt many polymers form a disordered structure called the amorphous state. Some of these materials, such as polymethyl methacrylate, polystyrene and rapidly cooled (melt-quenched) polyethylene terephthalate, have a comparatively high modulus at room temperature, but others, such as natural rubber and atactic polypropylene, have a low modulus. These two types of polymer are often termed *glassy* and *rubber-like*, respectively, and we shall see that the form of behaviour exhibited depends on the temperature relative to a glass–rubber transition temperature (T_g) that is dependent on the material and the test method employed. Although an amorphous polymer may be modelled as a random tangle of molecules (Figure 1.9(a)), features such as the comparatively high density [6] show that the packing cannot be completely random. X-ray diffraction techniques indicate no distinct structure, rather a broad diffuse maximum (the amorphous halo) that indicates a preferred distance of separation between the molecular chains.

When an amorphous polymer is stretched the molecules may be preferentially aligned along the stretch direction. In polymethyl methacrylate and polystyrene such molecular orientation may be detected by optical methods, which measure the small difference between the refractive index in the stretch direction and that in the perpendicular direction. X-ray diffraction methods still reveal no evidence of three-dimensional order, so the structure may be regarded as a somewhat oriented tangled skein (Figure 1.9(b)) that is oriented amorphous but not crystalline.

In polyethylene terephthalate, however, stretching produces both molecular orientation and small regions of three-dimensional order, termed crystallites, because the orientation processes have brought the molecules into adequate juxtaposition for regions of three-dimensional order to form.

Many polymers, including polyethylene terephthalate, also crystallize if they are cooled slowly from the melt. In this case we may say that they are crystalline but unoriented. Although such specimens are unoriented in the macroscopic sense, i.e. they possess isotropic bulk mechanical properties, they are not homogeneous in the microscopic sense and often show a spherulitic structure under a polarizing microscope.

(a)

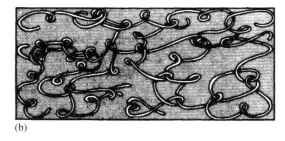

(b)

Figure 1.9 Schematic diagrams of (a) unoriented amorphous polymer and (b) oriented amorphous polymer

In summary, it may be said that for a polymer to crystallize the molecule must have a regular structure, the temperature must be below the crystal melting point and sufficient time must be available for the long molecules to become ordered in the solid state.

The structure of the crystalline regions of polymers can be deduced from wide-angle X-ray diffraction patterns of highly stretched specimens. When the stretching is uniaxial the patterns are related to those obtained from fully oriented single crystals. The crystal structure of polyethylene was determined by Bunn [7] as long ago as 1939 (Figure 1.10).

In addition to the discrete reflections from the crystallites, the diffraction pattern of a polymer shows diffuse scattering attributed to amorphous regions. Such polymers are said to be *semicrystalline*, with the crystalline fraction being controlled by molecular regularity. By comparing the relative amounts of crystalline and amorphous scattering of X-rays the crystallinity has been found to vary from more than 90 per cent for linear polyethylene to about 30 per cent for oriented polyethylene terephthalate.

Single crystals of many polymers may be grown from dilute solution [8]. Linear polyethylene, for example, forms single-crystal lamellae with lateral dimensions of the order of 10–20 μm and a thickness of about 10 nm. Electron diffraction shows that the molecular chains are oriented approximately normal to the lamellar

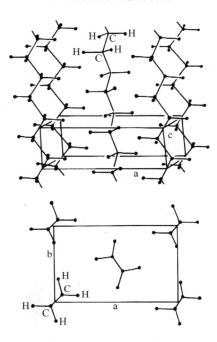

Figure 1.10 Arrangement of molecules in polyethylene crystallites. (From *Fibres from Synthetic Polymers* (ed. R. Hill), Elsevier, Amsterdam, 1953)

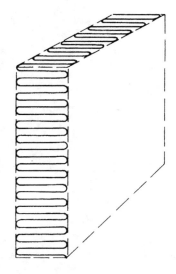

Figure 1.11 Diagrammatic representation of chain folding in polymer crystals with the folds drawn sharp and regular (Reproduced with permission from Keller, *Rep. Progr. Phys.*, **31**, 623 (1969))

surface, and as the molecules are typically about 1 μm in length they must be folded back and forth within the crystals. A diagrammatic representation in which the folds are sharp and regular, with adjacent re-entry, is given in Figure 1.11, but neutron scattering experiments suggest that such a picture may be oversimplified [9].

The crystallization of polymers from the melt is a more controversial process, as a single molecule is unlikely to be laid down on a crystalline substrate without interference from its neighbours, and it might be expected that the highly entangled topology of the chains that exists in the melt must be substantially retained in the semicrystalline state. There is, however, much evidence to support the existence of a lamellar morphology, with the separation between chain folds being greatest for material formed during the first stage of crystallization. Folded-chain lamellae grow outwards from initial nucleation centres, with the spaces between lamellae filled by material that crystallized later. Typically, spherulites

Figure 1.12 A photograph of typical spherulitic structure under a polarizing microscope

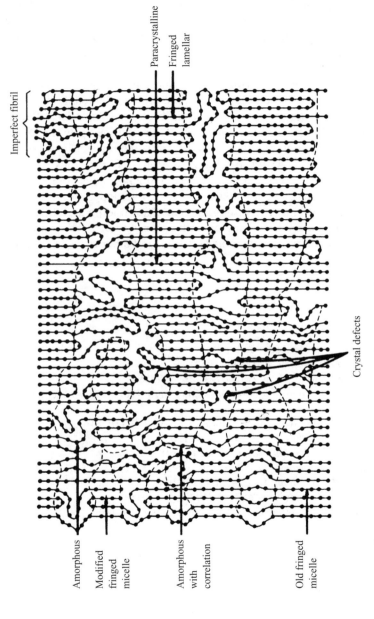

Figure 1.13 Schematic composite diagram of different types of order and disorder in oriented polymers. (Reproduced with permission from Hosemann, *Polymer*, **3**, 349 (1962))

1–10 μm in diameter are formed, which grow outwards until they impinge upon neighbouring spherulites (Figure 1.12). Although chain folding is predominant in the crystallization process there are still many chains that thread their way through the structure and provide continuity. Early models of spherulites are now considered to be oversimplified. For a good review, itself overtaken in some aspects, see the text by Bassett [10] and also more recent work directed by the same author.

Orientation through plastic deformation (*drawing*) destroys the spherulitic structure. What remains is determined to a large extent by the degree of crystallinity. Mechanical testing, described in the subsequent chapters, has helped to establish several models. At one extreme, some highly oriented, highly crystalline specimens of linear polyethylene behave as blocks or lamellae of crystalline material, connected together by tie molecules or crystalline bridges and separated by the amorphous component. Such materials in some respects can be treated as microscopic composites. At the other extreme one has materials such as polyethylene terephthalate in which the crystalline and amorphous components are so intermixed that a single-phase model appears to be more appropriate.

The current state of knowledge suggests that chain folding and the threading of molecules through the crystalline region both occur in typical polymers.

A schematic attempt to illustrate this situation, and other types of irregularity, is given in Figure 1.13.

References

1. Vaughan, M. F., *Nature*, **188**, 55 (1960).
2. Moore, J. C., *J. Polym. Sci. A*, **2**, 835 (1964).
3. Noël, C. and Navard, P., *Prog. Polym. Sci.*, **16**, 55 (1991).
4. Mizushima, S. I., *Structure of Molecules and Internal Rotation*, Academic Press, New York, 1954.
5. Grime, D. and Ward, I. M., *Trans. Faraday Soc.*, **54**, 959 (1958).
6. Robertson, R. E., *J. Phys. Chem.*, **69**, 1575 (1965).
7. Bunn, C. W., *Trans. Faraday Soc.*, **35**, 482 (1939).
8. Fischer, E. W., *Naturforschung*, **12a**, 753 (1957); Keller, A., *Philos. Mag.*, **2**, 1171 (1957); Till, P. H., *J. Polym. Sci.*, **24**, 301 (1957).
9. Keller, A., *Disc. Faraday Soc.*, **68**, 145 (1979).
10. Bassett, D. C., *Principles of Polymer Morphology*, Cambridge University Press, Cambridge, 1981.

Further reading

Billmeyer, F. W., *Textbook of Polymer Science*, Wiley, New York, 1963.
Bower, D. I., *An Introduction to Polymer Physics*, Cambridge University Press, Cambridge, 2002.

Cowie, J. M. G., *Polymers: Chemistry and Physics of Modern Materials* (2nd edn), Blackie Academic & Professional, London, 1997.

Gedde, U. W., *Polymer Physics*, Chapman and Hall, London, 1995.

Hamley, I. W., *The Physics of Block Copolymers*, Oxford University Press, Oxford, 1998.

Sperling, L., *Introduction to Physical Polymer Science* (3rd edn), Wiley, New York, 2001.

Tadokoro, H., *Structure of Crystalline Polymers*, Wiley, New York, 1979.

Ward, I. M., *Mechanical Properties of Solid Polymers* (2nd edn), Wiley, Chichester, 1983.

Ward, I. M., *Structure and Properties of Oriented Polymers* (2nd edn), Chapman and Hall, London, 1997.

Wunderlich, B., *Macromolecular Physics*, Vols 1 and 2, Academic Press, New York, 1973, 1976.

2

The Deformation of an Elastic Solid

In several of the following chapters we consider the behaviour of solid polymers subject to large deformations and show also that in general these materials are viscoelastic, which means that stress (or strain) varies with time. As a starting point, however, we need to consider a polymer as a linear elastic solid: when a load is applied the deformation is instantaneous, after which it remains constant until the load is removed, when the recovery is instantaneous and complete; linearity means that stress and strain are always proportional to one another.

2.1 The state of stress

The convention used in Figure 2.1 shows stresses designated as positive in the direction of the outward-facing normal. As a consequence of this definition inward-acting stresses, such as hydrostatic pressure above that of the surrounding atmosphere, are defined as negative quantities.

It is, however, customary when considering yield behaviour to envisage that hydrostatic pressure causes an increase in the yield stress. For this reason the hydrostatic pressure p in Chapters 10 and 11 is defined as $p = -(\sigma_{xx} + \sigma_{yy} + \sigma_{zz})$.

The components of stress in a body are defined by considering the forces acting on an infinitesimal cubical volume element (Figure 2.1) whose edges are parallel with coordinate axes x, y and z. In equilibrium the forces per unit area acting on the cube faces are P_1 on the yz plane and P_2 on the zx plane and P_3 on the xy plane. Equilibrium implies that similar forces must act on the directly opposite hidden faces of the cube in Figure 2.1.

The forces are then resolved into their nine components in the x, y and z directions as follows:

$$P_1: \sigma_{xx}, \sigma_{xy}, \sigma_{xz}$$

$$P_2: \sigma_{yx}, \sigma_{yy}, \sigma_{yz}$$

An Introduction to the Mechanical Properties of Solid Polymers I. M. Ward and J. Sweeney
© 2004 John Wiley & Sons, Ltd   ISBN: 0471 49625 1 (HB); 0471 49626 X (PB)

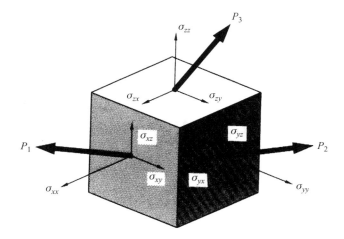

Figure 2.1   The components of stress

$$P_3: \sigma_{zx}, \sigma_{zy}, \sigma_{zz}$$

where the first subscript refers to the direction of the *normal* to the plane on which the stress acts, and the second subscript to the *direction* of the stress. As the cube is in equilibrium the net torque acting on it is zero, which implies the equalities:

$$\sigma_{xy} = \sigma_{yx}, \ \sigma_{xz} = \sigma_{zx}, \ \sigma_{yz} = \sigma_{zy}$$

The components of stress are therefore defined by six independent quantities: the normal stresses σ_{xx}, σ_{yy} and σ_{zz}, together with the shear stresses σ_{xy}, σ_{yz} and σ_{zx}. It is usual to write these components as the elements of a matrix, which is called the stress tensor σ_{ij} (for an explanation of tensors see Appendix 1)

$$\sigma_{ij} = \begin{bmatrix} \sigma_{xx} & \sigma_{xy} & \sigma_{xz} \\ \sigma_{xy} & \sigma_{yy} & \sigma_{yz} \\ \sigma_{xz} & \sigma_{yz} & \sigma_{zz} \end{bmatrix}$$

The state of stress at a point in a body is determined when we can specify the normal components and the shear components of estress acting on a plane drawn in any direction through the point. If we know these six components at a given point the stresses acting on any plane through the point can be calculated (see [1], section 74 and [2], section 47).

2.2   The state of strain

In elementary elasticity, which is concerned with the elastic behaviour of isotropic materials, it is usual to consider two types of strain only. First, there is extensional strain, which is defined as the fractional increase in length in the stretching direction (Figure 2.2(a)), and this definition of strain is an essential ingredient of the very familiar Hooke's law of elasticity. Secondly, there is a simple shear strain that is defined by the displacement of parallel planes, as shown in Figure 2.2(b). The lateral displacement divided by the perpendicular distance between the planes defines the 'engineering' shear strain, which is the angle θ in the figure. (For further discussion of shear strain, see Appendix 1.)

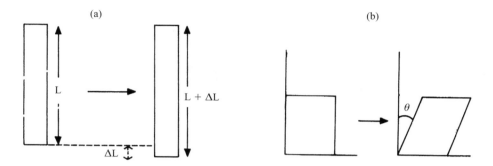

Figure 2.2   Illustration of (a) extensional strain and (b) simple shear strain

For deformation of a material, which involves extensions and shears superimposed in quite general directions, we require a more general starting point to define extensional and shear strains, i.e. components of extensional and shear strain, analogous to the components of tensile and shear stress.

2.2.1   The engineering components of strain

The displacement of any point P_1 (see Figure 2.3) in the body may be resolved into its components u, v and w parallel to x, y and z (Cartesian coordinate axes chosen in the undeformed state), so that if the coordinates of the point in the undisplaced

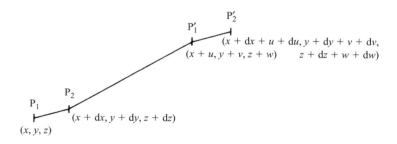

Figure 2.3   The displacement produced by deformation

position were (x, y, z) they become $(x + u, y + v, z + w)$ on deformation. In defining the strains we are not interested in the absolute displacement or rotation but in the deformation. The latter is the displacement of a point relative to adjacent points. Consider a point P_2, very close to P_1, that in the undisplaced position had coordinates $(x + dx, y + dy, z + dz)$ and let the displacement that it has undergone have components $(u + du, v + dv, w + dw)$. The quantities required are then du, dv and dw, the relative displacements.

If dx, dy and dz are sufficiently small, i.e. infinitesimal, then

$$du = \frac{\partial u}{\partial x} \, dx + \frac{\partial u}{\partial y} \, dy + \frac{\partial u}{\partial z} \, dz$$

$$dv = \frac{\partial v}{\partial x} \, dx + \frac{\partial v}{\partial y} \, dy + \frac{\partial v}{\partial z} \, dz$$

$$dw = \frac{\partial w}{\partial x} \, dx + \frac{\partial w}{\partial y} \, dy + \frac{\partial w}{\partial z} \, dz$$

Thus we need to define the nine quantities

$$\frac{\partial u}{\partial x}, \quad \frac{\partial u}{\partial y}, \ldots, \text{etc.}$$

For convenience these nine quantities are regrouped and denoted as follows:

$$e_{xx} = \frac{\partial u}{\partial x} \qquad e_{yy} = \frac{\partial v}{\partial y} \qquad e_{zz} = \frac{\partial w}{\partial z}$$

$$e_{yz} = \frac{\partial w}{\partial y} + \frac{\partial v}{\partial z} \qquad e_{zx} = \frac{\partial u}{\partial z} + \frac{\partial w}{\partial x} \qquad e_{xy} = \frac{\partial v}{\partial x} + \frac{\partial u}{\partial y}$$

$$2\bar{\omega}_x = \frac{\partial w}{\partial y} - \frac{\partial v}{\partial z} \qquad 2\bar{\omega}_y = \frac{\partial u}{\partial z} - \frac{\partial w}{\partial x} \qquad 2\bar{\omega}_z = \frac{\partial v}{\partial x} - \frac{\partial u}{\partial y}$$

The first three quantities e_{xx}, e_{yy} and e_{zz} correspond to the fractional expansions or contractions along the x, y and z axes of an infinitesimal element at P_1. The second three quantities e_{yz}, e_{zx} and e_{xy} correspond to the components of shear strain in the yz, zx and xy planes, respectively. The last three quantities $\bar{\omega}_x$, $\bar{\omega}_y$ and $\bar{\omega}_z$ do not correspond to a deformation of the element at P_1, but are the components of its rotation as a rigid body.

The concept of shear strain can be conveniently illustrated by a diagram showing the two-dimensional situation of shear in the yz plane (Figure 2.4). All strains are considered small, but must be depicted as large for clarity. ABCD is an infinitesimal square that has been displaced and deformed into the

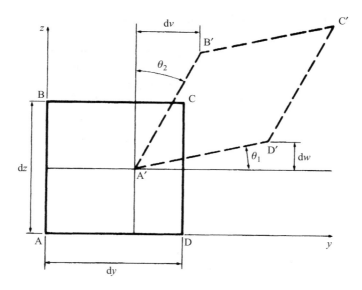

Figure 2.4 Shear strains. (Reproduced with permission from Kolsky, *Stress Waves in Solids*, Dover, New York, 1963)

rhombus A′B′C′D′, θ_1 and θ_2 being the (small) angles that A′D′ and A′B′ make with the y and z axes, respectively. Now

$$\tan\theta_1 = \theta_1 = \frac{dw}{dy} \to \frac{\partial w}{\partial y}$$

$$\tan\theta_2 = \theta_2 = \frac{dv}{dz} \to \frac{\partial v}{\partial z}$$

The shear strain in the yz plane is given by

$$e_{yz} = \frac{\partial w}{\partial y} + \frac{\partial v}{\partial z} = \theta_1 + \theta_2$$

Here $2\bar{\omega}_x = \theta_1 - \theta_2$ does not correspond to a deformation of ABCD but to twice the angle through which AC has been rotated. The deformation is therefore defined by the first six quantities e_{xx}, e_{yy}, e_{zz}, e_{yz}, e_{zx}, e_{xy}, which are called the components of strain. It is important to note that *engineering strains* have been defined (see below).

2.3 The generalized Hooke's law

The most general *linear* relationship between stress and strain is obtained by assuming that each of the six independent components of stress is linearly related to each of the six independent components of strain. Thus

$$\sigma_{xx} = ae_{xx} + be_{yy} + ce_{zz} + de_{xz} + \dots \text{ etc.}$$

and

$$e_{xx} = a'\sigma_{xx} + b'\sigma_{yy} + c'\sigma_{zz} + d'\sigma_{xz} + \dots \text{ etc.}$$

where $a, b \dots, a', b'$, are constants. This expression is the generalized Hooke's law for both isotropic and anisotropic solids. Most of this textbook is concerned with isotropic solids where there is no coupling between tensile stresses and shear stresses, or between shear stresses and extensional strains, so that these equations reduce to forms such as

$$\sigma_{xx} = ae_{xx} + be_{yy} + ce_{zz} \qquad \sigma_{xz} = fe_{xz}$$

and

$$e_{xx} = a'\sigma_{xx} + b'\sigma_{yy} + c'\sigma_{zz}$$

$$e_{xz} = f'\sigma_{xz}$$

To develop the generalized Hooke's law for isotropic materials it is convenient to construct equations for the strains e_{xx}, e_{yy}, etc. in terms of the applied stresses σ_{xx}, σ_{yy}, etc and so define Young's modulus E and Poisson's ratio ν. An applied stress σ_{xx} will produce a strain

$$e_{xx} = \frac{\sigma_{xx}}{E}$$

in the x direction and strains

$$e_{yy} = \frac{-\nu}{E}\sigma_{xx} \quad \text{and} \quad e_{zz} = \frac{-\nu}{E}\sigma_{xx}$$

in the y and z directions, respectively. (Note that Poisson's ratio ν, which defines the ratio of the contraction strain e_{yy} to the extensional strain e_{xx}, is conventionally positive, whereas the contraction e_{yy} is negative.)

A shear strain e_{xz} is related to the corresponding shear stress σ_{xz} by the relationship $e_{xz} = \sigma_{xz}/G$, where G is the shear modulus.

Thus we obtain the stress–strain relationships that are the starting point in many elementary textbooks of elasticity ([1], pp. 7–9):

$$e_{xx} = \frac{1}{E}\sigma_{xx} - \frac{\nu}{E}(\sigma_{yy} + \sigma_{zz})$$

$$e_{yy} = \frac{1}{E}\sigma_{yy} - \frac{\nu}{E}(\sigma_{xx} + \sigma_{zz})$$

$$e_{zz} = \frac{1}{E}\sigma_{zz} - \frac{\nu}{E}(\sigma_{xx} + \sigma_{yy})$$

$$e_{xz} = \frac{1}{G}\sigma_{xz} \qquad e_{yz} = \frac{1}{G}\sigma_{yz} \qquad e_{xy} = \frac{1}{G}\sigma_{xy}$$

A bulk modulus K, related to the fractional change in volume, can also be defined, but only two of the quantities E, ν, G and K are independent. For example

$$G = \frac{E}{2(1 + \nu)} \quad \text{and} \quad K = \frac{E}{3(1 - 2\nu)}$$

2.4 Finite strain elasticity: the behaviour of polymers in the rubber-like state

In the rubber-like state a polymer may be subjected to large deformation and still show complete recovery. The behaviour of a rubber band stretching to two or three times its original length and, when released, recovering instantaneously to its original shape is a matter of common experience. This is *elastic* behaviour at large strains. The first stage in developing an understanding of this behaviour is to consider how the fact that the strains are large affects the definition of stress and strain.

2.4.1 The definition of components of stress

In small strain elasticity theory, the components of stress in the deformed body are defined by considering the equilibrium of an elemental cube of material (Section 2.1 above). When the strains are small the areas of the cube faces are to a first approximation unaffected by the strain. It is therefore of no consequence whether the components of stress are referred to an elemental cube in the deformed body or to an elemental cube in the undeformed body. For finite strains this is no longer true, but for simplicity we will *choose* to define the components of stress with reference to the equilibrium of a cube in the *deformed* body. The components of stress can then be defined in exactly the same manner as for small strain elasticity.

In practice, the significance of this decision is that we must be careful to distinguish between true stresses σ, which relate to the stress tensor, and the convenience of using nominal stresses f, which are defined as the force per unit area of unstrained cross-section. This distinction will be exemplified further when the theory is developed for simple tension (Section 2.4.3 below).

2.4.2 The generalized definition of strain

In this introduction to finite elasticity it is only necessary to develop the most elementary definition of finite strains (for a more comprehensive discussion, see [3], chapter 3).

Consider a system of rectangular coordinate axes with an origin O (Figure 2.5) in which the point P has the coordinates x, y and z. When the body is deformed, consider that the origin of these coordinates is fixed at a point in the body, and that the point P moves to a point P′ with coordinates x', y' and z' where

$$x' = \lambda_1 x \qquad y' = \lambda_2 y \qquad z' = \lambda_3 z$$

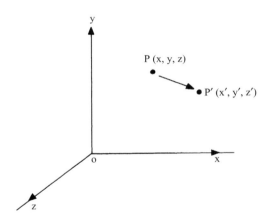

Figure 2.5   Displacement of point P to P′

A useful schematic way of representing the deformation, is to consider the change in the dimensions of a cube of unit dimensions (Figure 2.6). On deformation the cube deforms to a parallelepiped with sides of length λ_1, λ_2 and λ_3 in the x, y and z directions, respectively.

The quantities λ_1, λ_2 and λ_3 that define the deformation are called the deformation ratios, because they define the ratio of the length of lines in the x, y and z directions in the *deformed* body to their length in the *undeformed* body. Note

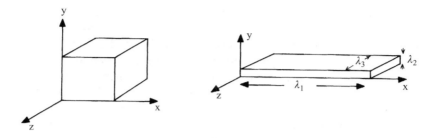

Figure 2.6 Deformation of a cube to a parallelepiped defines deformation ratios λ_1, λ_2 and λ_3

that, with the exception of bodies for which Poisson's ratio is negative, if one of the ratios is greater than unity at least one of the ratios must be less than unity.

Finite strain is most conveniently defined by these three deformation ratios, although we can equally well define the three components of extensional finite strain $\boldsymbol{\varepsilon}_{xx}$, $\boldsymbol{\varepsilon}_{yy}$ and $\boldsymbol{\varepsilon}_{zz}$ (in bold type to distinguish them from the small strain components defined in Section 2.4.1 above) as

$$\boldsymbol{\varepsilon}_{xx} = \tfrac{1}{2}(\lambda_1^2 - 1) \qquad \boldsymbol{\varepsilon}_{yy} = \tfrac{1}{2}(\lambda_2^2 - 1) \qquad \boldsymbol{\varepsilon}_{zz} = \tfrac{1}{2}(\lambda_3^2 - 1)$$

This generalized definition of strain, which is not limited to small strains, is compatible with our definition of strain in Section 2.2 above. For example, for small strain

$$\lambda_2^1 = (1 + e_{xx})^2 \rightarrow 1 + 2e_{xx} \quad \text{and} \quad \boldsymbol{\varepsilon}_{xx} = e_{xx}$$

A deformation in which lines of material along the three coordinate axes x, y and z in the undeformed state remain mutually perpendicular is called normal strain (often normal homogeneous strain, to include the idea that it is also uniform throughout the body) because the shear strain components are zero. We will always develop the theories of rubber elasticity with this simplification because it involves no loss of generality, as can be appreciated from the following.

Consider a sphere drawn in the undeformed solid (Figure 2.7). For a quite

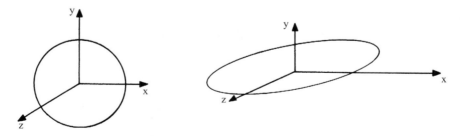

Figure 2.7 The strain ellipsoid

general deformation this sphere is transformed into an ellipsoid (called the strain ellipsoid). It can be seen that the principal axes of this ellipsoid do not coincide with the x, y and z axes that were chosen to refer to the undeformed state. However, without any loss of generality, we could have *chosen* to define the x, y and z axes in the undeformed state to coincide with the principal axes of the strain ellipsoid. We can therefore define the deformation in terms of normal strain (only extensional strain components defined by λ_1, λ_2 and λ_3) provided that we choose our axes carefully by reference to the deformed state. Such axes are referred to as *principal axes* or *principal directions*. In practice this is the most convenient way of proceeding because, as explained in Section 2.4.1 above, stress components also are defined in the deformed state.

2.4.3   The strain energy function

It has been shown that the finite strain deformation can be readily described by the deformation of a cube of unit dimensions in the undeformed state to the rectangular parallelepiped shown in Figure 2.6, which has edges λ_1, λ_2 and λ_3 in the x, y and z directions, respectively. In the deformed state the forces acting on the faces are f_1, f_2 and f_3, with $f =$ force per unit of undeformed cross-section, i.e. the forces are calculated in terms of the applied loads per unit cross-section in the undeformed state.

The corresponding stress components as defined in the deformed state are $\boldsymbol{\sigma}_{xx}$, $\boldsymbol{\sigma}_{yy}$ and $\boldsymbol{\sigma}_{zz}$, where

$$\boldsymbol{\sigma}_{xx} = \frac{f_1}{\lambda_2\lambda_3} = \lambda_1 f_1 \qquad \boldsymbol{\sigma}_{yy} = \frac{f_2}{\lambda_3\lambda_1} = \lambda_2 f_2 \qquad \boldsymbol{\sigma}_{zz} = \frac{f_3}{\lambda_1\lambda_2} = \lambda_3 f_3$$

and the bold type indicates the distinction between these components of stress and those defined for small strain elasticity. The above equations depend on the assumption of incompressibility, $\lambda_1\lambda_2\lambda_3 = 1$.

The work done (per unit of initial undeformed volume) in an infinitesimal displacement from the deformation state where λ_1, λ_2 and λ_3 change to $\lambda_1 + \mathrm{d}\lambda_1$, $\lambda_2 + \mathrm{d}\lambda_2$ and $\lambda_3 + \mathrm{d}\lambda_3$ is

$$\mathrm{d}W = f_1\,\mathrm{d}\lambda_1 + f_2\,\mathrm{d}\lambda_2 + f_3\,\mathrm{d}\lambda_3 \tag{2.1}$$

$$= \frac{\boldsymbol{\sigma}_{xx}}{\lambda_1}\,\mathrm{d}\lambda_1 + \frac{\boldsymbol{\sigma}_{yy}}{\lambda_2}\,\mathrm{d}\lambda_2 + \frac{\boldsymbol{\sigma}_{zz}}{\lambda_3}\,\mathrm{d}\lambda_3 \tag{2.1a}$$

For an elastic material the work done can be equated to a change in the stored elastic energy U. In the case of rubbers, it is usual to consider a reversible isothermal change of state at constant volume, so that the work done can be equated to the change in the Helmholtz free energy A, i.e. $\Delta U = \Delta A$. Here U is often called the strain energy function because it defines the energy stored as a result of the strain, i.e.

$$U = f(\lambda_1, \lambda_2, \lambda_3) \tag{2.2}$$

It would be unreasonable for a physical quantity such as energy to depend on the choice of axes. The use of *principal* extension ratios, with values independent of the axis set, goes some way to ensuring that this is not the case. However, the choice of subscripts 1, 2 and 3 is arbitrary, so the chosen form must be a symmetric function of λ_1, λ_2 and λ_3. For simplicity it should also become zero when $\lambda_1 = \lambda_2 = \lambda_3 = 1$, i.e. for zero strain. A further requirement is that for small strains we should obtain Hooke's law for simple tension and the equivalent equation for simple shear.

An equation that satisfies these requirements is

$$U = C_1(\lambda_1^2 + \lambda_2^2 + \lambda_3^2 - 3) \tag{2.3}$$

To obtain a stress–strain relationship from this equation we invoke Equation (2.1), together with the assumption that rubber is incompressible, i.e. there is no change in volume on deformation, which is true to a good approximation. For example, consider extension under a tensile force f in the x direction. This gives

$$\lambda_1 = \lambda \quad \text{and} \quad \lambda_2 = \lambda_3 = \lambda^{-1/2} \tag{2.4}$$

where we have used the incompressibility assumption $\lambda_1\lambda_2\lambda_3 = 1$.

Equation (2.3) becomes

$$U = C_1\left(\lambda^2 + \frac{2}{\lambda} - 3\right) \tag{2.5}$$

and from Equation (2.1) we have

$$f = \frac{\partial U}{\partial \lambda} = 2C_1\left(\lambda - \frac{1}{\lambda^2}\right) \tag{2.6}$$

This familiar equation is more usually represented as a consequence of the molecular theories of a rubber network. Here we see that it follows from purely phenomenological considerations as a simple constitutive equation for the finite deformation of an isotropic, incompressible solid. Materials that obey this relationship are sometimes called neo-Hookean.

Note that for small strain, where $\lambda = 1 + e$, Equation (2.6) reduces to

$$f = 2C_1\{1 + e - (1 - 2e)\} = 6C_1 e \qquad (2.7)$$

i.e. Hooke's law.

References

1. Timoshenko, S. and Goodier, J. N., *Theory of Elasticity*, McGraw-Hill International Editions, New York, 1970.
2. Love, A. E. H., *A Treatise on the Mathematical Theory of Elasticity* (4th edn), Macmillan, New York, 1944.
3. Ward, I. M., *Mechanical Properties of Solid Polymers* (2nd edn), Wiley, Chichester, 1983.

Further reading

Solecki, R. and Conant, R. J., *Advanced Mechanics of Materials*, Oxford University Press, Oxford, 2003.

3
Rubber-like Elasticity

3.1 General features of rubber-like behaviour

The most noticeable feature of natural rubber and other elastomers is the ability to undergo large and reversible elastic deformation. It is not unexpected that stress can cause polymeric molecules to adopt an extended configuration, but at first sight it may seem surprising that on removal of the stress the molecules retract, on average, to their initial coiled form. Simple theories of rubber-like elasticity assume, as an approximation, that both extension and retraction occur instantaneously, and neglect any permanent deformation. Natural rubber (*cis*-polyisoprene) in its native state does not satisfy this last criterion, as molecules in extended configurations tend to slide past one another and do not recover completely. Molecules need to be chemically cross-linked by sulphur bonds (vulcanization) to prevent any permanent flow, and we shall show that the degree of cross-linking determines the extensibility of the rubber for a given stress.

The application of stress is considered to cause molecules to change from a coiled to an extended configuration instantaneously. For this reason it is possible to apply equilibrium thermodynamics to determine how the stress is related to changes in both internal energy and entropy. The general nature of thermodynamics implies that this type of approach can give no direct information on molecular rearrangements but, when augmented by molecular theories of a statistical nature, it is possible to derive an equation of state that relates the force causing extension to molecular parameters. We will show that this equation of state is identical in form to Equation (2.6) above, which was derived as the equivalent of Hooke's law for finite deformations. It will be shown that the reason for this direct link between the behaviour at a molecular level and the mechanics of finite elasticity arises because of the applicability of the so-called 'affine deformation' assumption. Affine deformation in the molecular theory of rubber elasticity means that we can assume that the changes in the length and orientation of lines joining adjacent cross-links in the molecular network are identical to the changes in lines marked on the macroscopic rubber. It is also assumed that there is no change in volume on deformation. This assumption can be justified as a good approximation because the bulk modulus (K) is some 10^4 times greater than the

An Introduction to the Mechanical Properties of Solid Polymers I. M. Ward and J. Sweeney
© 2004 John Wiley & Sons, Ltd   ISBN: 0471 49625 1 (HB); 0471 49626 X (PB)

shear modulus (G): typical values are 10^{10} Pa and 10^6 Pa. As a consequence, at low strains, Poisson's ratio, given by

$$\nu = \frac{3K - 2G}{2(3K + 2G)}$$

is effectively 0.500, and deformation occurs essentially at constant volume.

A schematic force–extension curve for a typical rubber is shown in Figure 3.1, with the maximum extensibility varying between 500 and 1000 per cent, depending on the extent of cross-linking. The behaviour is Hookean, with a linear relationship between stress and strain only at strains of the order of 1 per cent of so. At larger strains the force–extension relation is non-linear, and we will show that its form is determined essentially by changes in configurational entropy rather than internal energy.

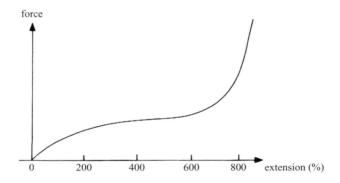

Figure 3.1 The force developed in uniaxial extension of a typical lightly cross-linked rubber

High extension results in a greatly reduced entropy, so that retraction is a consequence of the necessity for entropy to be maximized. A fully extended chain is a state of zero entropy because there is only one possible conformation of bonds through which it can occur. In contrast there are a very large number of ways of obtaining a given end-to-end distance for a contracted configuration of the chain. As all configurations have approximately the same internal energy, in the absence of external stress an extended chain will return to a more probable state. For this reason rubber is sometimes referred to as a 'probability or entropy spring', in contrast to the 'energy spring' characteristics of the elasticity of materials of low molecular mass, where extension causes an increase in internal energy. For a fuller discussion see Treloar [1].

3.2 The thermodynamics of deformation

The change in internal energy during deformation dU is given by

$$dU = dQ + dW \tag{3.1}$$

where dQ and dW are the heat absorbed and the work due to external forces, respectively. For a reversible process dQ relates to the entropy change dS by the relationship

$$dQ = T dS \tag{3.2}$$

Hence, for a reversible process we have

$$dU = T dS + dW \tag{3.3}$$

When an elastic solid of initial length l is extended uniaxially under a tensile force f, the work done on the solid in an infinitesimal displacement is

$$dW = f dl \tag{3.4}$$

By combining Equations (3.3) and (3.4) we have

$$dU = f dl + T dS \tag{3.5}$$

It is convenient to introduce the Helmholtz free energy A, which relates to changes that occur at constant volume

$$A = U - TS \tag{3.6}$$

Thus, for a change that occurs at constant temperature and constant volume we have

$$dA = dU - T dS \tag{3.7}$$

By combining Equations (3.3) and (3.7) we see that $dA = dW$ for isothermal changes at constant volume. Hence the tension f is given by

$$f = \left(\frac{\partial W}{\partial l}\right)_{T,V} = \left(\frac{\partial A}{\partial l}\right)_{T} = \left(\frac{\partial U}{\partial l}\right)_{T} - \left(\frac{\partial S}{\partial l}\right)_{T} \tag{3.8}$$

For any change in the Helmholtz free energy A (not necessarily an isothermal change)

$$dA = dU - TdS - SdT \tag{3.9}$$

But $dU = f\,dl + TdS$ from Equation (3.3), hence

$$dA = f\,dl - SdT \tag{3.10}$$

Then

$$\left(\frac{\partial A}{\partial l}\right)_T = f \quad \text{and} \quad \left(\frac{\partial A}{\partial T}\right)_l = -S \tag{3.11}$$

But

$$\frac{\partial}{\partial l}\left(\frac{\partial A}{\partial T}\right)_l = \frac{\partial}{\partial T}\left(\frac{\partial A}{\partial l}\right)_T$$

Substituting

$$\left(\frac{\partial S}{\partial l}\right)_T = -\left(\frac{\partial f}{\partial T}\right)_l \tag{3.12}$$

Hence Equation (3.8) becomes

$$\left(\frac{\partial U}{\partial l}\right)_T = f - T\left(\frac{\partial f}{\partial T}\right)_l \tag{3.13}$$

As long ago as 1935 Meyer and Ferri [2] showed that the tensile force at constant length was very nearly proportional to the absolute temperature, i.e. $f = \alpha T$. Differentiating this relationship we obtain

$$\left(\frac{\partial f}{\partial T}\right)_l = \alpha, \text{ a constant}$$

By substitution Equation (3.7) gives

$$\left(\frac{\partial U}{\partial l}\right)_T = 0$$

which demonstrates that elasticity arises entirely from changes in entropy to this good first approximation.

3.3 The statistical theory

The kinetic or statistical theory of rubber elasticity, originally proposed by Meyer, Susich and Valko [3], assumes that the very long molecular chains are each capable of assuming a wide variety of configurations in response to the thermal vibrations of their constituent atoms. Although the molecular chains are inter-linked to form a coherent network, the number of cross-links is assumed to be small enough not to interfere markedly with the motion of the chains. In the absence of external forces, the chain molecules will adopt configurations corre-sponding to a state of maximum entropy. When forces are applied, the chains will tend to extend in the direction of the force, thus reducing the entropy and producing a state of strain.

Quantitative evaluation of the stress–strain characteristics of the rubber network then involves calculation of the configurational entropy of the whole assembly of chains as a function of the state of strain. This calculation is considered in two stages: calculation of the entropy of a single chain and calculation of the change in entropy of a network of chains as a function of strain.

3.3.1 Simplifying assumptions

In reality, atoms are tightly packed along the length of a molecular chain. It is, however, convenient to represent molecular chains in terms of 'ball and stick' models, such as that shown for polyethylene in Figure 3.2. Here we show the fully

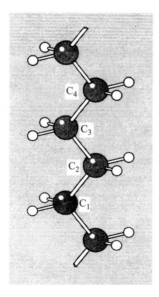

Figure 3.2   The polyethylene chain

extended molecule, which takes the form of a planar zigzag. If essentially free rotation from one conformation to another occurs, subject only to the limitation that the valence bond angle between carbon atoms must remain at 109.5°, the local situation $C_1C_2C_3C_4$ can change from the planar zigzag to a variety of conformations. In principle it is possible to calculate the number of molecular configurations that correspond to any chosen end-to-end length of the molecule: there will be only one fully extended configuration, but for a molecule containing possibly hundreds of backbone atoms the number of alternative contracted configurations will be very large. In practice it is more convenient to consider the 'freely jointed' chain, a mathematical abstraction in which the atoms are reduced to mere points, joined by one-dimensional equal links, with no restriction on the angle between adjacent links. It is assumed in this simple model that there is no difference in internal energy between the different molecular conformations along the chain.[†]

3.3.2   Average length of a molecule between cross-links

Consider a freely jointed chain with n links, each of length l. The length of a single chain is then given by

$$r = \sum_{i=1}^{n} l_i \tag{3.14}$$

For a large number of chains (q), or one chain considered at many different times, the mean length

$$r(q) = \frac{l}{q} \sum_{j=1}^{q} r_j = 0 \tag{3.15}$$

as the vector length is equally likely to be positive or negative.

We follow the procedure used, for example, with sinusoidally varying quantities such as alternating current and voltage, where the mean value is zero, and calculate the mean square chain length

$$\overline{r^2} = \frac{l}{q} \sum_{1}^{q} r_j^2 = \frac{l}{q} \sum_{1}^{q} \left(\sum_{1}^{n} l_i \right)_j^2 \tag{3.16}$$

[†]Conformation is used to denote differences in the immediate situation of a bond, e.g. *trans* and *gauche* conformations. Configuration is retained to refer to the arrangement of the whole molecular chain.

Expand, giving

$$\overline{r^2} = \frac{l}{q} \sum_{1}^{q} \left(l_1^2 + l_2^2 + \dots l_n^2 + \mathbf{l}_1 \cdot \mathbf{l}_2 + \mathbf{l}_1 \cdot \mathbf{l}_3 + \dots + \mathbf{l}_{n-1} \cdot \mathbf{l}_n \right)$$

but $l_1^2 = l_2^2 = l_n^2$ and $\mathbf{l}_m \cdot \mathbf{l}_n = l_m l_n \cos\theta$. In a freely jointed chain θ can have any value; hence $\Sigma \mathbf{l}_m \cdot \mathbf{l}_n = 0, m \neq n$. Therefore

$$\overline{r^2} = \frac{l}{q} \sum_{1}^{q} nl^2 = nl^2$$

The root mean square chain length is therefore

$$\sqrt{(r^2)} = l\sqrt{(n)} = r_{rms} \tag{3.17}$$

compared with the fully extended chain length of ln; i.e. for a chain of 100 bonds between cross-links the maximum extensibility is 10.

3.3.3 The entropy of a single chain

The expression just derived indicates the reason for the high extensibility of lightly cross-linked rubbers and serves to introduce the important concept of a mean square length, but yields no information on the probability of a chain having a particular end-to-end length. This latter problem was first analysed mathematically by Kuhn [4] and by Guth and Mark [5].

Consider a chain of n links each of length l, which has a configuration such that one end P is at the origin (Figure 3.3). The probability distribution for the position of the end Q is derived using approximations that are valid provided that the distance between the chain ends P and Q is much less than the extended chain length nl. The probability that Q lies within the elemental volume $dx \, dy \, dz$ at the point (x, y, z) can be shown to be

$$p(x, y, z) \, dx \, dy \, dz = \frac{b^3}{\pi^{3/2}} \exp(-b^2 r^2) \, dx \, dy \, dz \tag{3.18}$$

where $b^2 = 3/2nl^2$

This distribution has the form of the Gaussian error function and is spherically symmetrical about the origin, where the value is a maximum (Figure 3.4). The most probable end-to-end length is not, however, zero, as the probability that Q falls within an elemental volume situated between r and $(r + dr)$ from the origin, irrespective of direction, is the product of the probability distribution $p(r)$ and the volume of the concentric shell, $4\pi r^2 \, dr$. The overall probability is then

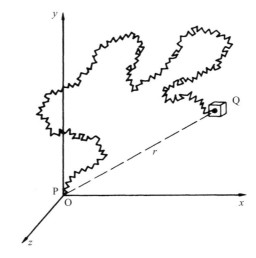

Figure 3.3   The freely jointed chain

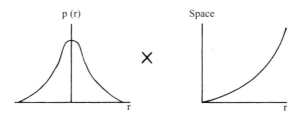

Figure 3.4   The Gaussian probability distribution for the free end of a chain must be multiplied by the volume in which that end can reside: $P(r) = p(r) \times 4\pi r^2 \, dr$

$$P(r)\,dr = p(r)4\pi r^2\,dr = (b^3/\pi^{3/2})\exp(-b^2 r^2) \cdot 4\pi^2\,dr \qquad (3.19)$$

$$= (4b^3/\pi^{1/2})r^2 \exp(-b^2 r^2)\,dr$$

which is illustrated in Figure 3.5.

It is seen that the most probable end-to-end distance, irrespective of direction, is not zero, but it is a function of b, i.e. of the length l of the links and the number n of links in the chain, as shown in Section 3.3.2 above.

Another important quantity is the root mean square chain length $(\overline{r^2})^{1/2}$

$$\overline{r^2} = \int_0^\infty r^2 P(r)\,dr$$

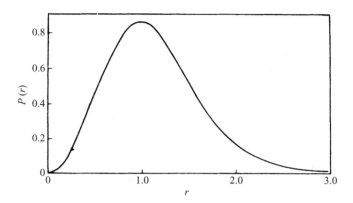

Figure 3.5 The distribution function $P(r) = \text{const} \, r^2 \exp(-b^2 r^2)$

Substitution of the above expression for P (r) gives

$$\overline{r^2} = 3/2b^2 = nl^2 \tag{3.20}$$

so that the root mean square length $(\overline{r^2})^{1/2} = l\sqrt{n}$, i.e. it is proportional to the square root of the number of links in the chain, as shown in Section 3.3.2 above.

The entropy of the freely jointed chain s is proportional to the logarithm of the number of configurations Ω so that

$$s = k \ln \Omega$$

where k is Boltzmann's constant. If $dx \, dy \, dz$ is constant, the number of configurations available to the chain is proportional to the probability per unit volume $p(x, y, z)$. The entropy of the chain is thus given by

$$s = c - kb^2 r^2 = c - kb^2(x^2 + y^2 + z^2) \tag{3.21}$$

where c is an arbitrary constant.

3.3.4 The elasticity of a molecular network

We wish to calculate the strain-energy function for a molecular network, assuming that this is given by the change in entropy of a network of chains as a function of strain.

The actual network is replaced by an ideal network in which each segment of a molecule between successive points of cross-linkage is considered to be a Gaussian chain.

Three additional assumptions are introduced:

1. In either the strained or unstrained state, each junction point may be regarded as fixed at its mean position.

2. The effect of the deformation is to change the components of the vector length of each chain in the same ratio as the corresponding dimensions of the bulk material (the 'affine' deformation assumption).

3. The mean square end-to-end distance for the whole assembly of chains in the unstrained state is the same as for a corresponding set of free chains and is therefore given by Equation (3.20).

In effect it is necessary to calculate the difference in probability between a spherical distribution of chain end-to-end vectors in the unstrained state and an ellipsoidal distribution for uniaxial extension (Figure 3.6). This difference is related to changes in entropy, and so to tensile force.

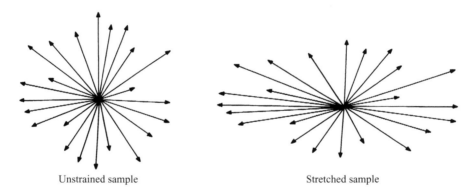

Unstrained sample   Stretched sample

Figure 3.6   Schematic representation of chain end-to-end vectors in the initial and final states

As discussed in Section 2.2, we can restrict our discussion to the case of normal strain without loss of generality. We choose principal extension ratios λ_1, λ_2 and λ_3 parallel to the three rectangular coordinate axes x, y and z. The affine deformation assumption implies that the relative displacement of the chain ends is defined by the macroscopic deformation. Thus, in Figure 3.7 we take a system of coordinates x, y and z in the undeformed body.

In this coordinate system a representative chain PQ has one end P at the origin. We refer any point in the deformed body to this system of coordinates. Thus the origin, i.e. the end of the chain P, could be moved bodily during the deformation. The other end Q (x, y, z) is displaced to the point Q$'(x', y', z')$ and from the affine deformation assumption we have

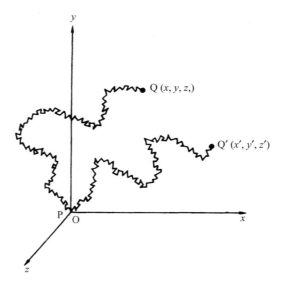

Figure 3.7   The end of the chain Q (x, y, z) is displaced to Q (x', y', z')

$$x' = \lambda_1 x \qquad y' = \lambda_2 y \qquad z' = \lambda_3 z$$

The entropy of the chain in the undeformed state is given by Equation (3.21) as

$$s = c - kb^2(x^2 + y^2 + z^2)$$

After deformation the entropy becomes

$$s' = c - kb^2(\lambda_1^2 x^2 + \lambda_2^2 y^2 + \lambda_2^3 z^2) \tag{3.22}$$

giving the change in entropy

$$\Delta s = s' - s = -kb^2\{(\lambda_1^2 - 1)x^2 + (\lambda_2^2 - 1)y^2 + (\lambda_3^3 - 1)z^2\} \tag{3.23}$$

Let there be N chains per unit volume in the network, with m of these having a given value of b (say b_p). The total entropy change for this particular group of chains is

$$\Delta s_b = \sum_1^m \Delta s = -kb_p^2\left\{(\lambda_1^2 - 1)\sum_1^m x^2 + (\lambda_2^2 - 1)\sum_1^m y^2 + (\lambda_3^2 - 1)\sum_1^m z^2\right\}$$
$$\tag{3.24}$$

where $\sum_1^m x^2$ is the sum of the squares of the x components for these m chains in the undeformed network. As there is no preferred direction for the chain vectors in the undeformed (isotropic) state, there is no preference for the x, y or z directions, so that

$$\sum_1^m x^2 = \sum_1^m y^2 = \sum_1^m z^2$$

but

$$\sum_1^m x^2 + \sum_1^m y^2 + \sum_1^m z^2 = \sum_1^m r^2$$

giving

$$\sum_1^m x^2 = \sum_1^m y^2 = \sum_1^m z^2 = \tfrac{1}{3}\sum_1^m r^2 \qquad (3.25)$$

From Equation (3.20)

$$\sum_1^m r^2 = mr^2 = m\left(\frac{3}{2b_p^2}\right) \qquad (3.26)$$

Combining Equations (3.25), (3.26) and (3.24)

$$\Delta s_b = -\tfrac{1}{2}mk\{\lambda_1^2 + \lambda_2^2 + \lambda_3^2 - 3\} \qquad (3.27)$$

We can now add the contribution of all the chains in the network (N per unit volume) and obtain the entropy change of the network ΔS, where

$$\Delta S = \sum_1^N \Delta s = -\tfrac{1}{2}Nk\{(\lambda_1^2 + \lambda_2^2 + \lambda_3^2) - 3\} \qquad (3.28)$$

Assuming no change in internal energy on deformation, this gives the change in the Helmholtz free energy

$$\Delta A = -T\,\Delta S = \tfrac{1}{2}NkT(\lambda_1^2 + \lambda_2^2 + \lambda_3^2 - 3)$$

If we assume that the strain-energy function U is zero in the undeformed state this gives

$$U = \Delta A = \tfrac{1}{2}NkT(\lambda_1^2 + \lambda_2^2 + \lambda_3^2 - 3) \qquad (3.29a)$$

Consider simple elongation λ in the x direction. The incompressibility relationship gives $\lambda_1 \lambda_2 \lambda_3 = 1$. Hence, by symmetry, $\lambda_2 = \lambda_3 = \lambda^{-1/2}$ and

$$U = \tfrac{1}{2} NkT \left(\lambda^2 + \frac{2}{\lambda} - 3 \right) \tag{3.29b}$$

From Equation (2.6) above

$$f = \frac{\partial U}{\partial \lambda} = NkT \left(\lambda - \frac{1}{\lambda^2} \right) \tag{3.30}$$

We have therefore obtained the neo-Hookean relationship of Section 2.4.3 above with a constant NkT. For small strain we can put $\lambda = 1 + e_{xx}$ and it follows from Equation (3.30) that

$$f = \sigma_{xx} = 3NkTe_{xx} = Ee_{xx}$$

where E is Young's modulus. Since for an incompressible material $E \equiv 3G$, we see that the quantity NkT in Equation (3.29) is equivalent to the shear modulus of the rubber, G. This term is sometimes written in terms of the mean molecular mass M_c of the chains, i.e. between successive points of cross-linkage. Then

$$G = NkT = \rho RT / M_c$$

where ρ is the density of the rubber and R is the gas constant.

3.4 Modifications of the simple molecular theory

The equation of state deduced above rests on many simplifying assumptions: freely jointed chains of negligible volume and with a Gaussian distribution of end-to-end lengths that give rise to no changes in internal energy on deformation. We mention briefly some of the investigations regarding those approximations, but direct the reader who requires further information to Treloar's monograph [1].

Even N, the number of chains per unit volume, is not a simple concept because junction points can be either (permanent) chemical cross-links or (temporary) physical entanglements. Not all cross-links will be effective, as seen in Figure 3.8, which shows (a) 'loose loops', where a chain folds back on itself, and (b) 'loose ends', where a chain does not contribute to the network [6].

Real polymer chains have fixed bond lengths and possibly hindered rotation. These effects are taken into account by the concept of 'the equivalent freely joined chain' [7]; for example, a paraffin-type chain with unhindered rotation will have r_{rms} $\sqrt{2}$ times greater than the simple model. More sophisticated treatments are possible in terms of random walk statistics.

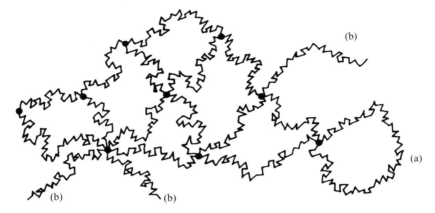

Figure 3.8   Types of network defect: (a) loose loop; (b) loose end

Kuhn and Grün showed that removing the restriction of a Gaussian distribution of chain end-to-end distances (but retaining other features of the simple model) gave a probability distribution $p(r)$ of the form

$$\ln p(r) = \text{const} - n\left[\frac{r}{nl}\beta + \ln(\beta/\sinh\beta)\right] \qquad (3.31)$$

where β is defined by

$$\frac{r}{nl} = \coth\beta - 1/\beta = \mathscr{L}(\beta)$$

where $\mathscr{L}$ is the Langevin function and $\beta = \mathscr{L}^{-1}\,(r/nl)$ is the inverse Langevin function.

The expression for probability can be expanded to give

$$\ln p(r) = \text{const} - n\left[\frac{3}{2}\left(\frac{r}{nl}\right)^2 + \frac{9}{20}\left(\frac{r}{nl}\right)^4 + \frac{99}{350}\left(\frac{r}{nl}\right)^6 + \dots\right] \qquad (3.32)$$

The Gaussian distribution is the first term of this series, and so is adequate when $r \ll nl$. James and Guth [8] subsequently used the inverse Langevin distribution function to give a revised expression for the force per unit unstrained area:

$$f = \frac{NkT}{3}n^{1/2}\left[\mathscr{L}^{-1}\left(\frac{\lambda}{n^{1/2}}\right) - \lambda^{-3/2}\mathscr{L}^{-1}\left(\frac{1}{\lambda^{1/2}n^{1/2}}\right)\right] \qquad (3.33)$$

Treloar's fit to the experimental data for natural rubber using Equation (3.33) and a suitable choice of parameters for N and n is shown in Figure 3.9. The maximum extension of the network is primarily determined by n (the number of chain links between successive cross-links), a result that is relevant for the cold-drawing and crazing behaviour discussed later (Section 11.6 and Section 12.5), where the basic deformation also involves the extension of a molecular network.

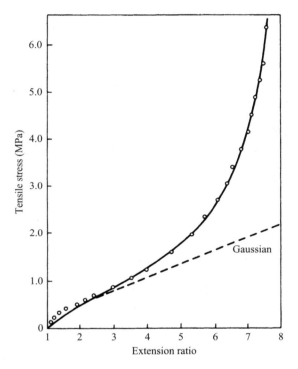

Figure 3.9 Theoretical non-Gaussian free extension curve obtained by fitting experimental data (0) to the James and Guth theory, with $NkT = 0.273$ MPa, $n = 75$. (Reproduced with permission from Treloar, *The Physics of Rubber Elasticity*, 3rd edn, Oxford University Press, Oxford, 1975)

Although we have seen how the extension to non-Gaussian statistics gives rise to a very large increase in tensile stress at large extensions, in the case of natural rubber it has been proposed that the observed increase in tensile stress occurs primarily because of strain-induced crystallization. The basic physical idea is that the melting point T_c of the rubber is increased due to extension: $T_c = \Delta H / \Delta S$, where ΔH and ΔS are the enthalpy and entropy of fusion, respectively. Because the entropy of the extended rubber is low, the change in entropy on crystallization is reduced and T_c correspondingly is increased. A higher degree of supercooling then gives rise to crystallization and the crystallites act to increase the modulus by forming additional physical cross-links.

Finally, the simple treatment of rubber elasticity given above makes two assumptions that require further consideration. First, it has been assumed that the internal energy contribution is negligible, which implies that different molecular conformations of the chains have identical internal energies. Secondly, the thermodynamic formulae that have been derived are, strictly, only applicable to measurements at constant volume, whereas most experimental results are obtained at constant pressure. For comprehensive elementary accounts of these complications, the reader is referred to the textbooks by Treloar [1] and Ward [9].

3.5 Recent developments in the molecular theory of rubber elasticity

There are two very clear discrepancies between the predictions of the simple Gaussian theory of rubber elasticity and experimental results. First, at comparatively low strains (~100 per cent) the stress does not rise with strain, but falls below the predicted values. This is usually called the Mooney softening. Secondly, the rise with strain at high extension is greater than that predicted and the rubber eventually fails at a limiting value of its extensibility. The classical approach to resolving these issues has been to modify Equation (2.5) and hence Equations (2.6) and (3.30) by introducing a more complex strain-energy function. This phenomenological approach is described in some detail in Appendix 2.

Recent attempts to explain these discrepancies on the basis of molecular models for the network deformation have come from three sources. First, there is the work of Edwards and co-workers who have advanced purely geometrical ideas to explain both the strain softening at low strains and the rapid strain hardening at high strains. Ball, Doi, Edwards and Warner [10] considered that in a rubber network there will be two types of junction points: the permanent cross-links where the chains are joined by chemical bonds; and entanglements, which are temporary junction points caused by chains becoming entangled. In an ingenious intuitive step these temporary junction points are replaced by slip links (Figure 3.10), which allow the chains to slide over one another. The change in free energy associated with the slip links is given by

$$\frac{F_s}{kT} = \frac{1}{2} N_s \sum_{i=1}^{3} \left[\frac{(1+\eta)\lambda_i^2}{1+\eta\lambda_i^2} + \log(1+\eta\lambda_i^2) \right] \tag{3.34}$$

where N_s is the number of slip links per unit volume, λ_i are the deformation ratios and η is the slipperiness factor, which in principle can vary between infinity for perfect sliding and zero for no sliding.

The total change in free energy is then

$$F = F_s + F_c$$

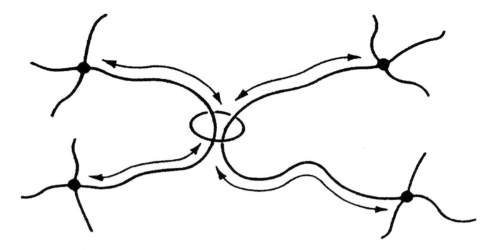

Figure 3.10 Slip-link model. (Reproduced with permission from Ball, Doi, Edwards and Warner, *Polymer*, **22**, 1010 (1981))

where

$$\frac{F_c}{kT} = \frac{1}{2} N_c \sum_{i=1}^{3} \lambda_i^2 \qquad (3.35)$$

and N_c is the number of permanent cross-links per unit volume.

Edwards and Vilgis [11] subsequently extended this theory to include the limitation of finite extensibility for the network that they regarded as arising when the network chains are fully extended, i.e. $\lambda_{max} \propto n^{1/2}$, where n is the number of links in the polymer chain between junction points (see Equation (3.17) above). This introduces another constant $\alpha = [(\lambda_1^2 + \lambda_2^2 + \lambda_3^2)^{-1/2}]_{max}$ into Equations (3.34) and (3.35) above. For example

$$\frac{F_c}{kT} = \frac{1}{2} N_c \left[\frac{\sum_{i=1}^{3} \lambda_i^2 (1 - \alpha^2)}{1 - \alpha^2 \sum_{i=1}^{3} \lambda_i^2} + \log\left(1 - \alpha^2 \sum_{i=1}^{3} \lambda_i^2\right) \right] \qquad (3.36)$$

Edwards and Vilgis showed that their theory could produce an excellent fit to data for natural rubber by assuming reasonable values of N_c, N_s, η and α (Figure 3.11).

The second approach, pioneered by Arruda and Boyce [12], involves the generalization of the non-Gaussian model of James and Guth [8]. The James and Guth model was essentially one of three mutually perpendicular molecular chains,

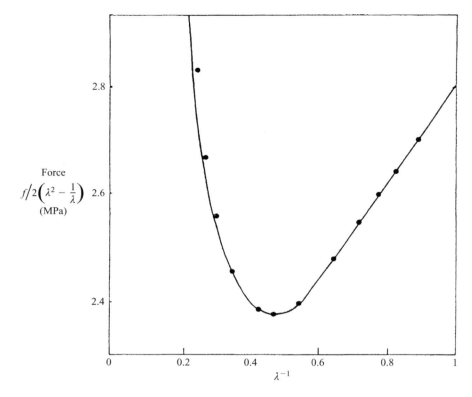

Force
$$f / 2\left(\lambda^2 - \frac{1}{\lambda}\right)$$
(MPa)

Figure 3.11   Stress–strain data for natural rubber according to Mullins. The solid line is calculated by the Edwards and Vilgis theory using the parameters $N_c k_B T = 1.2$ MPa, $N_s k_B T = 2.1$ MPa, $\eta = 0.2$, $\alpha^{-1} = \lambda_{max} = 7.5$. (Reproduced with permission from Edwards and Vilgis, *Polymer*, **27**, 486 (1986))

and this is generalized by increasing the number of chains. Thus, Arruda and Boyce introduced the eight-chain model, in which the chains are envisaged as being attached at one end to the corners of a cube of material and at the other end to the cube centre. On deformation of the cube into a rectangular element with extension ratios along its sides λ_1, λ_2 and λ_3, each of the eight chains develops the same extension ratio λ_{chain} defined by

$$\lambda_{chain} = \frac{1}{\sqrt{3}}\left(\lambda_1^2 + \lambda_2^2 + \lambda_3^2\right)$$

The analogue of the three-chain equation (Equation (3.33)) for uniaxial stretching becomes

$$f = \frac{NkT}{3} n^{1/2} \mathscr{L}^{-1} \left[\frac{\lambda_{\text{chain}}}{n^{1/2}} \right] \frac{\lambda - 1/\lambda^2}{\lambda_{\text{chain}}} \tag{3.37}$$

Arruda and Boyce [12] showed that the eight-chain model performed much better than three- and four-chain models in predicting the behaviour of vulcanized rubber in uniaxial and biaxial tension and shear, and also performed excellently in modelling uniaxial and plane strain compression of gum and neoprene rubber.

Wu and van der Giessen [13] produced a further generalization to an infinitely large number of chains – the full network model. Their expression for force is more complex than Equation (3.33) or Equation (3.37), but they showed that it could be approximated very well with a linear combination of the three- and eight-chain expressions. Contrary to expectations, the full network model performed less well than the eight-chain model for uniaxial and biaxial testing of natural gum rubber and silicone rubber. Wu and van der Giessen pointed out that both models were approximations, in that entanglement (slip link) effects and other intermolecular interactions are ignored, as is the existence of non-affine deformation. They raised the possibility that the eight-chain model was fortuitously compensating for these effects.

Sweeney [14] has compared the eight-chain, full network and Edwards–Vilgis models and shown that the first two can be approximated very well by the Edwards–Vilgis model provided that the chain extensibility limit is not approached too closely. This latter proviso is a result of the different forms of singularity in strain energy that control the approach to the extensibility limit. The additional feature of slip links in the Edwards–Vilgis model means that it is the most general of the three, and this is reflected in the greater number of fitting parameters.

A totally different approach to rubber elasticity has been developed by Stepto and co-workers [15, 16], which also accounts for the Mooney–Rivlin softening. Their approach is not phenomenological, but is based on structural considerations that give an accurate description of the molecular conformational states of the units in the polymer chains as the network is stretched. They have proposed a method for calculating the free energy of a stretched molecular network based on the rotational isomeric state of the network chains, with conformational energies determined from observations of conformational properties.

Using a series of Monte-Carlo calculations, the elastic properties of the network are derived from the network chain end-to-end distance distribution, and are assumed to arise solely as a result of allowed conformational changes in individual network chains. Figure 3.12 shows the calculation for the probability density functions $p(r)$ calculated from the simulated radial end-to-end distance distribution functions $P(r)$, where

$$p(r) = \frac{P(r)}{4\pi r^2}$$

as explained in Section 3.3.3 above.

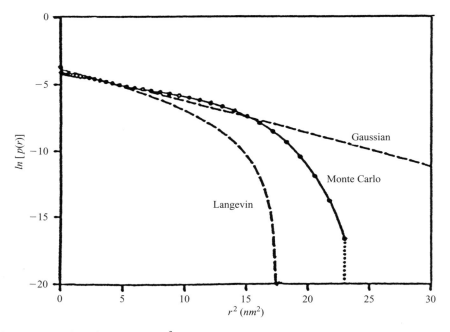

Figure 3.12 Plot of $\ln[\rho(r)]$ vs. r^2 at 298 K for polydimethyl siloxane chains of 40 bonds compared with Gaussian and Langevin treatments of the freely jointed chain. (Reproduced with permission from Stepto and Taylor, *J. Chem. Soc. Faraday Trans.*, **91**, 2639 (1995))

Comparison of the Monte-Carlo $p(r)$ with those predicted on the basis of the Gaussian distribution function (Equation (3.18) above) and the Langevin function (Equation above) show clear differences. In particular, the molecular structure-based Monte-Carlo $p(r)$ reflects clearly the limited extensibility of chains in the true network.

In the Stepto theory, the free energy of the network is given by

$$F = \frac{RT\rho}{M_c} s \left(\lambda^2 + \frac{2}{\lambda} - 3 \right)$$

where R, T, ρ and M_c have the same meaning as in Section 3.3.4 above but there is a new factor s, which is a function of λ and quantitatively accounts for the Mooney–Rivlin softening. In this theory, softening arises because some chains in the deformed network reach their full extensibility so that on further deformation these chains do not give a contribution to the reduction in entropy and hence to the network stress, which is correspondingly reduced. Stepto and co-workers have shown that their theory gives an exact quantitative fit to the stress–strain behaviour of polydimethyl siloxane networks [15], based only on the values for the rotational isomeric states obtained from the Flory–Crescenzi–Mark rotation isomeric state model for polydimethyl siloxane [16]. In addition, the optical birefringence (i.e. the stress-optical coefficient) for cross-linked polyethylenes was predicted quantitatively [18,19].

References

1. Treloar, L. R. G., *The Physics of Rubber Elasticity* (3rd edn), Clarendon Press, Oxford, 1975.
2. Meyer, K. H. and Ferri, C., *Helv. Chim. Acta*, **18**, 570 (1935).
3. Meyer, K. H., Von Susich, G. and Valko, E., *Kolloidzeitschrift*, **59**, 208 (1932).
4. Kuhn, W., *Kolloidzeitschrift*, **68**, 2 (1934); **76**, 258 (1936).
5. Guth, E. and Mark, H., *Lit. Chem.*, **65**, 93 (1934).
6. Flory, P. J., *Chem. Rev.*, **35**, 51 (1944).
7. Kuhn, W., *Kolloidzeitschrift*, **76**, 258 (1936); **87**, 3 (1939).
8. James, H. M. and Guth, E., *J. Chem. Phys.*, **11**, 455 (1943).
9. Ward, I. M., *Mechanical Properties of Solid Polymers* (2nd edn), Wiley, Chichester, 1983.
10. Ball, R. C., Doi, M., Edwards, S. F., *et al.*, *Polymer*, **22**, 1010 (1981).
11. Edwards, S. F. and Vilgis, Th., *Polymer*, **27**, 483 (1986).
12. Arruda, E. M. and Boyce, M. C., *J. Mech. Phys. Solids*, **41**, 389 (1993).
13. Wu, P. D. and van der Giessen, E., *J. Mech. Phys. Solids*, **41**, 427 (1993).
14. Sweeney, J., *Comput. Theor. Polym. Sci.*, **9**, 27 (1999).
15. Stepto, R. F. T. and Taylor, D. J. R., *Macromol. Symp.*, **93**, 261 (1995).
16. Stepto, R. F. T. and Taylor, D. J. R., *J. Chem. Soc. Faraday Trans.*, **91**, 2639 (1995)
17. Flory, P. J., Crescenzi, V. and Mark, J. E., *J. Am. Chem. Soc.*, **67**, 3202 (1971).
18. Taylor, D. J. R., Stepto, R. F. T., Jones, R. A., *et al.*, *Macromolecules*, **32**, 1978 (1999).
19. Cail, J. I., Taylor, D. J. R., Stepto, R. F. T., *et al.*, *Macromolecules*, **33**, 4966 (2000).

Problems for Chapters 2 and 3

1. The tensile stress σ in an *ideal* rubber when simply extended to a length λ times its initial length is given by

$$\sigma = NkT(\lambda^2 - \lambda^{-1})$$

Explain *without giving any mathematical details* the physical model that leads to this expression.

A sample of polyisoprene (density 1300 kg m^{-3}, monomer of relative molar mass 68) has a shear modulus of 4×10^5 Pa at room temperature. Calculate the average number of monomers between cross-links.

2. Rubber deforms at constant volume, and so has a Poisson's ratio of 0.5000 at small

strains. By how much must a rod of rubber be extended before Poisson's ratio falls to 0.4900?

3. What would be the root mean square end-to-end distance of a paraffinic chain consisting of 1000 carbon atoms? The length of the C—C bond is 1.53 Å and the chain can be considered to be freely jointed (i.e. without the restriction that valence angles should remain constant).

4. A non-Gaussian rubber has the strain-energy function

$$U = C_1(\lambda_1^{1.3} + \lambda_2^{1.3} + \lambda_3^{1.3} - 3)$$

where λ_1, λ_2 and λ_3 are the principal extension ratios and $C_1 = 4 \times 10^5$ Pa.

If a piece of such a rubber is initially 1 m long and has a cross-sectional area of 6×10^{-4} m^2, find the mass required to give a final extended length of 3 m.

5. The strain energy function for an ideal rubber is

$$U = C_1(\lambda_1^2 + \lambda_2^2 + \lambda_2^3 - 3)$$

where λ_1, λ_2, λ_3 are the principal extension ratios. Derive the stress–strain relations for the following:

(i) Simple extension $\lambda_1 = \lambda$ produced by a force applied in the 1 direction;

(ii) An equal two-dimensional extension $\lambda_1 = \lambda_2 = \lambda$, produced by the simultaneous application of equal forces in the 1 and 2 directions.

6. A non-Gaussian rubber has a strain-energy function

$$U = C_1(\lambda_1^2 + \lambda_2^2 + \lambda_3^2 - 3) + C_2(\lambda_1^2\lambda_2^2 + \lambda_2^2\lambda_3^2 + \lambda_1^2\lambda_3^2 - 3)$$

Derive the stress–strain relation for a simple extension $\lambda_1 = \lambda$ produced by a force applied in the 1 direction, and hence show that the low strain tensile modulus for this rubber is given by $E = 6(C_1 + C_2)$.

7. The application of a mass of 1 kg to a rubber strip of initial cross-sectional area 10 mm^2 causes a 100 per cent increase in length at 300 K. How many chains are there per unit volume? (Boltzmann's constant $k = 1.38 \times 10^{-23}$ J deg^{-1}).

8. A cross-linked rubber of undeformed cross-section 15 mm $\times$ 1.5 mm is stretched at 300 K to three times its initial length by suspending a mass of 1 kg. If the density of the rubber is 900 kg m^{-3}, what is the mean molecular mass of the network chains?

4

Principles of Linear Viscoelasticity

In this chapter we describe the common forms of viscoelastic behaviour and discuss the phenomena in terms of the deformation characteristics of elastic solids and viscous fluids. The discussion is confined to linear viscoelasticity, for which the Boltzmann superposition principle enables the response to multistep loading processes to be determined from simpler creep and relaxation experiments. Phenomenological mechanical models are considered and used to derive retardation and relaxation spectra, which describe the time-scale of the response to an applied deformation. Finally we show that in alternating strain experiments the presence of the viscous component leads to a phase difference between stress and strain.

4.1 Viscoelasticity as a phenomenon

The behaviour of materials of low relative molecular mass is usually discussed in terms of two particular types of ideal material: the elastic solid and the viscous liquid. The former has a definite shape and is deformed by external forces into a new equilibrium shape; on removal of these forces it reverts instantaneously to its original form. The solid stores all the energy that it obtains from the external forces during the deformation, and this energy is available to restore the original shape when the forces are removed. By contrast, a viscous liquid has no definite shape and flows irreversibly under the action of external forces.

One of the most interesting features of polymers is that a given polymer can display all the intermediate range of properties between an elastic solid and a viscous liquid depending on the temperature and the experimentally chosen time-scale. Bouncing putty, a silicone product, flows over a period of hours, fractures like a ductile solid when deformed rapidly and bounces like an elastomer when dropped. Of greater commercial importance are the rubber-like, and in extreme cases brittle, characteristics exhibited by molten polymers at high processing rates. This form of response, which combines both liquid-like and solid-like features, is termed viscoelasticity.

An Introduction to the Mechanical Properties of Solid Polymers I. M. Ward and J. Sweeney
© 2004 John Wiley & Sons, Ltd   ISBN: 0471 49625 1 (HB); 0471 49626 X (PB)

4.1.1   Linear viscoelastic behaviour

Newton's law of viscosity defines viscosity η by stating that stress σ is proportional to the velocity gradient in the liquid:

$$\sigma = \eta \frac{\partial V}{\partial y}$$

where V is the velocity and y is the direction of the velocity gradient. For a velocity gradient in the xy plane

$$\sigma_{xy} = \eta \left(\frac{\partial V_x}{\partial y} + \frac{\partial V_y}{\partial x} \right)$$

where $\partial V_x / \partial y$ and $\partial V_y / \partial x$ are the velocity gradients in the y and x directions, respectively (see Figure 4.1 for the case where the velocity gradient is in the y direction).

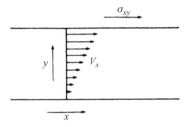

Figure 4.1   The velocity gradient

Since $V_x = \partial u / \partial t$ and $V_y = \partial v / \partial t$, where u and v are the displacements in the x and y directions, respectively, we have

$$\sigma_{xy} = \eta \left[\frac{\partial}{\partial y} \left(\frac{\partial u}{\partial t} \right) + \frac{\partial}{\partial x} \left(\frac{\partial v}{\partial t} \right) \right]$$

$$= \eta \frac{\partial}{\partial t} \left(\frac{\partial u}{\partial y} + \frac{\partial v}{\partial x} \right)$$

$$= \eta \frac{\partial e_{xy}}{\partial t}$$

It can be seen that the shear stress σ_{xy} is directly proportional to the rate of change of shear strain with time. This formulation brings out the analogy between Hooke's law for elastic solids and Newton's law for viscous liquids. In the former the stress is linearly related to the strain but in the latter the stress is linearly related to the rate of change of strain or strain rate.

Hooke's law describes the behaviour of a linear elastic solid and Newton's law that of a linear viscous liquid. A simple constitutive relation for the behaviour of a linear viscoelastic solid is obtained by combining these two laws:

1. For elastic behaviour $(\sigma_{xy})_E = Ge_{xy}$, where G is the shear modulus.

2. For viscous behaviour $(\sigma_{xy})_V = \eta(\partial e_{xy}/\partial t)$.

A simple possible formulation of linear viscoelastic behaviour combines these equations, making the assumption that the shear stresses related to strain and strain rate are additive:

$$\sigma_{xy} = (\sigma_{xy})_E + (\sigma_{xy})_V = Ge_{xy} + \eta \frac{\partial e_{xy}}{\partial t}$$

The equation represents one of the simple models for linear viscoelastic behaviour, the Kelvin or Voigt model, and is discussed in detail in Section 4.2.3 below.

As most of the experiments on linear viscoelasticity examine a single mode of deformation, usually corresponding to a measurement of Young's modulus or the shear modulus, our initial discussion of linear viscoelasticity will be confined to the one-dimensional situation.

For elastic solids Hooke's law is valid only at small strains, and Newton's law of viscosity is restricted to relatively low flow rates, as only when the stress is proportional either to the strain or the strain rate is analysis of the deformation feasible in simple form. A comparable limitation holds for viscoelastic materials: general quantitative predictions are possible only in the case of linear visco-elasticity, for which the results of changing stresses or strains are simply additive, but the time at which the change is made must be taken into account. For a single loading process there will be a linear relation between stress and strain at a given time. Multistep loading can be analysed in terms of the Boltzmann superposition principle (Section 4.2.1) because each increment of stress can be assumed to make an independent contribution to the overall strain.

In practice most useful plastics show some non-linearity at strains around 1 per cent, and so the following discussion may not be relevant for particular practical applications: for instance the fibres in a carpet pile suffer bending strains as high as 4 per cent.

4.1.2 Creep

Creep is the time-dependent change in strain following a step change in stress. Unlike the creep discussed by metallurgists, creep in polymers at low strains (1 per cent) is essentially recoverable after unloading, without the need for annealing at a raised temperature. The responses to two levels of stress for linear elastic and linear viscoelastic materials are compared in Figure 4.2. In the former case the strain follows the pattern of the loading programme exactly and in exact proportionality to the magnitude of the stresses applied. For the most general case of a linear viscoelastic solid the total strain e is the sum of three essentially separate parts: e_1 the immediate elastic deformation, e_2 the delayed elastic deformation and e_3 the Newtonian flow, which is identical with the deformation of a viscous liquid obeying Newton's law of viscosity.

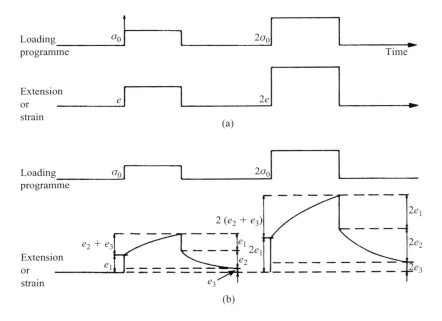

Figure 4.2 (a) Deformation of an elastic solid; (b) deformation of a linear viscoelastic solid

Because the material shows linear behaviour the magnitudes of e_1, e_2 and e_3 are exactly proportional to the magnitude of the applied stress, so that a creep compliance $J(t)$ can be defined, which is a function of time only:

$$J(t) = \frac{e(t)}{\sigma} = J_1 + J_2 + J_3$$

where J_1, J_2 and J_3 correspond to e_1, e_2 and e_3. Linear amorphous polymers show a significant J_3 above their glass transition temperatures, when creep may continue until the specimen ruptures, but at lower temperatures J_1 and J_2 dominate. Cross-linked polymers do not show a J_3 term, and to a very good approximation neither do highly crystalline polymers.

The separation of compliance into immediate and delayed terms is somewhat arbitrary, as J_1 is generally the limiting compliance observed at the shortest experimentally accessible times; for the purposes of analysis, however, we shall assume that there is a real distinction between the two phases of response. The initial elastic deformation is sometimes called the unrelaxed response to distinguish it from the relaxed response observed at times long enough for the various relaxation mechanisms to have occurred.

The maximum insight into the nature of creep is obtained by plotting the logarithm of creep compliance against the logarithm of time over a very wide time-scale (Figure 4.3). We shall show in Section 5.3.1 that an extended time-scale can be achieved through a series of short-term experiments at a range of temperatures. This diagram shows that at very short times the compliance (typically 10^{-9} Pa^{-1}) is that for a glassy solid and is independent of time; in contrast, at very long times the compliance (typically 10^{-5} Pa^{-1}) is that for a rubber-like solid and is again time independent. At intermediate times the compliance lies between these extremes and is time dependent, so that the behaviour is viscoelastic.

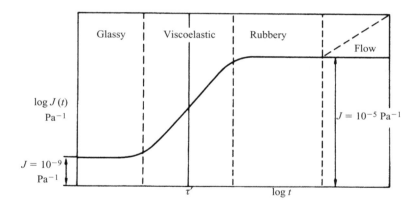

Figure 4.3   The creep compliance $J(t)$ as a function of time t; τ' is the characteristic time (the retardation time)

It is convenient to define a retardation time τ' in the middle of the viscoelastic region to characterize the time-scale for creep. The distinction between a rubber and a glassy plastic is then seen to be somewhat artificial, because it depends only on the value of τ' at room temperature. Compared with typical experimental response times, which can rarely be less than 1 s, τ' for a rubber is very small at room temperature, whereas the opposite is true for a glassy polymer. As the temperature is raised the frequency of molecular rearrangements increases, so reducing the value of τ'. Thus at sufficiently low temperatures a rubber behaves like a glassy plastic, and will shatter under impact conditions; correspondingly a glassy plastic will become rubber-like at a sufficiently high temperature.

Recovery curves from creep under constant stress are also illustrated in Figure 4.2, which indicates that at any selected time the extent of recovery (and so the unrecovered strain) is directly proportional to the stress that had been applied formerly. The relation between recovery and creep will be derived in Section 4.2.1.

4.1.3   Stress relaxation

When an instantaneous strain is applied to an ideal elastic solid a finite and constant stress will be recorded. For a linear viscoelastic solid subjected to a nominally instantaneous strain the initial stress will be proportional to the applied strain and will decrease with time (Figure 4.4), at a rate characterized by the relaxation time τ. This behaviour is called stress relaxation. For amorphous linear polymers at high temperatures the stress may eventually decay to zero. In the following discussion we shall ignore transient behaviour.

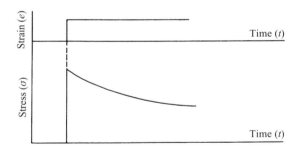

Figure 4.4   Stress relaxation (idealized)

Making the assumption of linear viscoelastic behaviour we can define the stress relaxation modulus $G(t) = \sigma(t)/e$. Where there is no viscous flow the stress decays to a finite value (Figure 4.5), to give an equilibrium or relaxed modulus G_r at finite time. As with creep we see that there are regions of glassy, viscoelastic, rubber-like and flow behaviour; similarly, changing temperature is equivalent to changing the time-scale. The relaxation time τ is of the same general magnitude as the retardation time τ', but the two are identical only for the simpler models.

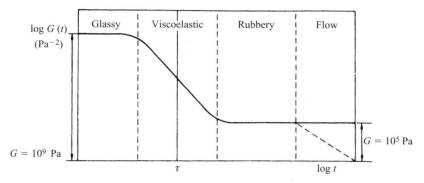

Figure 4.5   The stress relaxation modulus $G(t)$ as a function of time t; τ is the characteristic time (the relaxation time)

4.2 Mathematical representation of linear viscoelasticity

The models discussed here, which are phenomenological and have no direct relation with chemical composition or molecular structure, in principle enable the response to a complicated loading pattern to be deduced from a single creep (or stress-relaxation) plot extending over a long time interval. Interpretation depends on the assumption in linear viscoelasticity that the total deformation can be considered as the sum of *independent* elastic (Hookean) and viscous (Newtonian) components. In essence, the simple behaviour is modelled by a set of either integral or differential equations, which are then applicable in other situations.

4.2.1   The Boltzmann superposition principle

Boltzmann proposed, as long ago as 1876 [1], that:

1.   The creep is a function of the entire past loading history of the specimen.

2.   Each loading step makes an independent contribution to the final deformation, so that the total deformation can be obtained by the addition of all the contributions.

  Figure 4.6 illustrates the creep response to a multistep loading programme, in which incremental stresses $\Delta\sigma_1$, $\Delta\sigma_2$, ..., are added at times τ_1, τ_2, ..., respectively. The total creep at time t is then given by

$$e(t) = \Delta\sigma_1 J(t - \tau_1) + \Delta\sigma_2 J(t - \tau_2) + \Delta\sigma_3 J(t - \tau_3) + \ldots \qquad (4.1)$$

where $J(t - \tau)$ is the creep compliance function. For a particular loading step the

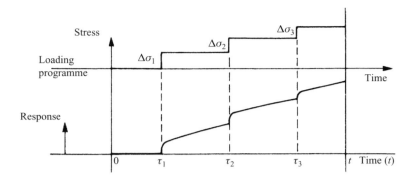

Figure 4.6 The creep behaviour of a linear viscoelastic solid

relevant form of the function is that for the time interval between the present instant and that at which the load increment was applied.

The summation of (4.1) can be generalized in integral form as

$$e(t) = \int_{-\infty}^{t} J(t - \tau)\, d\sigma(t) \tag{4.2}$$

It is usual to separate out the instantaneous elastic response in terms of the unrelaxed modulus G_u, giving

$$e(t) = \frac{\sigma}{G_u} + \int_{-\infty}^{t} J(t - \tau)\frac{d\sigma(\tau)}{d\tau}\, d\tau \tag{4.3}$$

where σ represents the total stress at the end of the experiment. Note that the integral extends from $-\infty$ to t, which implies that all previous elements of loading history must be taken into account and, in principle, the user must know the history of each specimen since its manufacture. In fact, when creep levels are low enough for linearity to apply, the deformation effectively levels off at sufficiently long times, so that only comparatively recent history is relevant, and this can be standardized by a conditioning treatment (see Section 5.1.1). For this reason viscoelastic solids are sometimes said to be materials with 'fading memory'.

The integral in Equation (4.3) is called a Duhamel integral, and it is a useful illustration of the consequences of the Boltzmann superposition principle to evaluate the response for a number of simple loading programmes. Recalling the development that leads to Equation (5.2) it can be seen that the Duhamel integral is most simply evaluated by treating it as the summation of a number of response terms. Consider two specific cases:

1. *Single-step loading of a stress σ_0 at time $\tau = 0$* (Figure 4.7). For this case

$$J(t - \tau) = J(t) \quad \text{and} \quad e(t) = \sigma_0 J(t)$$

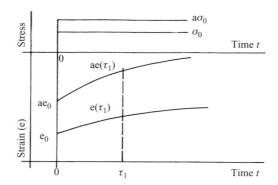

Figure 4.7 Creep for single step loading. For stress $a\sigma_0$ the strain at any time $(\tau \geqslant 0)$ is a times greater than that for stress σ_0

2. *Two-step loading of a stress σ_0 at time $\tau = 0$, followed by an additional stress σ_0 at time $\tau = t_1$* (Figure 4.8). For this case the creep deformation produced by the two loading steps are

$$e_1 = \sigma_0 J(t) \quad \text{and} \quad e_2 = \sigma_0 J(t - t_1)$$

so that

$$e(t) = e_1 + e_2 = \sigma_0 J(t) + \sigma_0 J(t - t_1)$$

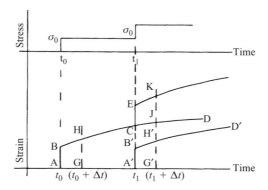

Figure 4.8 Creep for two equal loading steps. Additional instantaneous strain CE = AB. Additional total strain at $(t_1 + \Delta t)$: JK = GH

The 'additional creep' $e'_c(t - t_1)$ produced by the second loading step is given by

$$e'_c(t - t_1) = \sigma_0 J(t) + \sigma_0 J(t - t_1) - \sigma_0 J(t) = \sigma_0 J(t - t_1)$$

The above illustrates one consequence of the Boltzmann principle, viz. that the additional creep $e'_c(t - t_1)$ produced by adding the stress σ_0 is identical with the creep that would have occurred had this stress σ_0 been applied without any previous loading at the same instant in time t_1.

The principle is illustrated in Figure 4.8 where ABD represents creep under σ_0 alone. The response to an additional stress σ_0 at t_1 is found by sliding ABD along the time axis by t_1 to give curve A′B′D′, and at any time adding together the individual strains due to ABD and A′B′D′; e.g. at Δt after the initial loading step the deformation is GH; at Δt after t_1 the deformation due to the initial σ_0 is G′J. The total deformation at $(t_1 + \Delta t)$ is found by adding to G′J the strain JK = G′H′ = GH.

If the second load had been $a\sigma_0$, where a is a constant, then CE = A′B′ = aAB; JK = aG′H′ = aGH, etc.

3. *Creep and recovery.* In this case (Figure 4.9) the stress σ_0 is applied at time $\tau = 0$ and removed at time $\tau = t_1$. The deformation $e(t)$ at a time $t > t_1$ is given by the addition of two terms $e_1 = \sigma_0 J(t)$ and $e_2 = -\sigma_0 J(t - t_1)$, which express the application and removal of the stress σ_0, respectively. Thus $e(t) = \sigma_0 J(t) - \sigma_0 J(t - t_1)$.

The recovery $e_r(t - t_1)$ will be defined as the difference between the anticipated creep under the initial stress and the actual measured response. Thus

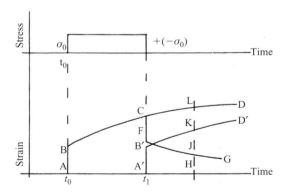

Figure 4.9 Recovery considered as the addition of a negative stress increment, i.e. subtract B′D′ from CD

$$e_r(t - t_1) = \sigma_0 J(t) - [\sigma_0 J(t) - \sigma_0 J(t - t_1)] = \sigma_0 J(t - t_1)$$

which is identical to the creep response to a stress σ_0 applied at a time t_1. This procedure demonstrates a second consequence of the Boltzmann super-position principle, that the creep and recovery responses are identical in magnitude.

The initial creep curve, ABD in Figure 4.9, is again moved along the time axis by t_1 to give A'B'D'. At any subsequent time the overall deformation is given by the difference between the two curves (CFG). It is important to realize that at a time $t_2 > t_1$ the residual deformation is obtained by subtracting A'B'D' from the deformation that would have occurred had unloading not taken place; e.g. residual deformation HJ = HL − HK, and not A'C − HK, where A'C is the maximum strain attained before unloading.

4.2.2 The stress relaxation modulus

Stress relaxation behaviour can be represented in an exactly complementary fashion using the Boltzmann superposition principle. Consider a stress relaxation programme in which incremental strains Δe_1, Δe_2, Δe_3, etc. are added at times τ_1, τ_2, τ_3, etc., respectively. The total stress at time t is then given by

$$\sigma(t) = \Delta e_1 G(t - \tau_1) + \Delta e_2 G(t - \tau_2) + \Delta e_3 G(t - \tau_3) + \ldots \quad (4.4)$$

where $G(t - \tau)$ is the stress relaxation modulus. Equation (4.4) may be generalized in an identical manner in which Equation (4.1) leads to Equations (4.2) and (4.3) to give

$$\sigma(t) = [G_r e] + \int_{-\infty}^{t} G(t - \tau) \frac{de(\tau)}{\Delta \tau} \, dt \quad (4.5)$$

where G_r is the equilibrium or relaxed modulus.

4.2.3 Mechanical models, retardation and relaxation time spectra

Linear viscoelasticity may be represented pictorially by models comprising massless Hookean springs and Newtonian dashpots, the latter being considered as oil-filled cylinders in which a loosely fitting piston moves at a rate proportional to the viscosity of the oil and to the applied stress. The models are used to establish differential equations that describe the deformation of the polymer under investigation. We start by considering the two possible combinations of a single spring and a single dashpot, and then discuss more realistic models. In the outcome the differential equation becomes the basis for further development, and the model itself is discarded.

The Kelvin or Voigt model

This model (Figure 4.10(a)) consists of a spring of modulus E_K, in parallel with a dashpot of viscosity η_K. If a constant stress σ is applied at time $t = 0$ there can be no instantaneous extension of the spring, as it is retarded by the dashpot. Deformation then occurs at a varying rate, with the stress shared between the two components until, after a time dependent on the dashpot viscosity, the spring approaches a finite maximum extension. When the stress is removed the reverse process occurs: there is no instantaneous retraction, but the initial unstretched length is eventually recovered (Figure 4.10(b)). The model does represent the time-dependent component of creep to a first approximation.

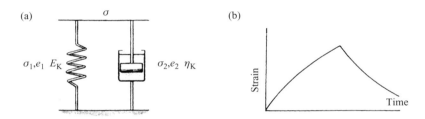

Figure 4.10   (a) The Kelvin, or Voigt, unit; (b) creep and recovery behaviour

The stress–strain relations are, for the spring, $\sigma_1 = E_K e_1$ and for the dashpot

$$\sigma_2 = \eta_K \frac{de_2}{dt}$$

The total stress σ is shared between spring and dashpot: $\sigma = \sigma_1 + \sigma_2$; but the strain in each component is the total strain: $e = e_1 = e_2$.

$$\therefore \quad \sigma = E_K e + \eta_K \frac{de}{dt} \tag{4.6}$$

Solving for $0 < t < t_1$, when the stress is σ

$$\frac{E_K}{\eta_K} \int_0^t dt = \int_0^e \frac{de}{(\sigma/E_K) - e}$$

where η_K/E_K has the dimensions of time, and represents the rate at which the deformation occurs: it is the retardation time τ'. Hence by integration

$$\frac{t}{\tau'} = \ln\left(\frac{\sigma/E_K}{(\sigma/E_K) - e}\right)$$

giving

$$\frac{\sigma}{E_K} = \left(\frac{\sigma}{E_K} - e\right) \exp(t/\tau')$$

Rearranging, we obtain

$$e = \frac{\sigma}{E_K} [1 - \exp(-t/\tau')] \tag{4.7}$$

It is convenient in creep experiments to replace E_K by $1/J$, where J is the spring compliance, to give

$$e = J\sigma[1 - \exp(-t/\tau')] \tag{4.8}$$

For $t > t_1$ after unloading, the solution becomes

$$e = e_{t_1} \exp\left(\frac{t_1 - t}{\tau'}\right), \text{ where } e_{t_1} = J\sigma[1 - \exp(-t_1/\tau')] \tag{4.9}$$

The retardation time τ' is the time after loading for the strain to reach

$$\left(1 - \frac{1}{\exp}\right)$$

of its equilibrium value; after stress removal the strain decays to $(1/\exp)$ of its maximum value in time τ'.

The Kelvin model is unable to describe stress relaxation, as at constant strain the dashpot cannot relax. In mathematical terms $(de/dt) = 0$, giving $\sigma = E_K e$.

The Maxwell model

The Maxwell model consists of a spring and dashpot in series as shown in Figure 4.11(a).

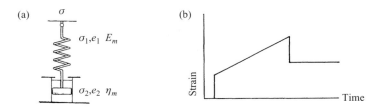

Figure 4.11   (a) The Maxwell unit; (b) creep and recovery behaviour

The equations for the stress–strain relations are

$$\sigma_1 = E_m e_1 \qquad\qquad (4.10\text{a})$$

relating the stress σ_1 and the strain e_1 in the spring, and

$$\sigma_2 = \eta_m \frac{de_2}{dt} \qquad\qquad (4.10\text{b})$$

relating the stress σ_2 and the strain e_2 in the dashpot. Because the stress is identical for the spring and the dashpot the total stress $\sigma = \sigma_1 = \sigma_2$. The total strain e is the sum of the strain in the spring and the dashpot, i.e. $e = e_1 + e_2$.

To find the relationship between total stress and total strain Equation (4.10a) can be written as

$$\frac{d\sigma}{dt} = E_m \frac{de_1}{dt}$$

and added to Equation (4.10b), giving

$$\frac{de}{dt} = \frac{1}{E_m} \frac{d\sigma}{dt} + \frac{\sigma}{\eta_m} \qquad\qquad (4.11)$$

The Maxwell model is of particular value in considering a stress relaxation experiment. In this case

$$\frac{de}{dt} = 0 \quad \text{and} \quad \frac{1}{E_m} \frac{d\sigma}{dt} + \frac{\sigma}{\eta_m} = 0$$

Thus

$$\frac{d\sigma}{\sigma} = -\frac{E_m}{\eta_m} dt$$

At time $t = 0$, $\sigma = \sigma_0$, the initial stress, and integrating we have

$$\sigma = \sigma_0 \exp\left(\frac{-E_m}{\eta_m}\right) t \tag{4.12}$$

This equation shows that the stress decays exponentially with a characteristic time constant $\tau = \eta_m / E_m$:

$$\sigma = \sigma_0 \exp\left(\frac{-t}{\tau}\right)$$

where τ is called the 'relaxation time'. There are two inadequacies of this simple model that can be understood immediately.

First, under conditions of constant stress, i.e

$$\frac{d\sigma}{dt} = 0, \qquad \frac{de}{dt} = \frac{\sigma}{\eta_m}$$

and Newtonian flow is observed. This is clearly not generally true for viscoelastic materials where the creep behaviour is more complex.

Secondly, the stress relaxation behaviour cannot usually be represented by a single exponential decay term, nor does it necessarily decay to zero at infinite time.

The standard linear solid

We have seen that the Maxwell model describes the stress relaxation of a viscoelastic solid to a first approximation, and the Kelvin model the creep behaviour, but that neither model is adequate for the general behaviour of a viscoelastic solid where it is necessary to describe both stress relaxation and creep.

A response closer to that of a real polymer is obtained by adding a second spring of modulus E_a in parallel with a Maxwell unit (Figure 4.12). This model is known as the 'standard linear solid' and is usually attributed to Zener [2]. It provides an approximate representation to the observed behaviour of polymers in their viscoelastic range. In creep, both springs extend, so that

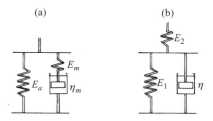

Figure 4.12   Three-element models; (a) is known as the standard linear solid (SLS)

$$\tau' = \eta_m \left[\frac{1}{E_a} + \frac{1}{E_m} \right]$$

but in stress relaxation E_a is unaffected, giving

$$\tau = \frac{\eta_m}{E_m}$$

The stress–strain relationship is

$$\sigma + \tau \frac{d\sigma}{dt} = E_a e + (E_a + E_m)\tau \frac{de}{dt} \tag{4.13}$$

This model describes both creep and stress relaxation and the transition from the glassy modulus E_m (for $E_m \gg E_a$) at short times, where the viscous dashpot has an infinite viscosity, to the rubbery modulus E_a at long times.

Multi-element models

For real materials a simple exponential response in creep or stress relaxation is not an adequate description of the time dependence. A good representation can be obtained by simulating creep with an array of Kelvin models in series and simulating stress relaxation with an array of Maxwell models in parallel (Figure 4.13). The dashpot in the first unit can be considered to be filled with a low-viscosity fluid, so that its response is effectively instantaneous. Figure 4.14 illustrates how such a system approaches reality, with successive units responding as time elapses.

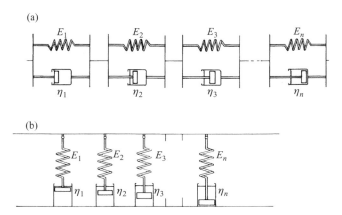

Figure 4.13   (a) Kelvin units in series for creep simulation; (b) Maxwell units in parallel for stress relaxation simulation

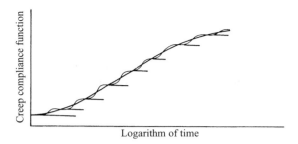

Figure 4.14 Simulation of creep function through combining Kelvin units in series (schematic)

Retardation and relaxation time spectra

Let the number of individual units in multi-element models tend to infinity. For creep, an infinite number of Kelvin units gives an infinite number of retardation times: this is called the spectrum of retardation times. The analogous development for stress relaxation leads to the spectrum of relaxation times.

For stress relaxation at constant strain e the Maxwell model gives

$$\sigma(t) = eE \exp\left(\frac{-t}{\tau}\right) \tag{4.14}$$

so that the stress relaxation modulus

$$G(t) = E \exp\left(\frac{-t}{\tau}\right)$$

For a series of Maxwell units all at strain e

$$\sigma(t) = e \sum^{n} E_n \exp\left(\frac{-t}{\tau_n}\right) \tag{4.15}$$

where E_n and τ_n refer to the nth Maxwell unit.

The summation can be written as an integral, giving

$$\sigma(t) = [G_r e] + e \int_0^\infty f(\tau)\exp\left(\frac{-t}{\tau}\right) d\tau \tag{4.16}$$

where the first term defines the eventual 'relaxed' stress, and the spring constant E_n is replaced by the weighting function $f(\tau)\,d\tau$, which defines the concentration of Maxwell units with relaxation times between τ and $(\tau + d\tau)$.

The stress relaxation modulus is then given by

$$G(t) = G_\mathrm{r} + \int_0^\infty f(\tau)\exp\left(\frac{-t}{\tau}\right)\mathrm{d}\tau \qquad (4.17)$$

As relaxation curves extend over many decades of time it is convenient to use a logarithmic time-scale, so that the stress relaxation modulus becomes

$$G(t) = G_\mathrm{r} + \int_{-\infty}^\infty H(\tau)\exp\left(\frac{-t}{\tau}\right)(\mathrm{d}\ln\tau) \qquad (4.18)$$

A new relaxation time spectra $H(\tau)$ is now defined where $H(\tau)\,\mathrm{d}(\ln\tau)$ gives the contribution to stress relaxation associated with relaxation times between $\ln\tau$ and $\ln\tau + \mathrm{d}(\ln\tau)$.

For creep under constant stress σ, the Kelvin model gives (from Equation (4.7))

$$J(t) = J\left[1 - \exp\left(\frac{-t}{\tau'}\right)\right]$$

Following a similar line of argument to that above, the creep compliance becomes

$$J(t) = J_\mathrm{u} + \int_{-\infty}^\infty L(\tau)\left[1 - \exp\left(\frac{-t}{\tau}\right)\right]\mathrm{d}(\ln\tau), \qquad (4.19)$$

where J_u is the instantaneous (unrelaxed) elastic compliance and $L(\tau)\,\mathrm{d}(\ln\tau)$, which gives the contributions to the creep compliance associated with retardation times between $\ln\tau$ and $\ln\tau + \mathrm{d}(\ln\tau)$, defines the retardation time spectrum $L(\tau)$. Note that when moving from a linear to logarithmic time-scale the lower limit of integration becomes minus infinity.

4.3 Dynamic mechanical measurements: the complex modulus and complex compliance

An alternative experimental procedure to creep and stress relaxation is to subject the specimen to an alternating strain and simultaneously measure the stress. For linear viscoelastic behaviour, when equilibrium is reached, the stress and strain will both vary sinusoidally, but the strain lags behind the stress. Thus we write

$$\text{strain } e = e_0 \sin\omega t$$

$$\text{stress } \sigma = \sigma_0 \sin(\omega t + \delta)$$

where ω is the angular frequency and δ the phase lag.

Expanding $\sigma = \sigma_0 \sin\omega t \cos\delta + \sigma_0 \cos\omega t \sin\delta$ we see that the stress can be considered to consist of two components: (1) of magnitude ($\sigma_0 \cos \delta$) in phase with the strain; (2) of magnitude ($\sigma_0 \sin \delta$) 90° out of phase with the strain.

The stress–strain relationship can therefore be defined by a quantity G_1 in phase with the strain and by a quantity G_2 90° out of phase with the strain, i.e.

$$\sigma = e_0 G_1 \sin \omega t + e_0 G_2 \cos \omega t \qquad (4.20)$$

where

$$G_1 = \frac{\sigma_0}{e_0} \cos \delta \quad \text{and} \quad G_2 = \frac{\sigma_0}{e_0} \sin \delta$$

A phasor diagram (Figure 4.15) then indicates that G_1 and G_2 define a complex modulus G^*. If $e = e_0 \exp(i\omega t)$, then $\sigma = \sigma_0 \exp[i(\omega t + \delta)]$, so that

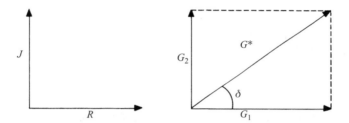

Figure 4.15 Phasor diagram for complex modulus $G^* = G_1 + iG_2$ and phase lag $\tan \delta = G_2/G_1$

$$G^* = \frac{\sigma}{e} = \frac{\sigma_0}{e_0} \exp(i\delta)$$

$$= \frac{\sigma_0}{e_0} (\cos \delta + i \sin \delta)$$

$$= G_1 + G_2 \qquad (4.21)$$

where G_1, which is in phase with the strain, is called the storage modulus because it defines the energy stored in the specimen due to the applied strain, and G_2, which is $\pi/2$ out of phase with the strain, defines the dissipation of energy that and is called the loss modulus, for a reason that becomes evident through calculating the energy (ΔE) dissipated per cycle:

$$\Delta E = \oint \sigma \, de = \int_0^{2\pi/\omega} \sigma \frac{de}{dt} \, dt.$$

Substituting for σ and e,

$$\Delta E = \omega e_0^2 \int_0^{2\pi/\omega} (G_1 \sin \omega t \cos \omega t + G_2 \cos^2 \omega t) \, dt \tag{4.22}$$

The integral is solved by using $\sin \omega t \cos \omega t = \frac{1}{2} \sin 2\omega t$ and $\cos^2 \omega t = \frac{1}{2}(1 + \cos 2\omega t)$, to give

$$\Delta E = \pi G_2 e_0^2 \tag{4.23}$$

If the integral for ΔE is evaluated over a quarter-cycle rather than over the complete period, the first term

$$G_1 \omega e_0^2 \int_0^{\pi/2\omega} \sin \omega t \cos \omega t \, dt \tag{4.24}$$

gives the maximum stored elastic energy (E).

Evaluating as before, we obtain

$$E = \tfrac{1}{2} G_1 e_0^2 \tag{4.25}$$

which, as expected, is independent of frequency. Equations (4.23) and (4.25) can be rewritten as

$$G_1 = \frac{2E}{e_0^2} \qquad G_2 = \frac{\Delta E}{\pi e_0^2}$$

Hence

$$\frac{G_2}{G_1} = \tan \delta = \frac{\Delta E}{2\pi E} \tag{4.26}$$

The ratio $\Delta E/E$ is called the specific loss

$$\frac{\Delta E}{E} = 2\pi \tan \delta \tag{4.27}$$

Typical values of G_1, G_2 and $\tan \delta$ for a polymer are 10^9 Pa, 10^7 Pa and 0.01, respectively. In such cases $|G^*|$ is approximately equal to G_1, and it is customary to define the dynamic mechanical behaviour in terms of the 'modulus' $G \approx G_1$ and the phase angle δ or $\tan \delta = G_2/G_1$.

A complementary treatment can be developed to define a complex compliance $J^* = J_1 - iJ_2$, which is directly related to the complex modulus, as $G^* = 1/J^*$.

4.3.1 Experimental patterns for G_1, G_2, etc. as a function of frequency

Consider the variation of G_1, G_2 and $\tan\delta$ with frequency for a viscoelastic solid that shows no flow (Figure 4.16). At very low frequencies the polymer is rubber-like and has a low modulus (G_1 probably about 10^5 Pa) that is independent of frequency. At the highest frequencies the rubber is glassy, with a modulus around 10^9 Pa, which is again independent of frequency. In the intermediate region, where the material behaves viscoelastically, the modulus will increase with increasing frequency. The loss modulus will be zero at low and high frequencies, where stress and strain are in phase for the rubbery and glassy states. In the intermediate viscoelastic region G_2 rises to a maximum value, close to the frequency at which G_1 is changing most rapidly with frequency. The loss factor $\tan\delta$ also has a maximum in the viscoelastic region, but this occurs at a slightly lower frequency than that in G_2, since $\tan\delta = G_2/G_1$, and G_1 is also changing rapidly in that frequency region.

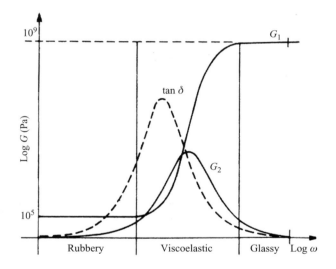

Figure 4.16 The complex modulus $G_1 + iG_2$ as a function of frequency ω

An analogous diagram (Figure 4.17) shows the variation of the compliances J_1 and J_2 with frequency.

4.3.2 The Alfrey approximation

Relaxation and retardation time spectra can be calculated exactly from stress relaxation, creep and dynamic mechanical measurements using Fourier or Laplace

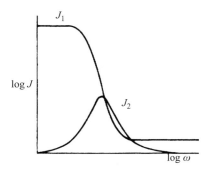

Figure 4.17   The complex compliance $J^* = J_1 - iJ_2$ as a function of frequency ω

transform methods. However, it is often adequate to use simple approximations due to Alfrey in which the exponential term for a single Kelvin or Maxwell unit is replaced by a step function, as shown schematically in Figure 4.18.

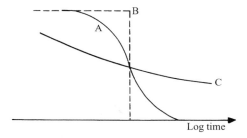

Figure 4.18   The Alfrey approximation: the stress relaxation of a Maxwell unit A is replaced by a step function B. The curve C represents relaxation of a typical viscoelastic polymer

Consider

$$G(t) = [G_r] + \int_{-\infty}^{\infty} H(\tau)\exp\left(\frac{-t}{\tau}\right)d(\ln \tau) \qquad (4.28)$$

If we assume that $e^{-t/\tau} = 0$ up to the time $\tau = t$, and $e^{-t/\tau} = 1$ for $\tau > t$, we can write

$$G(t) = [G_r] + \int_{\ln \tau}^{\infty} H(\tau)d(\ln \tau)$$

This gives the relaxation time spectrum

$$H(\tau) = \left[\frac{dG(t)}{d\ln t}\right]_{t=\tau} \tag{4.29}$$

which is known as the 'Alfrey approximation' [3].

The relaxation time spectrum can be expressed to a similar degree of approximation in terms of the real and imaginary parts of the complex modulus:

$$H(\tau) = \left[\frac{dG_1(\omega)}{d\ln\omega}\right]_{1/\omega=\tau} = \frac{2}{\pi}[G_2(\omega)]_{1/\omega=\tau} \tag{4.30}$$

These relationships are illustrated diagrammatically for the case of a single relaxation transition in Figure 4.19. To obtain the complete relaxation time spectrum the longer time part of $H(\tau)$ will be found from the stress relaxation

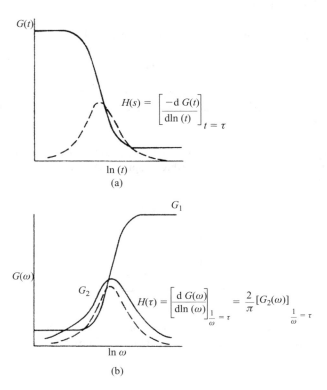

Figure 4.19 The Alfrey approximation for the relaxation time spectrum $H(\tau)$: (a) from the stress relaxation modulus $G(t)$; (b) from the real and imaginary parts G_1 and G_2, respectively, of the complex modulus $G(\omega)$

modulus data of Figure 4.19(a), and the shorter time part from the dynamic mechanical data of Figure 4.19(b).

Figure 4.19 shows that the relaxation time spectrum can be determined directly from the gradient of plots of the relaxation modulus or the dynamic modulus G_1 versus logarithm of time, or from G_2 even more directly.

Complementary relationships can be used to obtain the retardation time spectrum in terms of the complex compliances and the creep compliance.

References

1. Boltzmann, L., *Pogg. Ann. Phys. Chem.*, **7**, 624 (1876).
2. Zener, C., *Elasticity and Anelasticity of Metals*, Chicago University Press, Chicago, 1948.
3. Alfrey, T., *Mechanical Behaviour of High Polymers*, Interscience, New York, 1948.

Problems for Chapter 4

1.   The stress relaxation modulus of a certain polymer can be described approximately by

$$G(t) = G_0 e^{-t/\tau}$$

and has the values 2.0 GPa and 1.0 GPa at $t = 0$ and 10^4 s, respectively. Calculate the form of the creep compliance and so evaluate the strain 1000 s after the rapid application of a stress of 100 MPa.

2.   The creep deformation of a linear viscoelastic solid under constant stress can be represented by the model shown, in which a spring is in series with a Kelvin unit.

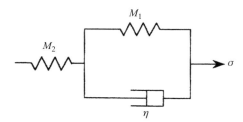

Here M_1 and M_2 represent Young's moduli, η the viscosity and σ the total constant stress. Show that the total strain is dependent on the relaxation time, $\tau_2 = \eta/M_1$, and is given by

$$\varepsilon = \frac{\sigma}{M_2} + \frac{\sigma}{M_1}[1 - \exp(-t/\tau_2)]$$

Immediately after applying the stress the strain is 0.002; after 1000 s the strain is

0.004; after a very long time the strain tends to 0.006. What is the retardation time τ_2?
NB It is not necessary to know the stress or the values of M_1, M_2 and η.

3. A strip of a linear viscoelastic polymer 200 mm long, 10 mm wide and 1 mm thick, and with an extensional Young's modulus of 2 GPa, is mounted in a dynamic testing apparatus. The specimen is initially extended by 1 mm, and then subjected to a sinusoidally varying strain with an amplitude of ± 1 mm.

    At 20 °C and 5 Hz the phase lag between stress and strain is 0.1 rad. Calculate the maximum stress developed, the elastic energy stored during the positive quarter-cycle and the work dissipated per cycle.

    Another specimen of the same polymer is tested in a simple torsion pendulum at 20 °C. The period of vibration is 2 s and the logarithmic decrement is 0.20. What would you expect the phase lag in a dynamic tester to be at 20 °C and 0.5 Hz? Comment on the result.

4. A long fibre is loaded by a mass of 0.1 kg attached to its lower end. The extension (%) is measured at various times after loading

Ext (%)	t (min)
0.300	0
0.328	10
0.350	20
0.390	40
0.428	60
0.462	80
0.490	100
0.514	120
0.535	140
0.555	160
0.572	180
0.585	200
0.593	220
0.600	240

    (The reading at $t = 0$ gives the immediate elastic extension.)

    Assuming the behaviour to be linear viscoelastic, calculate the extension under the following conditions:

    (i) Load with 0.1 kg at $t = 0$; remove load at $t = 40$; reload at $t = 80$; remove load at $t = 120$. Calculate the net residual extension at $t = 240$ min.

    (ii) Load with 0.1 kg at $t = 0$; add a further 0.2 kg at $t = 40$; unload completely at $t = 200$. What is the extension at $t = 80$; what is the immediate elastic recovery; what is the net residual extension at $t = 240$? NB A graphical solution is not necessary.

5. It is found that the stress relaxation behaviour of a certain polymer can be represented by a Maxwell model, where an elastic spring of modulus 10^8 Pa is in series with a viscous dashpot of viscosity 10^{10} Pa. State what is meant by the relaxation time and calculate its value for this polymer. If a strain of 1 per cent is applied at time $t = 0$, followed by a further additional strain of 2 per cent at time $t = 25$ s, calculate the stress at time $t = 50$ s.

6. For a polymer whose creep behaviour can be described by the Kelvin model, the creep compliance $J(t) = J_0 \, (1 - e^{-t/\tau'})$. If this polymer is subjected to an instantaneous stress σ_1 at time $t = 0$, which is increased to a value σ_2 at time $t = t_1$, find the strain at a time $t > t_1$.

7. A Maxwell element consists of an elastic spring of modulus $E_m = 10^9$ Pa and a dashpot of viscosity 10^{11} Pa. Calculate the stress at time $t = 100$ s in the following loading programme:

   (i)   At time $t = 0$ an instantaneous strain of 1 per cent is applied.

   (ii)   At time $t = 30$ s the strain is increased instantaneously from 1 per cent to 2 per cent.

8. The stress relaxation behaviour of a certain polymer can be represented by a Maxwell model of a spring and dashpot in series. The spring has the properties of an ideal rubber with a room temperature (293 K) modulus of 10^6 Pa. The dashpot has a room temperature viscosity η_{293} of 10^8 Pa, with a temperature dependence such that the viscosity at a temperature T is given by

$$\eta_T = \eta_{293} \exp\left(\frac{-\Delta H}{RT}\right)$$

where $\Delta H = 4 \, \text{kJ mol}^{-1}$. Calculate the stress relaxation time of the polymer at 100 °C. You may neglect the change in the density of the rubber with temperature. (The gas constant R may be taken as $8.3 \, \text{J K}^{-1} \, \text{mol}^{-1}$.)

5

The Measurement of Viscoelastic Behaviour

For a satisfactory understanding of the viscoelastic behaviour of polymers data are required over a wide range of frequency (or time) and temperature. The number of experiments required can sometimes be reduced by using either the equivalence of creep, stress relaxation and dynamic mechanical data (described in Chapter 4) or the equivalence of time and temperature as variables (to be discussed in Chapter 6). Nevertheless a variety of techniques need to be combined to cover a wide range of both time and temperature.

There are five main classes of experiment, which will be discussed in turn:

1. Transient measurements: creep and stress relaxation

2. Low-frequency vibrations: free oscillation methods

3. High-frequency vibrations: resonance methods

4. Forced vibration non-resonance methods

5. Wave propagation methods.

The approximate frequency scale for each technique is indicated in Figure 5.1.

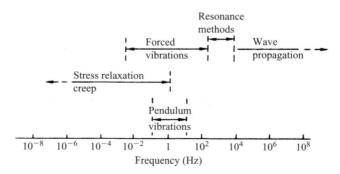

Figure 5.1 Approximate frequency scales for different experimental techniques. (Reproduced with permission from Becker, *Mater. Plast. Elast.*, **35**, 1387 (1969))

An Introduction to the Mechanical Properties of Solid Polymers I. M. Ward and J. Sweeney
© 2004 John Wiley & Sons, Ltd   ISBN: 0471 49625 1 (HB); 0471 49626 X (PB)

5.1 Creep and stress relaxation

Reliable creep and stress relaxation data are obtainable only if the specimens are well defined and strictly comparable. As deformations and deformation rates are usually quite small, if linearity is to hold then precision measurements are required: conditions that may be difficult to attain throughout the highly significant short-time regime.

5.1.1 Creep conditioning

Leaderman [1] was the first worker to emphasize that specimens must be cyclically conditioned at the highest temperature of measurement in order to obtain reproducible measurements in creep and recovery. Each cycle consists of application of the maximum load for the maximum period of loading, followed by a recovery period after unloading of about 10 times the loading period; cycling must be continued until reproducibility is obtained.

The conditioning procedure has two major effects on the creep and recovery behaviour. First, subsequent creep and recovery responses under a given load are identical, i.e. the sample has lost its 'long-term' memory and now only remembers loads applied in its immediate past history. Secondly, after the conditioning procedure the deformation produced by any loading programme is almost completely recoverable provided that the recovery period is about 10 times the period during which loads are applied. For tensile creep measurements over a wide range of temperature greater elaboration is required.

5.1.2 Specimen characterization

Many early experiments on the viscoelastic behaviour of polymers were unsatisfactory because the specimens were inadequately characterized, so that 'like' was not compared with 'like'.

Average molecular mass and its distribution are both critical parameters, and all polymers contain processing and stabilizing additives, which can sometimes have a significant effect on the response to stress and strain, particularly at temperatures well above that of the laboratory.

The physical structure of the polymer, in terms of morphology, crystallinity and molecular orientation, will also be important and should be well characterized.

5.1.3 Experimental precautions

Measurements made in the vicinity of a major relaxation region, for instance creep tests on isotactic polypropylene close to room temperature, are sensitive to small changes in temperature, so that a controlled temperature environment is essential. For some polymers, such as nylon, it is also essential to control the humidity, because the presence of moisture in the polymer has a dramatic effect on the mechanical behaviour [2] (in nylon by reducing the effect of interchain hydrogen bonding). Whenever possible, several nominally identical specimens should be examined, to confirm that inter-specimen variability is small compared with the effects under examination.

In the linear viscoelastic region strains are unlikely to exceed 1 per cent, so that the change of cross-section, and hence stress, with strain will be small. At larger strains the effective load should be reduced in a manner proportional to the decrease in cross-sectional area, to maintain a constant stress. Leaderman [3] used the device illustrated in Figure 5.2, in which a flexible tape (B) attached to the specimen (A) is wound round the periphery of a cylindrical drum C. A similar tape F, attached to a profiled cam D, supports a fixed mass E. As the specimen extends under load, the moment of E decreases according to the cam profile.

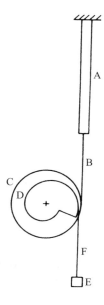

Figure 5.2 Cam arrangement for creep under constant stress. (Reproduced with permission from Leaderman, *Tran. Soc. Rheol.*, **6**, 361 (1962))

The creep and relaxation plots in Figures 4.2 and 4.4 are idealized because, as already mentioned, the immediate response is no more than that occurring before the first measurement after loading and it is evident that the relevant time interval should be no longer than necessary. For instance, if the load is applied very slowly during a stress relaxation experiment, a low value of maximum stress will be obtained (Figure 5.3). Conversely, too rapid a stress application will result in the complications of dynamic loading (Figure 5.4). This point is of particular importance, because some attempts to relate viscoelasticity to molecular parameters are particularly dependent on the short-time response. As explained in Section 6.2 below, some of the difficulties concerned with the early stages of deformation can be removed by employing time–temperature equivalence.

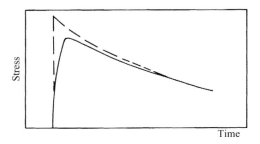

Figure 5.3   Effect on stress relaxation of the strain being applied slowly: (– – –) ideal; (—) actual

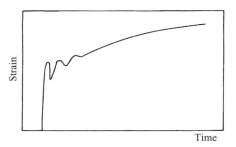

Figure 5.4   Damped vibrations resulting from rapid loading in a creep experiment

Specimens in extensional and torsional tests must be firmly clamped at their ends. However, the stresses in the clamp region will differ greatly from those in the bulk of the specimen and, if the complete length of the specimen is measured, end effects can be ignored only where the length is at least 10 times the diameter. For oriented samples, end effects are even more important. As an approximate guide it is reasonable to consider that the ratio of length to diameter should be greater than $10 \sqrt{E/G}$, where E and G are the Young's modulus (in the fibre direction) and the torsional modulus, respectively. A more satisfactory technique for high accuracy with robust specimens is to use an extensometer attached to the specimen away from its ends, strain being converted into an electrical signal by a displacement transducer [4]. For less robust specimens, non-contacting laser methods are available [5].

In stress relaxation measurements, as in standard mechanical testing devices, changing stress may be monitored using a strain gauge load cell. Since these devices rely on changes in strain, it must be confirmed that they are very stiff in comparison with the specimen so that the specimen strain is held effectively constant.

A range of measurement equipment is described and illustrated in the books by Turner [6] and Ward [7].

Where materials are being examined for their suitability for specific applications it is essential that creep measurements are performed over an extended time-scale. It is possible that a material that shows good short-term creep shows accelerated creep at longer times (Figure 5.5). As discussed later, comparable problems may occur for materials that are viscoelastically non-linear, so that the recovery response is very different from that during the early stages of creep.

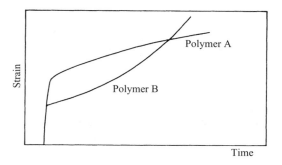

Figure 5.5 Short-term creep behaviour at a single temperature implies nothing about the eventual deformation

5.2 Dynamic mechanical measurements

A variety of techniques are required, capable of covering wide ranges of both time and temperature (Figure 5.1). Free oscillation pendulum methods, which have the advantage of simplicity, are confined to the frequency range $10^{-1}-10$ Hz. Forced vibration techniques, although more complicated, may yield higher reproducibility and can extend the frequency range by a further decade on either side, linking up with creep and stress relaxation at the lowest frequencies and resonance methods at the higher end. The latter, which are very sensitive to inter-specimen variability, are important above 10 kHz.

5.2.1   The torsion pendulum

The simplest device consists of a specimen of circular cross-section suspended vertically with its upper end rigidly clamped [8]. Its lower end supports a disc, or preferably a bar, fitted with adjustable weights (Figure 5.6), whose distance from the axis can be altered, thus changing the moment of inertia (I) and the period of oscillation. When the bar is twisted and released the oscillations gradually decrease in amplitude, and the logarithmic decrement Λ, the natural logarithm of the ratio of amplitude of successive oscillations, is recorded.

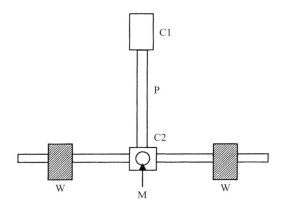

Figure 5.6   Free vibration torsion pendulum: P, polymer specimen; C1, fixed upper clamp; C2, lower clamp fixed to inertia bar; W, sliding masses to change moment of inertia; M, mirror to reflect light beam

Because the specimen is loaded by the inertia bar, the specimen is subjected to tensile as well as torsional stresses, which perturb the nominally free vibrations. For more precise work the specimen can be mounted as in Figure 5.7, with the inertia bar clamped at its upper end. The assembly is then suspended by an elastic wire or ribbon, which has a negligible effect on damping.

For an *elastic* rod the equation of motion is $I\ddot{\theta} + \tau\theta = 0$, where τ is the torsional rigidity of the rod, which is related to the shear modulus G through

$$\tau = \frac{\pi r^4 G}{2l}$$

where l is the length and r the radius of the rod. The elastic system executes simple harmonic motion with an angular frequency

$$\omega = \sqrt{\frac{\tau}{I}} = \sqrt{\frac{\pi r^4 G}{2lI}} \tag{5.1}$$

The high sensitivity to sample dimensions implies that inter-specimen comparisons can be subject to large uncertainties.

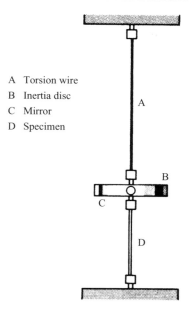

A Torsion wire
B Inertia disc
C Mirror
D Specimen

Figure 5.7 Apparatus for measuring torsional rigidity at low frequencies

When the vibrations are damped the amplitude decreases with time, but with light damping there is only a small effect on the period

$$\left(\frac{2\pi}{\omega}\right)$$

For linear viscoelastic solids the torsional modulus is complex, and may be written as $G^* = G_1 + iG_2$. When the damping is small, it is justified to replace G_1 for G in Equation (5.1), hence

$$\omega^2 = \frac{\pi r^4 G_1}{2lI} \tag{5.2}$$

The logarithmic decrement can then be related to the specific loss (Equation (4.33)) and hence tan δ.

$$\Lambda = \ln\left(\frac{A_n}{A_{n+1}}\right)$$

where A_n denotes the amplitude of the nth oscillation. For small damping

$$\Lambda = \ln\left(1 + \frac{\Delta A}{A_n}\right) = \frac{\Delta A}{A_n} - \frac{1}{2}\frac{\Delta A^2}{A_n^2}$$

Hence

$$\Lambda = \frac{1}{2}\left(\frac{A_n^2 - A_{n+1}^2}{A_n^2}\right)$$

But A^2 is proportional to stored energy, giving

$$\Lambda = \frac{1}{2}\frac{\Delta E}{E} = \pi\tan\delta \qquad (5.3)$$

from which $G_2 = G_1 \tan\delta$ can be obtained.

5.2.2 Forced vibration methods

Free vibration methods suffer from the disadvantage that the frequency of vibration depends on the stiffness of the specimen, which varies with temperature, so that forced vibration methods are to be preferred when the frequency and temperature dependence of viscoelastic behaviour are to be investigated.

As indicated in Section 4 above, when a sinusoidal strain is imposed on a linear viscoelastic body the strain lags behind the stress by the phase angle δ, which determines the degree of damping. The strains must be low enough for linearity to apply, and the strain must at all times remain positive. In practice the strain amplitude is typically ± 0.5 per cent, superposed on an initial extension slightly in excess of 0.5 per cent, to allow for some degree of stretch during the experiment. The specimen must be short enough for there to be no appreciable variation of stress along its length, i.e. the length must be short compared with the wavelength of the stress waves. Assuming that the lowest value of the modulus is 10^7 Pa for a specimen of density 10^3 kg m^{-3}, the longitudinal wave velocity is 10^2 ms^{-1}. At 100 Hz the wavelength of the stress waves is 1 m, which suggests that at that frequency the upper limit on specimen length is about 0.1 m. As the stress must never vanish, a lower limit to frequency is set by the stress relaxation time.

Typically strain and stress are measured by unbonded strain-gauge transducers, the signals from which are then fed to a phase meter, which provides a direct reading of the relative amplitudes and the phase difference, hence giving values of the modulus and $\tan\delta$ [9].

5.2.3 Dynamic mechanical thermal analysis (DMTA)

There are a number of commercial machines available for dynamic testing under varying temperature conditions. An actuator imposes an oscillatory (linear or angular) displacement, and typically a strain-gauge load cell measures force. Testing is frequently in bending mode, producing results that may be difficult to

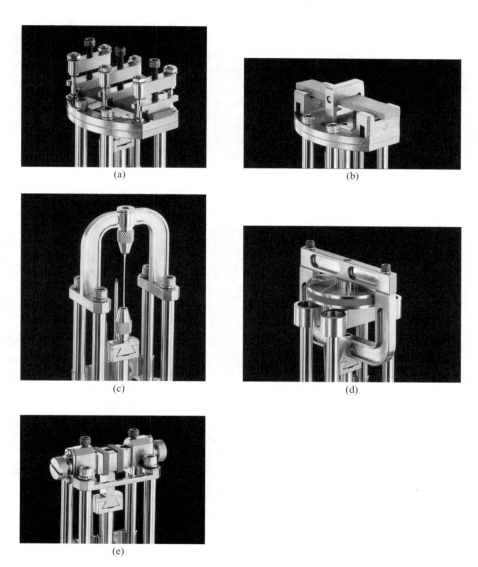

Figure 5.8 DMTA testing heads for various modes: (a) dual cantilever; (b) three-point bend; (c) tension; (d) compression; (e) shear sandwich (showing two specimens). Photographs by TA Instruments, Delaware USA

interpret in terms of viscoelasticity owing to the non-uniform nature of the stress field. However, such testing is useful in locating the temperatures of relaxation transitions (see Chapter 9). Temperature ranges are typically -150 to $600\,°C$, and frequency ranges are 10^{-6} to $200\,Hz$. In some cases a single machine can operate in a range of testing modes – bend, tension, compression, shear and torsion – by the use of different loading fixtures. Modulus and $\tan \delta$ data (see Chapter 4) are routinely derived using the manufacturer's proprietary software. Recent examples are due to: Damman and Buijs [10], investigating liquid-crystal polymer in tension; Beaudoir *et al.* [11], examining the properties of a polybutylene terephthalate composite in compression-tension; and Zhou and Chudnovsky [12], comparing drawn and undrawn polycarbonate in torsion. A commercial DMTA set-up is shown in Figure 5.8.

5.3 Wave-propagation methods

Wave-propagation methods are in three broad categories:

1. In the kilohertz frequency range.

2. In the megahertz frequency range: ultrasonic methods.

3. In the gigahertz frequency range: Brillouin spectroscopy.

5.3.1 The kilohertz frequency range

At frequencies in the kilohertz range the wavelength of the stress waves is of the order of the length of a viscoelastic specimen. Typically [13], a thin monofilament is stretched longitudinally, with one end attached to a stiff massive diaphragm, such as a loudspeaker. A piezoelectric crystal pickup then detects the changes in signal amplitude and phase along the length of the specimen. As shown in Figure 5.9 for low-density polyethylene, a plot of the phase (θ) against distance (l) takes the form of damped oscillations about the line $\theta = kl$, where k is the propagation constant.

Where the attenuation coefficient α is small it has been shown [14] that $V_{max}/V_{min} = (\alpha l + \beta)$, where V is the signal amplitude. A plot of

$$\tanh^{-1}\left(\frac{V_{max}}{V_{min}}\right)$$

against l then gives a line of slope α.

It is possible to relate α and k to the storage and loss moduli E_1 and E_2, and to $\tan \delta$. For a filament of density ρ, in which c is the longitudinal wave velocity

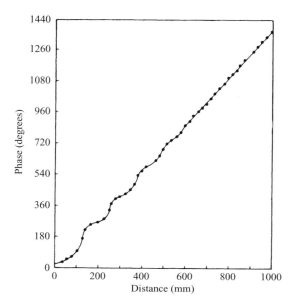

Figure 5.9 The variation of phase angle with distance along a polyethylene monofilament for transmission of sound waves at 3000 Hz. (Reproduced with permission from Hillier and Kolsky, *Proc. Phys. Soc. B*, **62**, 111 (1949))

$$E_1 = \frac{\omega^2}{k^2}\rho \qquad E_2 = \frac{2ac^3\rho}{\omega}$$

and

$$\tan\delta = \frac{2ac}{\omega}$$

For further information the reader is referred to review articles by Kolsky [15].

5.3.2 The megahertz frequency range: ultrasonic methods

Measurements of the velocity and attenuation of elastic waves at ultrasonic frequencies are important, especially for oriented polymers and composites. Compact solid specimens with dimensions of the order of 10 mm are required.

In a typical application of this technique, Chan *et al.* [16] measured the elastic constants of a uniaxially oriented rod, 12 mm in diameter, by cutting discs of thickness 4–8 mm parallel, perpendicular and at 45° to the axis of the rod (Figure 5.10). Quartz transducers were bonded to the discs, so that longitudinal and

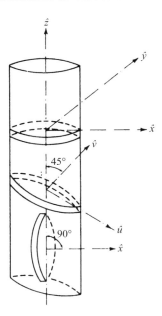

Figure 5.10   Schematic diagram illustrating the sample discs employed in ultrasonic measurements. (Reproduced with permission from Chan *et al.*, *J. Phys. D*, **11**, 481 (1975))

transverse waves were propagated along the geometrical axes of each disc. In principle nine different velocities v_{ab} can be measured, where a refers to the direction of polarization and b to the direction of wave propagation. For a specimen of density ρ we can then define $Q_{ab} = \rho v_{ab}^2$, where Q_{ab} is either an elastic stiffness constant or a linear combination of such constants. Velocities were measured using the pulse echo-overlap technique [17], and tan δ was obtained by making attenuation measurements [18].

An alternative approach [19, 20] is to immerse a specimen, thickness d, in a water-filled tank fitted with both a transmitter and a receiver of ultrasonic waves, and measure the change (τ) in transit time with and without the specimen in the beam. If v is the velocity in the polymer and v_w that in the water we have

$$\frac{1}{v} = \frac{1}{v_w} - \frac{\tau}{d}$$ (5.6)

The various wave velocities, which can be derived from measurements made over a range of incident angles, are related to the elastic stiffness constants.

In a variation of the above method Wright *et al.* [21] detected the component of the incident beam that was reflected from the immersed specimen and hence measured the critical angle of incidence.

More recently this technique has been developed to measure the anisotropy of uniaxial composites [22]. A specimen of uniform thickness, placed between the transmitting and receiving heads in a water-filled container, could be rotated about a vertical axis to change the angle of incidence and hence the direction of the beam in the sample (Figure 5.11). The velocity V and angle of refraction r of the wave are then calculated following the method of Markham [19]. Let X_1 and X_2 axes define the isotropic plane perpendicular to the fibre axis. It can then be shown [23] that for quasi-tensile waves propagating in the X_1–X_3 plane at an angle r to the X_1 axis, with the specimen axis X_2 vertical,

$$V_t^2 = \frac{B_{11} + B_{33} + [(B_{33} - B_{11})^2 + 4B^2{}_{13}]^{1/2}}{2\rho}$$

and for quasi-shear waves

$$V_s^2 = \frac{B_{11} + B_{33} - [(B_{33} + B_{11})^2 + 4B^2{}_{13}]^{1/2}}{2\rho}$$

where ρ is the specific gravity of the specimen.

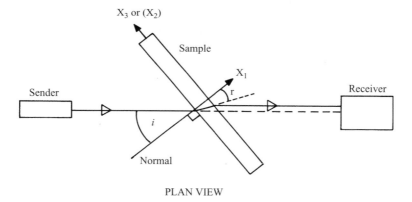

Figure 5.11 Schematic diagram showing layout of ultrasonic apparatus used for measurement of elastic constants. (From Dyer *et al.*, *J. Phys. D* (1992))

The elastic stiffness constants C_{ij} are obtained from

$$B_{11} = C_{11} \cos^2 r + C_{44} \sin^2 r$$

$$B_{33} = C_{33} \sin^2 r + C_{44} \cos^2 r$$

$$B_{13} = (C_{44} + C_{13}) \sin r \cos r$$

5.3.3 The hypersonic frequency range: Brillouin spectroscopy

Brillouin spectroscopy enables the elastic constants of polymers to be determined at frequencies of several gigahertz, i.e. three orders of magnitude higher than those pertaining to ultrasonic measurements, which are known as hypersonic frequencies.

The principle of the method is to use Fabry-Perot spectroscopy to measure the frequency shift in laser light scattered through 90° after passage through a parallel polymer sheet.

The hypersonic velocity V_s is obtained from the equation

$$V_s = \frac{f_B \lambda_i}{\sqrt{2}}$$

where λ_i is the laser wavelength and f_B is the measured Brillouin shift.

For an isotropic polymer the hypersonic sound velocity is determined as a function of direction, and the elastic constants are obtained by fitting the data to a set of equations known as the Christoffel equations, which essentially relate the values of V_s to the stiffness constants C_s through $C_s = \rho V_s^2$, where ρ is the density [24]. For a detailed discussion of this technique the reader is referred to papers by Kruger, Pietralla and co-workers [25, 26].

References

1. Leaderman, H., *Elastic and Creep Properties of Filamentous Materials and Other High Polymers*, Textile Foundation, Washington, DC, 1943.
2. Hadley, D. W., Pinnock, P. R. and Ward, I. M., *J. Mater. Sci.*, **4**, 152 (1969).
3. Leaderman, H., *Trans. Soc. Rheol.*, **6**, 361 (1962).
4. Ward, I. M., *Polymer*, **5**, 59 (1964).
5. Spathis, G. and Kontou, E., *J. Appl. Polym. Sci.*, **71**, 2007, (1999).
6. Turner, S., *Mechanical Testing of Plastic* (2nd edn), G. Godwin, Harlow, 1983.
7. Ward, I. M., *Mechanical Properties of Solid Polymers*, Wiley, Chichester, 1983.
8. Schmeider, K. and Wolf, K., *Kolloidzeitschrift*, **127**, 65 (1952).
9. Pinnock, P. R. and Ward, I. M., *Proc. Phys. Soc.*, **81**, 261 (1963); Takayanagi, M., in *Proceedings of Fourth International Congress on Rheology*, Part 1, Interscience, New York, 1965, p. 161; Becker, G. W., *Mat. Plast. Elast.*, **35**, 1387 (1969).
10. Damman, S. B. and Buijs J. A. H. M., *Polymer*, **35**, 2359 (1994).
11. Beaudoin, O., Bergeret, A., Quantin, J. C., *et al.*, *Compos. Interfaces*, **5**, 543 (1998).
12. Zhou, Z. and Chudnovsky, A., *Polym. Eng. Sci.*, **35**, 304 (1995).
13. Hillier, K. W. and Kolsky, H., *Proc. Phys. Soc. B*, **62**, 111 (1949); Kolsky, H., *Structural Mechanics*, Pergamon Press, Oxford, 1960.
14. Hillier, K. W., in *Progress in Solid Mechanics* (eds I. N. Sneddon and R. Hill), North-Holland, Amsterdam, 1961, pp. 199–243.

15. Kolsky, H., *Appl. Mech. Rev.*, **11**, 9 (1958); *Structural Mechanics*, Pergamon Press, Oxford, 1960.

16. Chan, O. K., Chen, F. C., Choy, C. L., *et al.*, *J. Phys. D*, **11**, 481 (1975).

17. Papadakis, E. P., *J. Appl. Phys.*, **35**, 1474 (1964); Kwan, S. F., Chen, F. C. and Choy, C. L., *Polymer*, **16**, 481 (1975).

18. Roderick, R. L. and Truell, R., *J. Appl. Phys.*, **23**, 267 (1952).

19. Markham, M. F., *Composites*, **1**, 145 (1970).

20. Rawson, F. F. and Rider, J. G., *J. Phys. D*, **7**, 41 (1974).

21. Wright, H., Faraday, C. S. N., White, E. F. T., *et al.*, *J. Phys. D*, **4**, 2002 (1971).

22. Dyer, S. R. A., Lord, D., Hutchinson, I. J., *et al.*, *J. Phys. D*, **25**, 66 (1992).

23. Musgrove, M. J. P., *Proc. R. Soc.*, **A226**, 339 (1954).

24. Auld, B. A., *Acoustic Fields and Waves in Solids*. Wiley, New York, 1973, p. 211.

25. Kruger, J. K., Marx, A., Peetz, L., *et al.*, *Colloid Polym. Sci.*, **264**, 403 (1986).

26. Krbecek, H., Kruger, J. K. and Pietralla, M., *J. Polym. Sci. Phys. Ed.*, **37**, 1477 (1993).

6

Experimental Studies of Linear Viscoelastic Behaviour as a Function of Frequency and Temperature: Time–Temperature Equivalence

6.1 General introduction

An introduction to the extensive experimental studies of linear viscoelastic behaviour in polymers falls conveniently into three parts, in which amorphous polymers, crystalline polymers and temperature dependence are discussed in turn.

6.1.1 Amorphous polymers

Many of the earlier investigations of linear viscoelastic behaviour in polymers were confined to amorphous polymers. During the early 1950s R. S. Marvin [1, 2] of the National Bureau of Standards, Washington, DC, coordinated the assembly of data from many laboratories who had measured the complex shear modulus and complex shear compliance of a specimen of polyisobutylene $(CH_2—CCH_3CH_3)_n$ of high relative molecular mass over a wide range of frequencies. The results, redrawn in Figure 6.1, show clearly the four regions, i.e. the glassy, the viscoelastic, rubbery and flow regions that are characteristic for amorphous high polymers. At high frequencies the complex modulus has a value around 10^9 Pa. For material of high molecular mass a plateau in modulus occurs, and appreciable molecular flow occurs only at frequencies below 10^{-5} Hz (i.e. period >1 day).

The reader may use the *Alfrey* approximation (see Section 4.3.2) to derive relaxation and retardation time spectra from the data of Figure 6.1. These spectra can be approximated by a 'wedge and box' distribution [3], shown by the dotted lines in Figure 6.2.

The observed plateau in the rubbery region is a consequence of high molecular

An Introduction to the Mechanical Properties of Solid Polymers I. M. Ward and J. Sweeney
© 2004 John Wiley & Sons, Ltd   ISBN: 0471 49625 1 (HB); 0471 49626 X (PB)

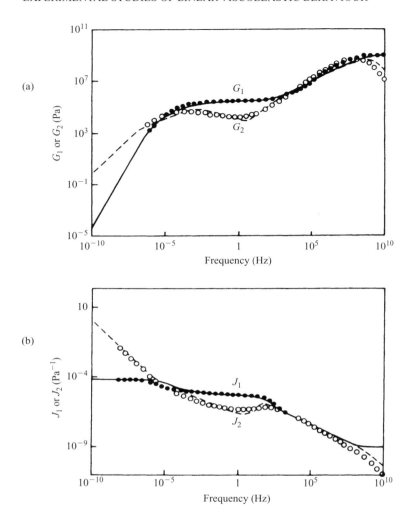

Figure 6.1 Complex shear modulus (a) and complex shear compliance (b) for 'standard' polyisobutylene reduced to 25 °C. Points from averaged experimental measurements; curves from a theoretical model for viscoelastic behaviour. (Reproduced with permission from Marvin and Oser, *J. Res. Natl. Bur. Stand. B*, **66**, 171 (1962))

mass, as the long molecules tend to entangle, with the formation of physical cross-links that restrict molecular flow through the formation of temporary networks. At long times such physical entanglements are usually labile and lead to some irreversible flow, in contrast with the situation for permanent chemical cross-links, such as those introduced when rubber is vulcanized. It follows directly from the theory of rubber elasticity (Chapter 3) that the value of the modulus in the

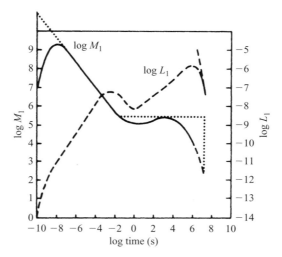

Figure 6.2 Approximate distribution functions of relaxation (M_1) and retardation (L_1) times for polyisobutylene. (Reproduced with permission from Marvin, *Proceedings of the 2nd International Congress of Rheology*, Butterworths, London 1954)

rubber-like plateau region is directly related to the number of effective cross-links per unit volume.

The influence of molecular entanglements is illustrated by Figure 6.3, which shows the stress relaxation behaviour for two samples of polymethyl methacrylate. It is seen that the lower molecular mass sample does not show a rubbery plateau

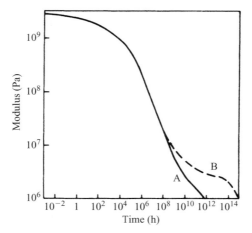

Figure 6.3 Master stress-relaxation curves for low molecular mass (molecular mass 1.5×10^5 daltons, curve A) and high molecular mass (molecular mass 3.6×10^6 daltons, curve B) polymethyl methacrylate. (Reproduced with permission from McLoughlin and Tobolsky, *J. Colloid Sci.*, **7**, 555 (1952))

region of modulus but passes directly from the viscoelastic region to the region of permanent flow.

6.1.2   Temperature dependence of viscoelastic behaviour

Previously we have referred only indirectly to the effect of temperature on viscoelastic behaviour. From a practical viewpoint, however, the temperature dependence of polymer properties is of paramount importance because plastics and rubbers show very large changes in properties with changing temperature.

In purely scientific terms, the temperature dependence has two primary points of interest. In the first place, as we have seen in Chapter 5, it is not possible to obtain from a single experimental technique a complete range of measuring frequencies to evaluate the relaxation spectrum at a single temperature. It is therefore a matter of considerable experimental convenience to change the temperature of the experiment, and so bring the relaxation processes of interest within a time-scale that is readily available. This procedure, of course, assumes that a simple interrelation exists between time-scale and temperature, and we will discuss shortly the extent to which this assumption is justified.

Secondly, there is the question of obtaining a molecular interpretation of the viscoelastic behaviour. In most general terms polymers change from glass-like to rubber-like behaviour as either the temperature is raised or the time-scale of the experiment is increased. In the glassy state at low temperatures we would expect the stiffness to relate to changes in the stored elastic energy on deformation that are associated with small displacements of the molecules from their equilibrium positions. In the rubbery state at high temperatures, on the other hand, the molecular chains have considerable flexibility, so that in the undeformed state they can adopt conformations that lead to maximum entropy (or, more strictly, minimum free energy). The rubber-like elastic deformations are then associated with changes in the molecular conformations.

The molecular physicist is interested in understanding how this conformational freedom is achieved in terms of molecular motions, for example to establish which bonds in the structure become able to rotate as the temperature is raised. One approach, which has proved successful to some degree, has been to compare the viscoelastic behaviour with dielectric relaxation behaviour and more particularly with nuclear magnetic resonance behaviour.

We have tacitly assumed that there is only one viscoelastic transition, corresponding to the change from the glassy low-temperature state to the rubbery state. In practice there are several relaxation transitions. For a typical amorphous polymer the situation is summarized in Figure 6.4. At low temperatures there are usually several secondary transitions involving comparatively small changes in modulus. These transitions are attributable to such features as side-group motions, e.g. methyl ($—CH_3$) groups in polypropylene

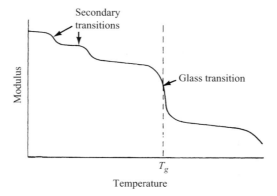

Figure 6.4   Temperature dependence of modulus in a typical polymer

$$\left[\begin{array}{c} CH_3 \\ | \\ CH_2{-}CH{-} \end{array}\right]_n$$

In addition, there is one primary transition, usually called the 'glass transition',

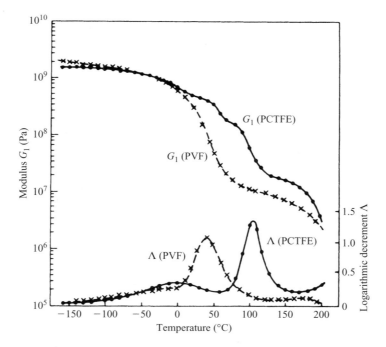

Figure 6.5   Shear modulus G_1 and logarithmic decrement Λ of polychlorotrifluoroethylene (PCTFE) and polyvinyl fluoride (PVF) as a function of temperature at ~ 3 Hz. (Reproduced with permission from Schmieder and Wolf, *Kolloidzeitschrift*, **134**, 149 (1953))

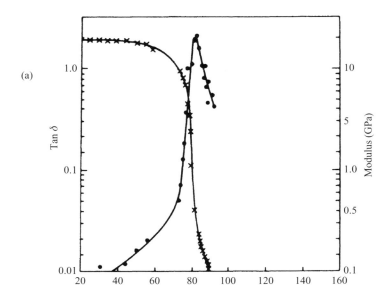

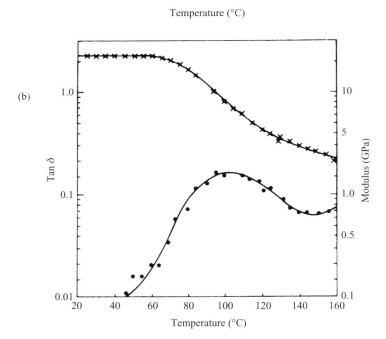

Figure 6.6 Tensile modulus and loss factor $\tan \delta$ for unoriented amorphous polyethylene terephthalate (a) and unoriented crystalline polyethylene terephthalate (b) as a function of temperature at ~ 1.2 Hz: (x) modulus; ($\bullet$) $\tan \delta$. (Reproduced with permission from Thompson and Woods, *Trans. Faraday Soc.*, **52**, 1383 (1956))

that involves a large change in modulus. The temperature at which it occurs is commonly denoted by T_g.

6.1.3   Crystallinity and inclusions

Although the viscoelastic behaviour of semicrystalline polymers gives some indication of the four characteristic regions that can be identified for amorphous polymers, they are much less clearly defined, as is illustrated in Figure 6.5, which shows data for polychlorotrifluoroethylene $(CClF\text{—}CF_2)_n$ and polyvinyl fluoride $(CH_2CHF)_n$ obtained by Schmeider and Wolf [4]. The fall in modulus over the glass transition region for semicrystalline materials is, at between one and two orders of magnitude, much less than for amorphous polymers, and the change in modulus or loss factor with temperature or frequency is much more gradual, indicating a broader relaxation time spectrum. At high temperatures (or low frequencies) molecular mobility is severely curtailed by the crystalline regions, so it is no longer correct to regard the polymer as rubber-like. These differences are clearly illustrated by the data of Thompson and Woods [5] for polyethylene terephthalate

$$\left[-O\text{—}CH_2\text{—}CH_2\text{—}O\text{—}\overset{\displaystyle O}{\underset{\displaystyle \|}{C}}\text{—}\hspace{-4pt}\bigcirc\hspace{-4pt}\text{—}\overset{\displaystyle \|}{\underset{\displaystyle O}{C}}\text{—} \right]_n$$

a material that is amorphous when quenched rapidly from the melt (Figure 6.6(a)), but semicrystalline when slowly cooled or subsequently heat treated (Figure 6.6(b)).

Molecular mobility may be restricted by other factors, such as the addition of nanometre-scale inclusions. This and other factors have been reviewed recently by Aharoni [6], with particular reference to changes in the glass transition temperature T_g.

6.2   Time–temperature equivalence and superposition

Time–temperature equivalence in its simplest form implies that the viscoelastic behaviour at one temperature can be related to that at another temperature by a change in the time-scale only. Consider the idealized double logarithmic plots of creep compliance versus time shown in Figure 6.7(a). The compliances at temperatures T_1 and T_2 can be superimposed exactly by a horizontal displacement $\log a_t$, where a_t is called the shift factor. Similarly (Figure 6.7(b)), in dynamic

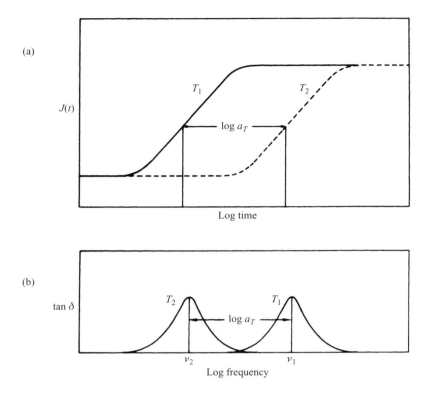

Figure 6.7 Schematic diagrams illustrating the simplest form of time–temperature equivalence for: (a) compliance, $J(t)$; (b) loss factor, $\tan \delta$

mechanical experiments, double logarithmic plots of $\tan \delta$ versus frequency show an equivalent shift with temperature.

The experimental procedure is illustrated in Figures 6.8 and 6.9. A series of creep compliance curves each typically extending over 2 h, so that individual tests

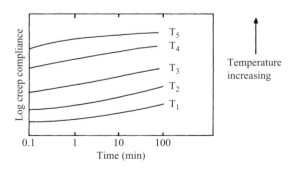

Figure 6.8   Creep plots at different temperatures (schematic)

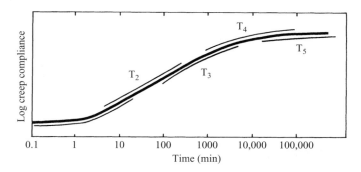

Figure 6.9　Master curve of creep from superposing plots of Figure 6.8

can be performed on successive days, is plotted using a specimen that has been mechanically conditioned at the highest temperature used. The individual plots are then transposed along the logarithmic time axis until they coincide, using any required temperature within the experimental range as the reference value. The variation of shift factor with temperature should also be recorded, for comparison with the predictions of theoretical interpretations to be discussed shortly.

Ferry and co-workers [7], on the basis of the molecular theory of viscoelasticity, proposed that superposition should incorporate a small vertical shift factor $T_0\rho_0/T\rho$, where ρ is the density at the experimental temperature T and ρ_0 relates to the reference temperature T_0. Further corrections have been suggested by McCrum and Morris [8] to deal with the changes in unrelaxed and relaxed compliances with temperature.

The situation is illustrated schematically in Figure 6.10. When we compare the

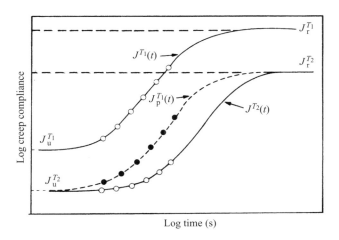

Figure 6.10　Schematic diagram illustrating McCrum's reduction procedure for superposition of creep data: $J_r^{T_1}$ and $J_u^{T_1}$ are the relaxed and unrelaxed compliances, respectively, at the temperature T_1; $J_r^{T_2}$ and $J_u^{T_2}$ are the corresponding quantities at the temperature T_2

creep compliance curves at the two temperatures T_1 and T_2 we see that the relaxed and unrelaxed compliances are both changing with temperature. McCrum and Morris [8] propose a scaling procedure for obtaining a modified or 'reduced' compliance curve at the temperature T_1, to give the dashed curve $J_\rho^T(t)$ in Figure 6.10. The shift factor is now obtained by a horizontal shift of $J_\rho^{T_1}(t)$ to superimpose $J^{T_2}(t)$.

6.3 Molecular interpretations of time–temperature equivalance

6.3.1 Molecular rate processes with a constant activation energy: the site nodel theory

The simplest theories that attempt to deal with the temperature dependence of viscoelastic behaviour are the transition state or barrier theories [9, 10]. The site model was originally developed to explain the dielectric behaviour of solids [11, 12], but was later applied to mechanical relaxations in polymers [13].

In its simplest form there are two sites, separated by an equilibrium free energy difference ΔG_1–ΔG_2, the barrier heights being ΔG_1 and ΔG_2 per mole, respectively (Figure 6.11).

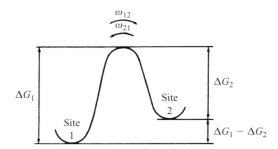

Figure 6.11   The two-site model

The transition probability for a jump from site 1 to site 2 is given by

$$\omega_{12}^0 = A' \exp(-\Delta G_1/RT) \tag{6.1}$$

and for a jump from site 2 to site 1 by

$$\omega_{21}^0 = A' \exp(-\Delta G_2/RT) \tag{6.2}$$

where A' is a constant.

To give rise to a mechanical relaxation process, the energy difference between the two sites must be changed by the application of the applied stress. There is then a change in the populations of site 1 and site 2, and it is assumed that this relates directly to the strain. It is not difficult to imagine how this might arise at a molecular level if, for example, the uncoiling of a molecular chain involved internal rotations. Locally, the chain conformations could be changing from crumpled *gauche* conformations to extended *trans* conformations (see Section 1.2.1).

Assume that the applied stress σ causes a small linear shift in the free energies of the sites such that

$$\delta G_1' = \lambda_1 \sigma \tag{6.3}$$

and

$$\delta G_2' = \lambda_2 \sigma \tag{6.4}$$

for sites 1 and 2, respectively, where λ_1 and λ_2 are constants with the dimensions of volume. The transition probabilities ω_{12} and ω_{21} in the presence of the applied stress are related to those in the absence of stress by expanding the exponential

$$\omega_{12} \approx \omega_{12}^0 \left[1 - \frac{\delta G_1'}{RT} \right] = \omega_{12}^0 \left[1 - \frac{\lambda_1 \sigma}{RT} \right] \tag{6.5}$$

where ω_{12}^0 is the transition probability in the absence of the stress. Similarly

$$\omega_{21} \approx \omega_{21}^0 \left[1 - \frac{\lambda_2 \sigma}{RT} \right] \tag{6.6}$$

The rate equations for sites 1 and 2 are then

$$\frac{dN_1}{dt} = -N_1 \omega_{12} + N_2 \omega_{21} \tag{6.7}$$

$$\frac{dN_2}{dt} = -N_2 \omega_{21} + N_1 \omega_{12} \tag{6.8}$$

where we can write the occupation number N_1 of state 1 as $N_1 = N_1^0 + n$ and similarly $N_2 = N_2^0 - n$, where N_1^0 and N_2^0 are the occupation numbers at zero stress, $N_1^0 + N_2^0 = N_1 + N_2 = N$.

Combining these equations and making suitable approximations gives a rate equation

$$\frac{dn}{dt} + N(\omega_{12}^0 + \omega_{21}^0) = N_1^0 \omega_{12}^0 \left[\frac{\lambda_1 - \lambda_2}{RT}\right] \sigma \tag{6.9}$$

which describes the change in the site population n as a function of time. Assuming that the observed strain e is a consequence of the change in site population it is given by

$$e = e_u + n\bar{e} \tag{6.10}$$

In this equation e_u is the instantaneous or unrelaxed elastic deformation and it is considered that each change in site population produces a proportionate change in strain by an amount $\bar{e}$.

Equation (6.9) can then be seen to have the form

$$\frac{de}{dt} + Be = C$$

where B and C are constants, which is formally identical to the equation of a Kelvin unit (Section 4.2.3 above), with a characteristic retardation time given by $\tau' = 1/B$, i.e.

$$\tau' = \frac{1}{(\omega_{12}^0 + \omega_{21}^0)} = \frac{\exp(\Delta G_2/RT)}{A'[\exp\{-(\Delta G_1 - \Delta G_2)/RT\} + 1]} \tag{6.11}$$

Since RT is usually small compared with the equilibrium free energy difference, we may approximate to

$$\tau' = \frac{1}{A'} \exp(\Delta G_2/RT) \tag{6.12}$$

Since

$$2\pi\nu = \frac{1}{\tau'}$$

where ν is the frequency of molecular jumps between the two rotational isomeric states of the chain molecule, we can write

$$\nu = \frac{A'}{2\pi} \exp(-\Delta G_2/RT) \tag{6.13}$$

This equation states that the frequency of molecular conformational changes depends on the *barrier height* ΔG_2 and not on the free energy difference between the equilibrium sites. Equation (6.13) may also be written as

$$\nu = \frac{A''}{2\pi} \exp(\Delta S/R) \exp(-\Delta H/RT) = \nu_0 \exp(-\Delta H/RT) \qquad (6.14)$$

This form of Equation (6.14) emphasizes the way in which temperature affects ν primarily through the activation energy ΔH. To a good approximation the activation energy for the process (actually an enthalpy) is thus given by

$$\Delta H = -R \left[\frac{\partial(\ln \nu)}{\partial(1/T)}\right]_p \qquad (6.15)$$

Equation (6.14) is known as the 'Arrhenius equation', because it was first shown by Arrhenius [14] that it describes the influence of temperature on the velocity of chemical reactions.

As an example of the application of the Arrhenius equation, consider the $\tan \delta$ curve of Figure 6.7(b). At temperatures T_1 and T_2 the peak value of $\tan \delta$ occurs at frequencies ν_1 and ν_2, respectively. Using

$$\nu = \nu_0 \exp(-\Delta H/RT) \qquad (6.16)$$

we obtain

$$\frac{\nu_1}{\nu_2} = \frac{\exp(-\Delta H/RT_1)}{\exp(-\Delta H/RT_2)}$$

giving the shift factor as

$$\frac{\log \nu_1}{\log \nu_2} = \log a_T = \frac{\Delta H}{R} \left\{ \frac{1}{T_2} - \frac{1}{T_1} \right\} \qquad (6.17)$$

The activation energy for the process can then be obtained by plotting $\log a_T$ versus the reciprocal of the absolute temperature, and for large values of ΔH changes in temperature are equivalent to very large changes in frequency.

It can be shown [15] that the magnitude of the relaxation is proportional to

$$S \left[\frac{\exp(-\Delta G_1 - \Delta G_2)/RT}{\{\exp[-(\Delta G_1 - \Delta G_2)/RT] + 1\}^2} \right] \frac{(\lambda_1 - \lambda_2)^2}{RT} \qquad (6.18)$$

where S is the number of species per unit volume. Thus the intensity of the relaxation is predicted to be low at both high and low temperatures, and passes through a maximum when the free energy difference $(\Delta G_1 - \Delta G_2)$ and RT are of the same order of magnitude.

It must be emphasized that the site model is applicable only to relaxation processes showing a constant activation energy, examples being those associated with localized motions in the crystalline regions of semicrystalline polymers. The temperature dependence of the glass transition relaxation behaviour of polymers does not fit a constant activation energy model, and where this has appeared to be true it is probably a consequence of the limited range of experimental frequencies that were available.

6.3.2   The Williams–Landel–Ferry (WLF) equation

In considering the time–temperature equivalence of the glass transition behaviour in amorphous polymers, we will follow a treatment very close to that given by Ferry [16]. To fix our ideas, consider the storage compliance J_1 of an amorphous polymer (poly-*n*-octyl methacrylate) as a function of temperature and frequency (Figure 6.12). It can be seen that there is an overall change in the shape of the compliance–frequency curve as the temperature changes. At high temperatures there is an approximately constant high compliance: the rubbery compliance. At low temperatures the compliance is again approximately constant but at a low

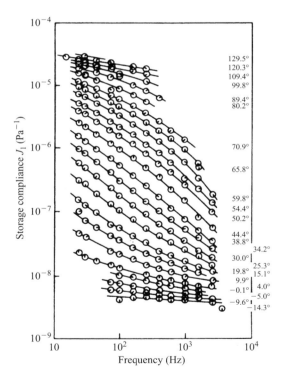

Figure 6.12   Storage compliance of poly-*n*-octyl methacrylate in the glass transition region plotted against frequency at 24 temperatures as indicated. (Reproduced from Ferry, *Viscoelastic Properties of Polymers* (3rd edn), Wiley, New York, 1980, Ch. 11)

value: the glassy compliance. At intermediate temperatures there is the frequency-dependent viscoelastic compliance.

Applying time–temperature equivalence (as discussed in Section 6.2 above) gives the storage compliance as a function of frequency over a very wide range of frequencies, as shown in Figure 6.13. Thus it is now possible to calculate the retardation time spectrum, and compare this with any theoretical models that may be proposed.

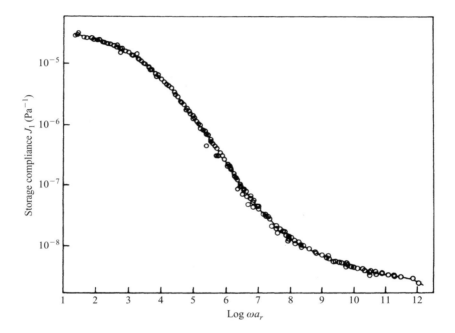

Figure 6.13 Composite curve obtained by plotting the data of Figure 6.12 with suitable shift factors, giving the behaviour over an extended frequency scale at temperature T_0. (Reproduced from Ferry, *Viscoelastic Properties of Polymers* (3rd edn), Wiley, New York, 1980, Ch. 11)

The horizontal shift on a logarithmic time-scale is shown in Figure 6.14. Remarkably, Williams, Landel and Ferry [7] found an approximately identical shift factor–temperature relation for all amorphous polymers, which could be expressed as

$$\log a_T = \frac{C_1(T - T_S)}{C_2 + (T - T_S)}$$

where C_1 and C_2 are constants. This relation, known as the WLF equation, applies over a temperature range $(T_S \pm 50)\,°C$, where T_S is a reference temperature

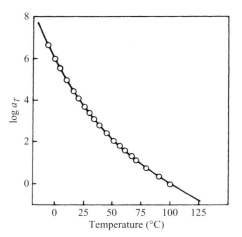

Figure 6.14 Temperature dependence of the shift factor α_T used in plotting Figure 6.13. Points are chosen empirically; curve is WLF equation with a suitable choice of T_g (or T_s). (Reproduced from Ferry, *Viscoelastic Properties of Polymers* (3rd edn), Wiley, New York, 1980, Ch. 11)

specific to a particular polymer. The constants C_1 and C_2 were originally determined by arbitrarily choosing $T_S = 243$ K for polyisobutylene.

Following this empirical discovery there was naturally some speculation as to whether the WLF equation has a fundamental interpretation. To proceed further we must consider the dilatometric glass transition, and its interpretation in terms of free volume.

The glass transition can be defined on the basis of dilatometric measurements. As shown in Figure 6.15, if the specific volume of the polymer is measured against temperature, a change of slope is observed at a characteristic temperature that we may call T_g. In practice the change in slope is somewhat less sharp than the diagram suggests and T_g is determined by extrapolation. When dilatometric measurements are carried out at very slow rates of temperature change, T_g approaches a constant value and will vary by only 2–3 K when the heating rate is decreased from 1 K min^{-1} to 1 K day^{-1}. Thus it appears possible to define a rate-independent value of T_g to at least a very good approximation.

It has subsequently been shown that the original WLF equation can be rewritten in terms of this dilatometric transition temperature such that

$$\log a_T = \frac{C_1^g(T - T_g)}{C_g^2 + (T - T_g)}$$

where C_1^g and C_g^2 are new constants and $T_g = T_s - 50\,°\text{C}$.

Moreover, it is now possible to give a plausible theoretical basis to the WLF

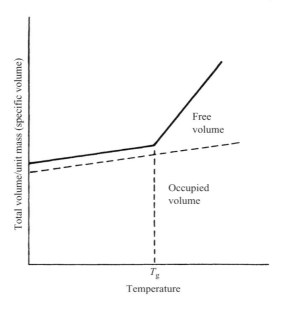

Figure 6.15   The volume–temperature relationship for a typical amorphous polymer

equation in terms of free volume [7]. In liquids, this concept has proved useful in discussing transport properties such as viscosity and diffusion, which are considered to relate to the difference $v_f = v - v_0$, where v is the total macroscopic volume, v_0 is the actual molecular volume of the liquid molecules, the 'occupied volume', and v_f is the proportion of holes or voids, the 'free volume'.

Figure 6.15 shows the schematic division of the total volume of the polymer into both occupied and free volumes. It is argued that the occupied volume increases uniformly with temperature. The discontinuity in the expansion coefficient at T_g then corresponds to a sudden onset of expansion in the free volume, which suggests that certain molecular processes that control the viscoelastic behaviour commence at T_g, and not merely that T_g is the temperature when their time-scale becomes comparable with that of the measuring time-scale. This behaviour would seem to imply that T_g is a genuine thermodynamic second-order transition temperature. The point is not, however, completely resolved, and it has been shown by Kovacs [17] that the T_g measured dilatometrically is still sensibly dependent on the time-scale, i.e. the rate of heating. However, as already mentioned, this time dependence is small. Thus to a good approximation it can be assumed that the free volume is constant up to T_g and then increases linearly with increasing temperature.

The fractional free volume $f = v_f/v$ can therefore be written as

$$f = f_g + \alpha_f(T - T_g) \tag{6.19}$$

where f_g is the fractional free volume at the glass transition T_g and α_f is the coefficient of expansion of the free volume.

The WLF equation can now be obtained in a simple manner. The model representations of linear viscoelastic behaviour all show that the relaxation times are given by expressions of the form $\tau = \eta/E$ (see the Maxwell model in Section 4.2.3 above), where η is the viscosity of a dashpot and E the modulus of a spring.

If we ignore the changes in the modulus E with temperature compared with changes in the viscosity η, this suggests that the shift factor a_T for changing temperature from T_g to T will be given by

$$a_T = \frac{\eta_T}{\eta_{T_g}} \tag{6.20}$$

At this juncture, we introduce Doolittle's viscosity equation [18], based on experimental data for monomeric liquids, that relates the viscosity to the free volume through

$$\eta = a \exp(bv/v_f) \tag{6.21}$$

where a and b are constants. Using Equations (6.20) and (6.21) it can be shown that the Doolittle equation becomes

$$\ln a_T = b\left\{\frac{1}{f} - \frac{1}{f_g}\right\} \tag{6.22}$$

Substituting $f = f_g + \alpha_f(T - T_g)$ we have

$$\log a_T = -\frac{(b/2.303f_g)(T - T_g)}{f_g/\alpha_f + T - T_g} \tag{6.23}$$

which is the WLF equation.

The fractional free volume at the glass transition temperature, f_g, is 0.025 ± 0.003 for most amorphous polymers. The thermal coefficient of expansion of free volume α_f is a more variable quantity, but has the physically reasonable 'universal' average value of 4.8×10^{-4} K^{-1}.

Substituting for a_T from Equation (6.20) the WLF equation may be rewritten as

$$\log \eta_T = \log \eta_{T_g} + \frac{C_1^g(T - T_g)}{C_2^g + (T - T_g)} \tag{6.24}$$

which implies that at a temperature $T = T_g - C_2^g$, i.e. $T = T_g - 51.6$, the viscosity of the polymer becomes infinite. This feature has suggested that at the molecular level the WLF equation should be related to the temperature $T_2 = T_g - 51.6$, rather than to the dilatometric glass transition T_g.

There have been two basic approaches along these lines:

1. The free volume theory is modified so that the changes in free volume with temperature relate to a discontinuity that occurs at T_2 rather than T_g.

2. It is considered that T_2 represents a true thermodynamic transition temperature. Adam and Gibbs [19] have developed a modified transition state theory in which the frequency of molecular jumps relates to the cooperative movement of a group of segments of the chain. The number of segments acting cooperatively is then calculated from statistical thermodynamic considerations.

Despite the successes mentioned above there have been objections to free volume theories [13, 20] based on observations that a few polymers behave similarly under both constant pressure and constant volume conditions, although in the latter case free volume must decrease with increasing temperature.

6.4 Flexible molecular chain models

Condensed matter physicists calculate many properties of crystalline solids in terms of a model, due initially to Debye, in which massive point atoms are connected by linear elastic springs. The dynamics of molecular chains can be considered from this starting point. The theories discussed below, although initially derived for polymer solutions, can be used to predict relaxation spectra and time–temperature equivalence for amorphous solid polymers. As a full treatment involves quite advanced mathematics, we shall discuss the theories only in outline.

6.4.1 Normal mode theories

If a sinusoidal vibration is applied to a longitudinal rod, and both ends of the rod are then fixed, the wave will be reflected at these fixed points or nodes. A series of discrete standing waves, known as the *normal modes*, can be established in the rod, with wavelengths given by

$$\lambda_n = \frac{2l}{n}$$

where l is the length of the rod and n represents a series of integers. It is more usual to express normal modes in terms of the wave vector,

$$k = \frac{2\pi}{\lambda}, \quad \text{so that} \quad k_n = \frac{n\pi}{l}$$

To calculate the normal modes of vibration in a crystal it is necessary to take into account interatomic forces and crystal structure. The starting point for such calculations is a model of an infinite one-dimensional linear chain in which point atoms of identical mass are joined by equal springs with a common force constant. It can be seen from Figure 6.16 that the equation of motion of the nth atom must take into account the displaced positions of both the $(n-1)$th and $(n+1)$th atoms. A full analysis of this type of model can be found in a university-level text on solid state physics, for example that by Guinier and Jullien [21].

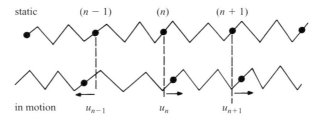

Figure 6.16 Vibrations in a linear chain. The displacement of the nth atom is U_n. The equation of motion for atom n must take into account U_{n-1} and U_{n+1}

The similarities between a one-dimensional lattice analogue and a polymer chain are evident, and related models have been used to set up a series of linear differential equations that can represent the motion of polymer chains in a viscous solvent. In the simplest model, due to Rouse [22], each polymer chain is considered as built from submolecules, represented as beads, which are linked by springs whose behaviour is that of a freely jointed chain on the Gaussian theory of rubber elasticity. The molecular chains between beads are of equal length and are assumed to be long enough for the end-to-end separation to approximate to a Gaussian probability distribution. It is assumed that only the beads and not the chains interact directly with the solvent molecules. When a bead is displaced from its equilibrium position there are two types of forces acting on it: those that result from the viscous interaction with the solvent molecules, and those that represent the tendency of the molecular chains to return to a state of maximum entropy by Brownian diffusional movements.

Consideration of the restoring force when a bead is displaced from its equilibrium position leads to the expression

$$\eta \dot{x}_i + \frac{3kT}{zl^2}(2x_i - x_{i-1} - x_{i+1}) = 0$$

where η is the coefficient defining the viscous interaction between the beads and the solvent, l is the length of each link in a chain, z is the number of links in a submolecule and $\dot{x}_i$ is the time differential of the displacement of a bead situated between the $(i-1)$th and $(i+1)$th submolecules. For m submolecules there are $3m$ of these equations, each of which is equivalent to the equation of a kelvin unit (cf. $\eta\dot{e} + Ee = 0$).

The $3m$ equations have to be uncoupled using a normal coordinate transformation, to obtain eigenfunctions that are linear combinations of the positions of the submolecules. Each eigenfunction then describes a single viscoelastic element with characteristic time-dependent properties.

It can be shown, for example, that the stress relaxation modulus is given by

$$G(t) = NkT \sum_{p=1}^{m} \exp\left(-\frac{t}{\tau_p}\right) \qquad (6.25)$$

and the real part of the complex modulus in dynamic experiments is

$$G_1(\omega) = NkT \sum_{p=1}^{m} \left(\frac{\omega^2\tau_p^2}{1+\omega^2\tau_p^2}\right) \qquad (6.26)$$

where there are N molecules per unit volume and τ_p is the relaxation time of the pth mode, given by

$$\tau_p = zl^2\eta[24kT \sin^2\{p\pi/2(m+1)\}]^{-1}, \qquad p = 1, 2 \ldots, m \qquad (6.27)$$

The moduli are thus determined by a discrete spectrum of relaxation times, each of which characterizes a given normal mode of motion. These normal modes are shown schematically in Figure 6.17. In the first mode, corresponding to $p = 1$, the ends of the molecule move, with the centre of the molecule remaining stationary. In the second mode there are two nodes in the molecule, and in the general case the pth mode has p nodes, with the motion of the molecule occurring in $(p+1)$ segments.

For the Rouse model the submolecule is the shortest length of chain that can undergo relaxation, and the motion of segments within a submolecule is ignored. This limitation implies that the theory is applicable only when $m \gg 1$, and means that the equation for τ_p reduces to

$$\tau_p = \frac{n^2 l^2 \eta_0}{6\pi^2 p^2 kT} \qquad (6.28)$$

where $\eta_0 = \eta/z$ is the friction coefficient per random link. The relaxation times depend on temperature through $1/T$, through nl^2 (which defines the equilibrium mean-square separation of the chain ends) and through changes in η_0. This last parameter varies rapidly with temperature and is primarily responsible for the changes in τ_p. As each τ_p has the same temperature dependence, the theory

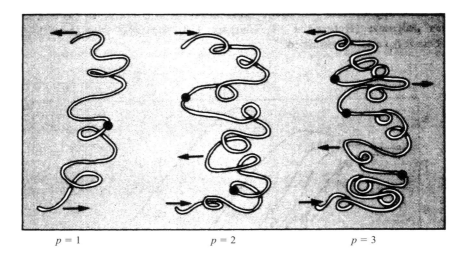

$$p = 1 \qquad\qquad p = 2 \qquad\qquad p = 3$$

Figure 6.17   Illustration of the first three normal modes of a chain molecule

satisfies the requirements of thermorheological simplicity and gives justification for time–temperature equivalence.

A later theory due to Zimm [23] gives a modified relaxation spectrum by taking into account the way in which the solvent is affected by the movement of the polymer molecules, and by considering the hydrodynamic interaction between the moving submolecules.

6.4.2   The dynamics of highly entangled polymers

In a concentrated polymer solution, a melt or a solid polymer the molecular chains cannot pass through one another, a constraint that effectively confines each chain within a tube [24]. The centre line of this tube defines the overall path of the chain in space, and has been called by Edwards the primitive chain (Figure 6.18). Each chain 'sees' its environment as a tube because, although all the other chains are moving, there are so many entanglements that at any one time the tube is well defined. De Gennes [25] has described the possible motions of a polymer chain confined to a tube as snake-like, and has called the phenomenon 'reptation'. He considered two distinct forms of motion. First, the comparatively short-term wriggling motions that correspond to the migration of a molecular kink along the chain, for which the longest relaxation time is proportional to the square of the molecular mass. Second, there is the much longer time associated with the movement of the chain as a whole through the polymer. This motion corresponds to an overall movement of the centre of gravity of the chain, and has a characteristic time proportional to the cube of the molecular mass.

Figure 6.18 Chain segment AB in dense rubber. The points A and B denote the cross-linked points and the dots represent other chains that, in this drawing, are assumed to be perpendicular to the paper. Due to entanglements the chain is confined to the tube-like region denoted by the broken line. The bold line shows the primitive path. (Reproduced with permission from Doi and Edwards, *J. Chem. Soc. Faraday Trans.*, **74** 1802 (1978))

Doi and Edwards [24] have extended the work of de Gennes, and have derived mathematical expressions for features such as the stress relaxation that occurs after a large strain. Their explanation for the physical situation is illustrated in Figure 6.19, in which the hatched area indicates the deformed part of the tube. Here (a) represents the tube before deformation, when the conformation of the primitive chain is in equilibrium. The deformation is considered to be affine, so that each molecule deforms to the same extent as the macroscopic body. In (b) the situation immediately after the step deformation is given, with the primitive chain

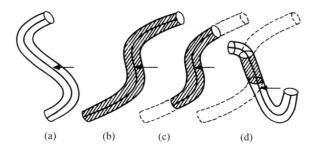

(a)   (b)   (c)   (d)

Figure 6.19 Explanation of the stress relaxation after a large step strain. (a) Before deformation the conformation of the primitive chain is in equilibrium ($t = -0$). (b) Immediately after deformation, the primitive chain is in the affinely deformed conformation ($t = +0$). (c) After time τ_R, the primitive chain contracts along the tube and recovers the equilibrium contour length ($t \approx \tau_R$). (d) After the time τ_d, the primitive chain leaves the deformed tube by reptation ($t \approx \tau_d$). The oblique lines indicate the deformed part of the tube. (Reproduced from Doi and Edwards, *The Theory of Polymer Dynamics*, Oxford University Press, 1986)

in the affinely deformed conformation. In (c) the situation after a characteristic time τ_R is given, with the primitive chain recovering to its equilibrium contour length by contracting along the tube. For these very short-term motions the constraints of the tube are not felt, so the motions are the same as those for the Rouse chain without constraints. The relaxation time τ_R is therefore the longest Rouse relaxation time. After a longer characteristic time τ_d (Figure 6.19(d)) the primitive chain leaves the deformed tube by reptation. Hence τ_d is called the disengagement or reptation time and is the terminal relaxation time. If N is the number of segments in the chain, τ_d is proportional to N^3 (i.e. to the cube of the molecular mass) whereas τ_R is proportional to N^2. For a fuller discussion, refer to the advanced text by Doi and Edwards [24].

Recent research has addressed the shortcomings of the original Doi–Edwards exposition of reptation theory, firstly for low shear rate linear behaviour and secondly for non-linear behaviour at large shear rates. Doi–Edwards linear reptation theory predicts thatar τ_d scales with N as N^3, whereas there is a large body of experimental evidence that viscosity scales with molecular mass M as $M^{3.4}$. Secondly, linear theory predicts that the dynamic loss modulus $G_2(\omega)$ is proportional to $\omega^{-1/2}$ in the intermediate frequency range whereas experiment gives a much weaker frequency dependence, with a power law between 0 and $-1/4$ depending on chain length.

The physical origin of the 3.4 scaling was recognized to be due to the relaxation of a part of the stress by a faster process than reptation, this process becoming less important as N becomes large, so that the viscosity increases faster than the asymptotic N^3 relationship: hence the apparent 3.4 relationship with molecular weight. Doi [26] identified this faster process with what he called contour-length fluctuations. These fluctuations involve the chain contracting within the tube and then stretching out again, so that orientation of the ends of the tube is forgotten and part of the stress is relaxed. Milner and McLeish [27] recently have proposed a quantitative theory for the stress relaxation in monodisperse linear polymer melts that incorporates reptation and contour-length fluctuations. The basis for the theory is that linear chains can be regarded as two-armed stars. In star polymers reptation cannot occur because the arms do not have a single tube. The stress relaxes by arm retraction, where the star contracts by fluctuations down its tube towards the branch point, i.e. similar to the contour-length fluctuations of Doi.

The stress relaxation modulus $G(t)$ then has three components:

1. High-frequency modes due to contour-length fluctuations at the ends of the tubes.

2. Lower frequency modes associated with reptation.

3. Very high-frequency (short-time) Rouse modes where the tube constraints are not felt.

It was shown that the Milner–McLeish theory predicted the loss modulus $G_2(\omega)$ for monodisperse polystyrene melts, in addition to achieving the 3.4 molecular weight relationship for the viscosity, which converges to 3 for high molecular weight. More recently, Likhtman and McLeish [28] have produced a more satisfactory expression for the stress relaxation modulus, which incorporates contour-length fluctuations, longitudinal stress relaxation along the tube and constraint release, where the effect of tube breakdown is taken into account in a less *ad hoc* fashion. The previous predictions of Milner and McLeish are modified to some extent, but the main advance is that the Likhtman–McLeish theory is more self-consistent and avoids many of the approximations in the original theory.

The Doi–Edwards model also has some shortcomings for describing non-linear behaviour at high shear rates where the viscosity is not linearly proportional to the shear rate. The Doi–Edwards theory predicts that the viscosity $\eta(\dot{\gamma})$ reduces with $\dot{\gamma}$ (shear thinning) as $\dot{\gamma}^{-3/2}$, which would imply that there is a maximum in the shear stress as a function of $\dot{\gamma}$, which is not observed. Shear thinning is due to the orientation of the entangled chain network under flow. Marucci [29] proposed that the severity of the orientation is reduced due to what he termed convective constraint release, which relates to the relaxation (i.e. contraction) of chains extended affinely by the flow, by retraction to their equilibrium length in their tubes. Marucci and others have developed theories for the non-linear behaviour based on this idea [30, 31].

An alternative approach is that convective constraint release can be regarded as due to a hopping motion of the tube, most simply following Rouse dynamics [32]. This has been developed analytically by Milner, McLeish and Likhtman [33], whose theory predicts a monotonically increasing shear stress with increasing shear strain rate, with no stress maximum for polymer melts.

References

1. Marvin, R. S., in *Proceedings of the Second International Congress of Rheology*, Butterworths, London, 1954, pp. 156–164.
2. Marvin, R. S. and Oser, H., *J. Res. Natl Bur. Stand. B*, **66**, 171 (1962).
3. Marvin, R. S. and Berger, J. T., *Viscoelasticity: Phenomenological Aspects*, Academic Press, New York, 1960, p. 27.
4. Schmieder, K. and Wolf, K., *Kolloidzeitschrift*, **134**, 149 (1953).
5. Thompson, A. B. and Woods, D. W., *Trans. Faraday Soc.*, **52**, 1383 (1956).
6. Aharoni, S. M., *Polym. Adv. Technol.*, **9**, 169 (1998).
7. Williams, M. L., Landel, R. F. and Ferry, J. D., *J. Am. Chem. Soc.*, **77**, 3701 (1955).
8. McCrum, N. G. and Morris, E. L., *Proc. R. Soc.*, **A281**, 258 (1964).
9. Glasstone, S., Laidler, K. J. and Eyring, H., *The Theory of Rate Processes*, McGraw-Hill, New York, 1941.
10. Glasstone, S., *Textbook of Physical Chemistry* (2nd edn), Macmillan, London, 1953.
11. Debye, P., *Polar Molecules*, Dover, New York, 1945.
12. Fröhlich, H., *Theory of Dielectrics*, Oxford University Press, Oxford, 1949.

13. Hoffman, J. D., Williams, G. and Passaglia, E., *J. Polym. Sci., C*, **14**, 173 (1966).
14. Arrhenius, Z., *J. Phys. Chem.*, **4**, 226 (1889).
15. Wachtman, J. B., *Phys. Rev.*, **131**, 517 (1963).
16. Ferry, J. D., *Viscoelastic Properties of Polymers* (3rd edn), Wiley, New York, 1980, Ch. 11.
17. Kovacs, A., *J. Poym. Sci.*, **30**, 131 (1958).
18. Doolittle, A. K. *J. Appl. Phys.*, **22**, 1471 (1951).
19. Adam, G. and Gibbs, J. H., *J. Chem. Phys.*, **43**, 139 (1965).
20. Williams, G., *Trans. Faraday Soc.*, **60**, 1556 (1964).
21. Guinier, A. and Jullien, R., *The Solid State*, Oxford University Press, Oxford, 1989, p. 13.
22. Rouse, P. E., *J. Chem. Phys.*, **21**, 1272 (1953).
23. Zimm, B. H., *J. Chem. Phys.*, **24**, 269 (1956).
24. Doi, M. and Edwards, S. F., *The Theory of Polymer Dynamics*, Oxford University Press, 1986.
25. de Gennes, P. G., *J. Chem. Phys.*, **55**, 572 (1971).
26. Doi, M., *J. Polym. Sci., Polym. Phys. Ed.*, **21**, 667 (1983).
27. Milner, S. T. and McLeish, T. C. B., *Phys. Rev. Lett.*, **81**, 725 (1998).
28. Likhtman, A. E. and McLeish, T. C. B., *Macromolecules*, **35**, 6532 (2002).
29. Marucci, G., *J. Non-Newtonian Fluid Mech.*, **62**, 279 (1996).
30. Ianniruberto, G. and Marucci, G., *J. Non-Newtonian Fluid Mech.*, **65**, 241 (1996).
31. Mead, D. W., Larson, R. G. and Doi, M., *Macromolecules*, **31**, 7895 (1998).
32. Viovy, J. L., Rubinstein, M. and Colby, R. A., *Macromolecules*, **24**, 3587 (1991).
33. Milner, S. T., McLeish, T. C. B. and Likhtman, A. E., *J. Rheol.*, **45**, 539 (2001).

7

Anisotropic Mechanical Behaviour

In many practical applications of synthetic polymers molecular orientation is produced by the fabrication process to give improved properties, especially with regard to stiffness and strength. Well known examples are textile fibres such as Terylene or nylon, polypropylene packaging films and polyester bottles. Most natural materials such as silk and cotton fibres, muscle and bone also show significant molecular orientation. All these synthetic and natural materials are anisotropic, i.e. their properties are different in different directions.

7.1 Elastic constants and polymer symmetry

The mechanical properties of an anisotropic elastic solid where the stresses are linearly related to the strains are defined by the generalized Hooke's law where each component of stress can relate to all six independent components of strain, and equivalently each component of strain can relate to all six independent components of stress. In the former case we have, for example

$$\sigma_1 = c_{11}e_1 + c_{12}e_2 + c_{13}e_3 + c_{14}e_4 + c_{15}e_5 + c_{16}e_6$$

where $\sigma_1 = \sigma_{xx}$, and $e_1 = e_{xx}$, $e_2 = e_{yy}$, $e_3 = e_{zz}$, $e_4 = e_{yz}$, $e_5 = e_{xz}$ and $e_6 = e_{xy}$ are the six engineering components of strain; the equivalent equations for all six components of stress are $\sigma_1 = \sigma_{xx}$, $\sigma_2 = \sigma_{yy}$, $\sigma_3 = \sigma_{zz}$, $\sigma_4 = \sigma_{yz}$, $\sigma_5 = \sigma_{xz}$ and $\sigma_6 = \sigma_{xy}$. In abbreviated notation we have

$$\sigma_p = c_{pq}e_q$$

and, equivalently

$$e_p = s_{pq}\sigma_q$$

where the constants c_{pq} are called the stiffness constants and form the elements of a six by six symmetric matrix

An Introduction to the Mechanical Properties of Solid Polymers I. M. Ward and J. Sweeney
© 2004 John Wiley & Sons, Ltd   ISBN: 0471 49625 1 (HB); 0471 49626 X (PB)

$$\begin{bmatrix} c_{11} & c_{12} & c_{13} & c_{14} & c_{15} & c_{16} \\ c_{12} & c_{22} & c_{23} & c_{24} & c_{25} & c_{26} \\ c_{13} & c_{23} & c_{33} & c_{34} & c_{35} & c_{36} \\ c_{14} & c_{24} & c_{34} & c_{44} & c_{45} & c_{46} \\ c_{15} & c_{25} & c_{35} & c_{45} & c_{55} & c_{56} \\ c_{16} & c_{26} & c_{36} & c_{46} & c_{56} & c_{66} \end{bmatrix}$$

and the constants s_{pq} are called the compliance constants and form the matrix

$$\begin{bmatrix} s_{11} & s_{12} & s_{13} & s_{14} & s_{15} & s_{16} \\ s_{12} & s_{22} & s_{23} & s_{24} & s_{25} & s_{26} \\ s_{13} & s_{23} & s_{33} & s_{34} & s_{35} & s_{36} \\ s_{14} & s_{24} & s_{34} & s_{44} & s_{45} & s_{46} \\ s_{15} & s_{25} & s_{35} & s_{45} & s_{55} & s_{56} \\ s_{16} & s_{26} & s_{36} & s_{46} & s_{56} & s_{66} \end{bmatrix}$$

In practical applications of oriented polymers we are concerned with films or fibres, which reduces the number of independent elastic constants to nine or six, respectively. It is also convenient at this stage to limit the discussion to the compliance constants because these can be related directly to the readily measured engineering elastic constants, such as Young's moduli, shear moduli and Poisson's ratios.

7.1.1   Specimens possessing orthorhombic symmetry

Polymer films prepared by commercial drawing or rolling processes generally possess orthorhombic symmetry, which means that there are three orthogonal planes of symmetry. As shown in Figure 7.1, the initial drawing or rolling direction is normally taken as the z axis of a system of rectangular Cartesian

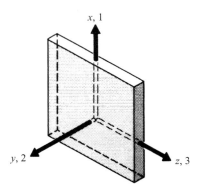

Figure 7.1   Choice of axes for a polymer sheet possessing orthorhombic symmetry

coordinates; the x axis then lies in the plane of the sheet, with the y axis normal to that plane. Figure 7.1 illustrates the convention $1 \sim x$, $2 \sim y$, $3 \sim z$.

The compliance matrix for these axes involves nine independent elastic constants:

$$S_{pq} = \begin{bmatrix} s_{11} & s_{12} & s_{13} & 0 & 0 & 0 \\ s_{12} & s_{22} & s_{23} & 0 & 0 & 0 \\ s_{13} & s_{23} & s_{33} & 0 & 0 & 0 \\ 0 & 0 & 0 & s_{44} & 0 & 0 \\ 0 & 0 & 0 & 0 & s_{55} & 0 \\ 0 & 0 & 0 & 0 & 0 & s_{66} \end{bmatrix}$$

There are three Young's moduli:

$$E_1 = \frac{1}{s_{11}}, \qquad E_2 = \frac{1}{s_{22}}, \qquad E_3 = \frac{1}{s_{33}}$$

and six Poisson's ratios:

$$v_{21} = -\frac{s_{21}}{s_{11}}, \qquad v_{31} = -\frac{s_{31}}{s_{11}}, \qquad v_{32} = -\frac{s_{32}}{s_{22}}$$

$$v_{12} = -\frac{s_{12}}{s_{22}}, \qquad v_{13} = -\frac{s_{13}}{s_{33}}, \qquad v_{23} = -\frac{s_{23}}{s_{33}}$$

Note that in each of these expressions the denominator indicates the direction x, y, z (or 1, 2, 3) in which the tensile stress is applied.

There are also three independent shear moduli:

$$G_1 = \frac{1}{s_{44}}, \qquad G_2 = \frac{1}{s_{55}}, \qquad G_3 = \frac{1}{s_{66}}$$

corresponding to shear in the yz, zx and xy planes, respectively. Torsion experiments where the sheet is twisted about the x, y or z axis will in general involve a combination of shear compliances (see Section 7.2).

7.1.2 Specimens possessing uniaxial symmetry, often termed transverse isotropy

A fibre or a uniaxially drawn film will usually show no preferred orientation in the plane normal to the fibre axis or the draw direction (see Figure 7.2). The compliance matrix then reduces to a form with only five independent constants:

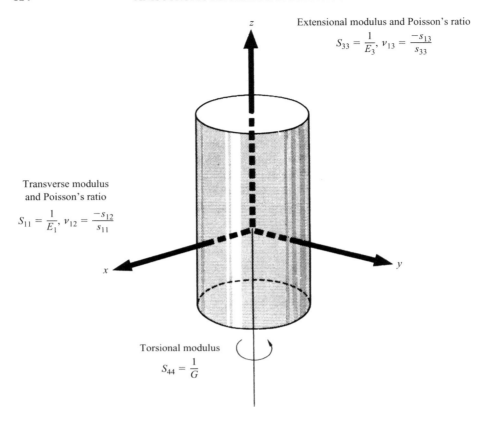

Figure 7.2 The fibre compliance constants

$$
\begin{bmatrix}
s_{11} & s_{12} & s_{13} & 0 & 0 & 0 \\
s_{12} & s_{22} & s_{23} & 0 & 0 & 0 \\
s_{13} & s_{23} & s_{33} & 0 & 0 & 0 \\
0 & 0 & 0 & s_{44} & 0 & 0 \\
0 & 0 & 0 & 0 & s_{55} & 0 \\
0 & 0 & 0 & 0 & 0 & 2(s_{11} - s_{12})
\end{bmatrix}
$$

and we have two Young's moduli:

$$
E_1 = \frac{1}{s_{11}}, \; E_3 = \frac{1}{s_{33}}
$$

two Poisson's ratios:

$$
\nu_{13} = -\frac{s_{13}}{s_{33}}, \; \nu_{23} = -\frac{s_{23}}{s_{33}}
$$

and one independent shear modulus:

$$G = \frac{1}{s_{44}}$$

7.2 Measuring elastic constants

Here we describe only the principal features of methods of measuring the elastic constants, and refer the reader either to the references cited or to Ward's more advanced text [3] for details of the experimental arrangements.

The experimental methods employed for filaments are different from those used for flat sheets, and the two cases will therefore be discussed separately.

7.2.1 Measurements on films or sheets

In all cases the draw or roll direction is taken as the z axis, with the x axis in the plane of the sheet and the y axis perpendicular to that plane.

Extensional moduli

The principle here is to measure the Young's modulus for long narrow strips cut at various angles to the z axis. End effects that arise from the non-uniform stresses near the clamps are very severe [4, 5], so that it is desirable to measure specimens with the highest possible aspect ratio, i.e. length to width and thickness ratios. A good general rule is that the aspect ratio, i.e. ratio of length of specimen to the longest lateral dimension, should be much greater than

$$10 \times \sqrt{\frac{E}{G}}$$

where E is the relevant Young's modulus and G is the appropriate shear modulus (i.e. either G_1 or G_2).

It is convenient to measure strips cut at 0°, 45° and 90° to the z axis. As shown in Appendix 1 the required relations are as follows:

$$E_0 = \frac{1}{s_{33}}, \qquad E_{90} = \frac{1}{s_{11}}, \quad \text{and} \quad \frac{1}{E_{45}} = \frac{1}{4}[s_{11} + s_{33} + (2s_{13} + s_{55})]$$

For a transversely isotropic sheet $s_{55} = s_{44}$.

Transverse stiffness

The stiffness c_{22} normal to the plane of the sheet has been determined by compressing narrow strips in a compressional creep apparatus [6]. A lever device is used to determine the deformation, and precision can be improved by compressing a sandwich of sheets separated by effectively rigid spacers. Wilson *et al.* [6] found that for sheets of polyethylene terephthalate frictional constraints prevented strain in the x or z directions. In this case $e_x = e_z = 0$ and $\sigma_y = c_{22}e_y$, or

$$e_y = \left[s_{22} + s_{12}(s_{13}s_{23} - s_{12}s_{33}) - \frac{s_{23}(s_{11}s_{23} - s_{12}s_{23})}{(s_{11}s_{33} - s_{13}^2)} \right] \sigma_y$$

Lateral compliances and Poisson's ratio

For a polymer film possessing orthorhombic symmetry there are three lateral compliances s_{12}, s_{13} and s_{23}, which relate to the six Poisson's ratios previously defined.

The compliance s_{13}, which defines the contraction in the x direction for a stress applied along the z direction, has been determined from the change in shape of a grid of straight lines parallel with the x and z axes vacuum- deposited on the surface of the long thin specimen [7]. The grid was photographed under the smallest load that held the specimen taut, and again at a fixed time after the application of further loads.

Compliances s_{12} and s_{23} relate to contraction in the thickness (y) direction for stresses along the x and z axes, respectively. For sufficiently thick specimens the changes in thickness, claimed accurate to within 0.25 μm, can be measured by extensometers, fitted with lever arms, which bear against the faces of the polymer sheet. Thin cover slides are inserted between the extensometer elements and the specimen faces to remove any possibility of indentation [8]. For thinner specimens, or materials of high optical clarity, the sample can be inserted in one arm of a Michelson interferometer and immersed in turn in two fluids of different refractive index. The fringe shift Δm can be related to the change in thickness Δt by

$$\Delta m = \frac{2}{\lambda} [(n_i - 1)\Delta t + t \Delta n_i]$$

where λ is the wavelength, n_i the refractive index and Δn_i the change in refractive index [9].

A Hall effect extensometer (Figure 7.3), due to Richardson and Ward [10], is capable of recording strains down to 10^{-3} in sheets of 0.5 mm thickness. Two permanent magnets A_1 and A_2 are mounted in tubes T with their like poles adjacent, so that the intermediate field has a null point and the field gradient is twice that of a single magnet. Springs P_1 and P_2 enable the polymer sheet S to be

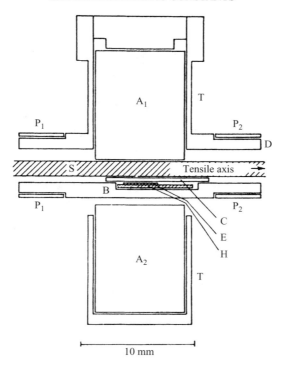

Figure 7.3 Scale diagram of the Hall effect lateral extensometer. (Reproduced with permission from Richardson and Ward, *J. Polym. Sci., Polym. Phys. Ed.*, **16**, 667 (1978))

held against the face of A_1 by the stainless steel plate B and its cover slip C, which contain the Hall effect device H. The Hall plate is positioned so that its sensing element E lies initially along the null field axis.

Measurement in torsion

For sheets of orthorhombic symmetry a solution to the elastic torsion problem is possible only when the sheets are cut as rectangular prisms with their surfaces normal to the three axes of symmetry, and where the torsion axis coincides with one of these three axes. For example, torsion about the z axis involves compliances s_{44} in the yz plane and s_{55} in the xz plane.

In this case, for a specimen of length l, thickness a and width b (Figure 7.4), the torsional rigidity, which is the ratio of the torque Q_z to the angle of twist T, is given by theory due to St Venant (see [11] p. 283) as

$$\frac{Q_z}{T} = \frac{ab^3}{s_{55}l}\beta(c_z) = \frac{a^3b}{s_{44}l}\beta(\overset{+}{c}_z)$$

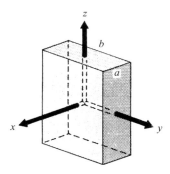

Figure 7.4   The orthorhombic sheet

where

$$c_z = \frac{1}{\overset{+}{c}_z} = \frac{a}{b}\left(\frac{s_{55}}{s_{44}}\right)^{1/2}$$

and $\beta(c_z)$ is a rapidly converging function of c_z, which for $c_z > 3$ can be approximated to

$$\beta(c_z) = \frac{1}{3}\left\{1 - \frac{0.630}{c_z}\right\}$$

For a sheet with isotropy perpendicular to the z axis, $s_{44} = s_{55}$, so that the torque Q_z about an axis perpendicular to the symmetry axis is given by

$$Q_z = \frac{bc^3 T}{s_{66} l}\beta(\overset{*}{c}) \frac{cb^3 T}{s_{44} l}\beta(\overset{+}{c})$$

where

$$\overset{+}{c} = \frac{1}{\overset{*}{c}} = \frac{c}{b}\left(\frac{s_{44}}{s_{66}}\right)^{1/2}$$

with b the thickness, c the width and l the length of the specimen. Both compliances have been obtained from measurements on sheets of a range of aspect ratios [12].

For torsion about the symmetry axis z, the torque for transversely isotropic sheets is given by

$$Q_z = \left(\frac{ab^3}{s_{44} l}\right)\beta(c),$$

where $c = a/b$; $\beta(c)$ is now the same function of $c = a/b$ only.

Practical difficulties arise because the specimen is normally deformed additionally by an axial stress, and warping of the twist axes is restrained in the vicinity of the end clamps. The former effect can be compensated for by carrying out experiments over a range of axial stresses and extrapolating to zero stress. Folkes and Arridge [5] have considered end effects as confined to a block with compliance s' and length p at each end of the specimen of total length l, the central region of which has the true compliance s_0. The overall sample compliance s is then

$$s = s_0 + \left(\frac{2p}{l}\right)(s' - s_0)$$

which means that s_0 can be found by extrapolating measurements on samples of different length to zero reciprocal length.

Because of these complicating effects it can be advantageous to determine s_{44} and s_{66} from experiments in simple shear. Lewis, Richardson and Ward [13] have used a Hall effect transducer to measure this type of deformation and obtain shear compliances that agree with those measured in torsion.

7.2.2   Measurements on filaments

The extensional modulus

$$E = \frac{1}{s_{33}}$$

has been measured statically and dynamically over a wide range of frequencies. The shear modulus can be obtained either from free torsional vibrations, as described in Chapter 5, or from a forced vibration pendulum. The Poisson's ratio

$$\nu_{13} = -\frac{s_{13}}{s_{33}}$$

has been measured directly, using a microscope to record the radial contraction corresponding to a known change in length of a marked section of a much longer filament [14]. For strains of $0.01-0.02$ the uncertainty was in the region of 10 per cent. The transverse modulus

$$\frac{1}{s_{11}}$$

and the transverse Poisson's ratio

$$\nu_{12} = -\frac{s_{12}}{s_{11}}$$

have both been measured for filaments compressed under known loads between parallel glass flats (Figure 7.5(a)) using a simple device attached to a microscope stage [14, 15].

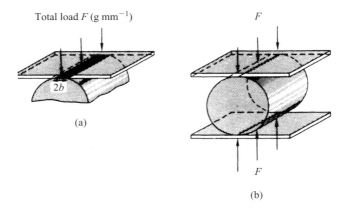

(a)

(b)

Figure 7.5   The contact zone in the compression of a fibre monofilament (a); for consideration of deformation in the central zone of the compressed monofilament it is sufficient to assume line contacts (b); F is the force acting on unit length of the filament

The filament's transverse isotropy implies that under compressive loading normal to the fibre axis the stresses in the transverse plane are identical in form to those for the compression of an isotropic cylinder and, provided that the length of the filament under compression is long compared with the width of the contact strip $2b$, friction ensures that compression occurs under plane strain conditions. As $e_{zz} = 0$ only a normal stress acts along the filament axis, which can be found in terms of the normal stresses σ_{xx} and σ_{yy} in the perpendicular plane, i.e.

$$\sigma_{zz} = -\frac{s_{13}}{s_{33}}(\sigma_{xx} + \sigma_{yy})$$

The stresses can therefore be obtained from the solution to the problem of compression of an isotropic cylinder, and the corresponding strains derived from $e_p = s_{pq}\sigma_q$.

   Through an extension of Hertz's solution for the compression of an *isotropic* cylinder Ward was able to derive an analytical solution for the contact width $2b$, of the form

$$b^2 = \frac{4FR}{\pi}\left(s_{11} - \frac{s_{13}^2}{s_{33}}\right)$$ (7.1)

where F is the compressive force per unit length of filament of radius R (Figure 7.5(a)). By rewriting Equation (7.1) as

$$b^2 = \frac{4FR}{\pi}(s_{11} - v_{13}^2 s_{33})$$

we can consider the relative importance of the two terms inside parentheses. Poisson's ratio is typically close to 0.5, and s_{33} is normally small compared with s_{11}, as highly oriented polymers are usually much stiffer along the z axis than transverse to it. Hence the contact problem provides a good method for determining s_{11}.

Further, it can be shown that the diametral compression u_1 parallel to the direction of the applied load is [16]

$$u_1 = -\frac{4F}{\pi}\left(s_{11} - \frac{s_{13}^2}{s_{33}}\right)\left[0.19 + \sinh^{-1}\left(\frac{R}{b}\right)\right]$$ (7.2)

Provided that the width of the contact zone is small compared with the filament radius, the change in diameter parallel with the plane of contact u_2 can be calculated by considering the deformation of a cylinder under concentrated loads (Figure 7.5). The stresses here correspond exactly with those for the isotropic case, whose solution can be found in texts on elasticity (e.g. [17] p. 414). When the strains have been calculated it is possible to derive the diametral expansion as

$$u_2 = F\left[\left(\frac{4}{\pi} - 1\right)\left(s_{11} - \frac{s_{13}^2}{s_{33}}\right) - \left(s_{12} - \frac{s_{13}^2}{s_{33}}\right)\right]$$ (7.3)

Thus measurement of the diametral expansion provides a method of determining s_{12}, once s_{11} has been derived from a measurement of the contact width.

7.3 Experimental studies of mechanical anisotropy in oriented polymers

The review of experimental studies has two principal aims. First, we will describe early studies on low-density polyethylene and oriented monofilaments because they highlighted several unexpected features of the anisotropy, indicating the differences between individual polymers and providing the basis for the aggregate model against which many subsequent measurements have been tested. Secondly, we will provide very brief summaries of further work in this area, and indicate its relevance to our present understanding.

7.3.1   Sheets of low-density polyethylene

Raumann and Saunders [18] uniaxially stretched isotropic sheets of low-density (i.e. branched) polyethylene to varying final extensions, and measured the tensile modulus in directions over a range of angles to the initial draw direction. For a highly oriented sample the plot of Young's modulus ($1/s_\theta$) against angle with the draw direction (Figure 7.6) shows the lowest stiffness at an angle close to 45° to that direction.

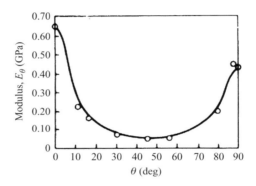

Figure 7.6   Comparison of the observed variation of modulus E_0 with angle θ to draw direction and the theoretical relation (i.e. full curve) calculated from E_0; E_{45} and E_{90} for low-density polyethylene sheet drawn to a draw ratio of 4.65. (Reproduced with permission from Raumann and Saunders, *Proc. Phys. Soc.*, **77**, 1028 (1961))

The general compliance equation for transverse isotropy (see Appendix 1, Section A1.7) is

$$'s_\theta = s_{11} \sin^4 \theta + s_{33} \cos^4 \theta + (2s_{13} + s_{44}) \sin^2 \theta \cos^2 \theta$$

The experimental result implies that $(2s_{13} + s_{44})$ is much greater than either s_{11} or s_{33}, because when $\theta = 45°$ the terms will be equally weighted.

Replotting to obtain the modulus at a given angle as a function of draw ratio (Figure 7.7), the results are again somewhat unexpected: E_0 initially falls with increasing draw ratio, so that at low draw ratios $E_{90} > E_0$. Subsequently, Gupta and Ward [19] showed that this unusual behaviour was specific to room-temperature measurements, and at a sufficiently low temperature the behaviour resembled that of most other polymers (Figure 7.8).

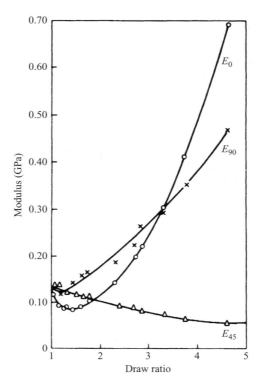

Figure 7.7 The variation of E_0, E_{45} and E_{90} with draw ratio in cold-drawn sheets of low-density polyethylene. Modulus measurements taken at room temperature (Reproduced with permission from Raumann and Saunders, *Proc. Phys. Soc.*, **77**, 1028 (1961))

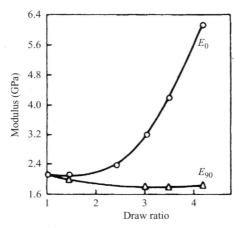

Figure 7.8 The variation of E_0 and E_{90} with draw ratio in cold-drawn sheets of low-density polyethylene. Modulus measurements taken at $-125\,^\circ$C.

7.3.2 Filaments tested at room temperature

In a comprehensive study at room temperature, Hadley, Pinnock and Ward [20] determined the five independent elastic constants for oriented filaments of polyethylene terephthalate, nylon 6:6, low and high-density polyethylene and polypropylene. The orientation was determined in terms of draw ratio and optical birefringence. Subsequent studies indicated that it would have been appropriate to record not only the overall orientation, as derived from birefringence, but also the crystal orientation, obtainable from X-ray measurements. The results are summarized in Table 7.1 and Figures 7.9–7.13 (see Section 7.5 for discussion of the aggregate theory predictions).

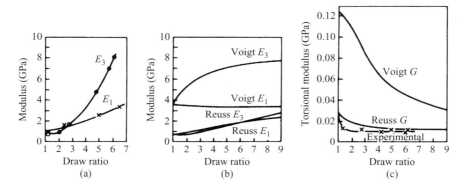

Figure 7.9 Low-density polyethylene filaments: extensional (E_3), transverse (E_1) and torsional moduli (G); comparison between experimental results and simple aggregate theory for E_3 and E_1 ((a) and (b)) and for G (c)

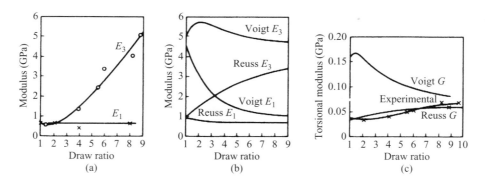

Figure 7.10 High-density polyethylene filaments: extensional (E_3), transverse (E_1) and torsional moduli (G); comparison between experimental results and simple aggregate theory for E_3 and E_1 ((a) and (b)) and for G (c)

Table 7.1 Elastic compliances of oriented fibres (units of compliance are 10^{-10} Pa^{-1}; errors quoted are 95% confidence limits) [1]

Material	Birefringence (Δn)	s_{11}	s_{12}	s_{33}	s_{13}	s_{44}	ν_{13}	ν_{12}
Low-density polyethylene film 14		22	-15	14	-7	680	0.50	0.68
Low-density polyethylene 1	0.0361	40 ± 4	-25 ± 4	20 ± 2	-11 ± 2	878 ± 56	0.55 ± 0.08	0.61 ± 0.20
Low-density polyethylene 2	0.0438	30 ± 3	-22 ± 3	12 ± 1	-7 ± 1	917 ± 150	0.58 ± 0.08	0.73 ± 0.20
High-density polyethylene 1	0.0464	24 ± 2	-12 ± 1	11 ± 1	-5.1 ± 0.7	34 ± 1	0.46 ± 0.15	0.52 ± 0.08
High-density polyethylene 2	0.0594	15 ± 1	-16 ± 2	2.3 ± 0.3	-0.77 ± 0.3	17 ± 2	0.33 ± 0.12	1.1 ± 0.14
Polypropylene 1	0.0220	19 ± 1	-13 ± 2	6.7 ± 0.3	-2.8 ± 1.0	18 ± 1.5	0.42 ± 0.16	0.68 ± 0.18
Polypropylene 2	0.0352	12 ± 2	-17 ± 2	1.6 ± 0.04	-0.73 ± 0.3	10 ± 2	0.47 ± 0.17	1.5 ± 0.3
Polyethylene terephthalate 1	0.153	8.9 ± 0.8	-3.9 ± 0.7	1.1 ± 0.1	-0.47 ± 0.05	14 ± 0.5	0.43 ± 0.06	0.44 ± 0.09
Polyethylene terephthalate 2	0.187	16 ± 2	-5.8 ± 0.7	0.71 ± 0.04	-0.31 ± 0.03	14 ± 0.2	0.44 ± 0.07	0.37 ± 0.06
Nylon 6:6	0.057	7.3 ± 0.7	-1.9 ± 0.4	2.4 ± 0.3	-1.1 ± 0.15	15 ± 1	0.48 ± 0.05	0.26 ± 0.08

$\nu_{13} = -\dfrac{s_{13}}{s_{33}}$ $\qquad \nu_{12} = -\dfrac{s_{12}}{s_{11}}$

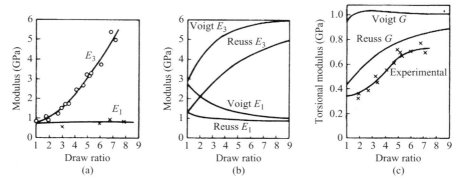

Figure 7.11 Polypropylene filaments; extensional (E_3), transverse (E_1) and torsional moduli (G); comparison between experimental results and simple aggregate theory for E_3 and E_1 ((a) and (b)) and for G (c)

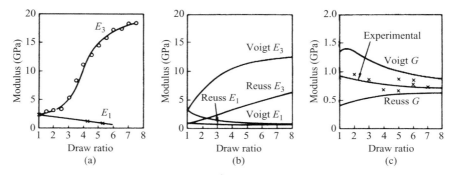

Figure 7.12 Polyethylene terephthalate filaments: extensional (E_3), transverse (E_1) and torsional moduli (G); comparison between experimental results and simple aggregate theory for E_3 and E_1 ((a) and (b)) and for G (c)

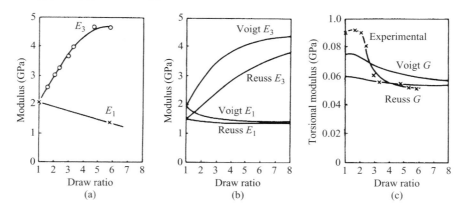

Figure 7.13 Nylon filaments: extensional (E_3), transverse (E_1) and torsional moduli (G); comparison between experimental results and simple aggregate theory for E_3 and E_1 ((a) and (b)) and for G (c)

Although the detailed development of mechanical anisotropy in these particular filaments must depend on their exact chemical composition and subsequent processing, several general features can be distinguished. The principal effect of drawing (i.e. increasing molecular orientation) is to increase the Young's modulus E_3 measured along the filament axis. In nylon 6:6 and polyethylene terephthalate there is a corresponding but small decrease in the transverse modulus E_1, for polypropylene and high-density polyethylene E_1 is almost independent of draw ratio and for low-density polyethylene E_1 increases significantly, in agreement with the results of Raumann and Saunders. Note here, too, the anomalous behaviour of this polymer at low draw ratios. Overall, E_3 for highly oriented filaments is greater than E_1, with the anisotropy being greatest for polyethylene terephthalate:

$$\frac{E_3}{E_1} = \frac{s_{11}}{s_{33}} \sim 27$$

The shear modulus G of low-density polyethylene, and its change with orientation, provides another striking contrast with the other materials examined. A decrease of G by more than a factor of 3 over the range of orientation used compares with only small changes for the other filaments. In polyethylene terephthalate, high-density polyethylene and polypropylene $s_{44} \sim s_{11}$; in nylon $s_{44} \sim 2s_{11}$. By contrast, low-density polyethylene, at least as regards room-temperature behaviour, is exceptional, with the extensional compliance s_{33} having the same order of magnitude as the transverse compliance s_{11}, and with the shear compliance s_{44} being more than an order of magnitude greater than either s_{11} or s_{33}. Such measurements provide the basis for the discussion of relaxation transitions in Chapter 9.

In all cases the compliance s_{13} is low and appears to decrease fairly rapidly with increasing draw ratio, in a manner comparable with s_{33}. Hence the extensional Poisson's ratio $\nu_{13} = s_{13}/s_{33}$ is rather insensitive to draw ratio and, with the exception of high-density polyethylene, does not differ significantly from 0.5. It is thus generally a valid approximation to assume that the filaments are incompressible. NB For anisotropic bodies ν_{13} is not confined to a maximum value of 0.5, but is limited solely by the inequalities necessary for a positive strain energy [2]

$$s_{12}^2 < s_{11}^2; \; s_{13}^2 < \tfrac{1}{2}s_{33}(s_{11} + s_{12})$$

There have been no similar attempts to determine comprehensive sets of elastic constants for oriented fibres or monofilaments. Kawabata [21] devised an apparatus that used a linear differential transformer to measure diametral changes of 0.05 μm in single fibres of diameter 5 μm subjected to transverse compression. Equation (7.2) above was then used to calculate the transverse modulus $E_1 = 1/S_{11}$. Results were obtained for poly(p-phenylene terephthalamide) (Kevlar) and high-modulus polyethylene (Tekmilon) fibres. Values of E_1 were in the range 2.31–2.59 GPa for Kevlar and a value of 1–2 GPa was found for Tekmilon.

Further work in this area has been undertaken by Ward and co-workers [22], combining measurements of contact width and diametral compression on mono-filaments of polyethylene terephthalate (PET), polyethylene and a thermotropic liquid crystalline polymer based on hydroxybenzoic and hydroxynaphthoic acid in the ratio 73:27 (Vectra). Values for E_1 were in the range 1.94–2.34 GPa for PET, 0.63–1.50 GPa for polyethylene and 0.96–1.01 GPa for Vectra.

Recently Hine and Ward [23] have used the ultrasonic immersion method (Section 5.3.2 above) to determine a full set of elastic constants for a range of fibres, by making uniaxially oriented fibre composites. In the first method the oriented composites were produced by the Leeds hot compaction process. Here fibres are compacted under suitable conditions of temperature and pressure to form a homogeneous-oriented material in which only a small fraction of the original fibre is melted and recrystallized to form the matrix of the fibre composite. This matrix fraction can be removed and the results extrapolated to 100 per cent fibre fraction. The measured elastic properties for a range of fibres are shown in Table 7.2. The overall patterns of anisotropy determined by these ultrasonic measurements are similar to those found earlier by Hadley, Pinnock and Ward. It is particularly notable that ν_{12} is greater than 0.5, which is consistent with the view that in highly oriented fibres the high axial stiffness means that transverse compression becomes close to pure shear in the transverse plane.

Table 7.2   The elastic properties of the plates of compacted fibres

Fibre type	E_{33} (GPa)	E_{11} (GPa)	ν_{13}	ν_{12}	G_{13} (GPa)
Tenfor polyethylene	57.7	4.68	0.45	0.55	1.63
Dyneema polyethylene	74.3	4.31	0.47	0.57	1.36
Polyethylene terephthalate	14.9	3.70	0.39	0.65	1.62
Polypropylene	11.0	2.41	0.39	0.58	1.52
Liquid crystal polymer	97.2	3.24	0.48	0.73	1.30

Standard deviation: $E_{33} \pm 3\%$; all others, $\pm 2\%$.

More recently, Wilczynski, Ward and Hine [24] have proposed an 'inverse calculation' method where the elastic constants of a fibre can be estimated from fibre resin composite and the elastic constants of the resin. The method was confirmed by measurements on polyethylene/epoxy and carbon fibre/epoxy resin composites. It has been applied [25] to the determination of the elastic constants of a new organic fibre, poly{2,6-dimidazo[4,5-6: 4'5'-e]pyridinylene – 1,4(2,5-dihydroxy)pheny-lene} (PIPD). This fibre is a lyotropic liquid crystalline fibre with a very high Young's modulus of 285 GPa and a much higher tensile strength (5.21 GPa) and compressive strength (500MPa) than other polyaramid fibres such as Kevlar.

The technique of Brillouin spectroscopy (Section 5.3.3 above) has been applied to determine the elastic constants of oriented polymer fibres. Early studies of this nature were undertaken by Krüger and co-workers [26, 27] on oriented polycarbonate films, also determining the third-order constants that define the elastic non-linear behaviour. Wang and co-workers [28, 29] have described measurements on oriented polyvinylidene fluoride and polychlorotrifluoroethylene films. In the latter case the results were interpreted using an aggregate model differing in detail from that of Ward discussed in Section 7.5 below.

Recent Brillouin spectroscopy measurements include those of Kumar, Renisch and Grimsditch [30] on uniaxially and biaxially stretched polypropylene films. Using the Ward aggregate model with a modified definition of molecular alignment enabled elastic anisotropy and refractive index data to be related quantitatively to molecular orientation.

Choy, Leung and colleagues, initially in collaboration with Ward, have undertaken very extensive measurements of the elastic constants of oriented polymers using the ultrasonic techniques described in Section 5.3.2 above. Results have been obtained for hydrostatically extruded polypropylene and polyethylene terephthalate [31], die-drawn polyethylene [32], hot-rolled nylon 6:6 [33] and polyoxymethylene [34], high-modulus polyethylene [35] and polypropylene [36] produced by tensile drawing. The results have been analysed in terms of the aggregate model, which has proved a fair approximation for the low-temperature data, and the Takayanagi model with tie molecules or crystalline bridges, which is more satisfactory for high-temperature data.

7.4 Interpretation of mechanical anisotropy: general considerations

The mechanical anisotropy of oriented polymers is determined by the following factors, which will be discussed in turn: (1) the structure of the molecular chain and, where the polymer crystallizes, the crystal structure; (2) the molecular orientation and, in a crystalline polymer, the morphology; (3) thermally activated relaxation processes in both the crystalline and non-crystalline regions.

7.4.1 Theoretical calculations of elastic constants

In recent years there has been increased interest in the theoretical calculation of the elastic constants for ideal and fully oriented polymers based on knowledge of their crystal structures. This increased interest arises from the major developments in computational methods and from the success achieved in producing very highly oriented polymers with reasonably high stiffness.

The theoretical calculations require knowledge of the force constants for deformation of the structure. These fall into two categories, those relating to intramolecular deformation of the polymer chains, which are primarily bond stretching and bond bending, and those concerned with intermolecular deformations, which relate to dispersion forces and can be represented, for example, by Lennard-Jones potentials.

Estimation of the intramolecular deformations of polymer chains is comparatively straightforward, provided that the chain structure is well defined, because the force constants for bond stretching and bond bending can be determined by infrared and Raman spectroscopy. This means that an estimate of the so-called chain modulus for important crystalline polymers such as polyethylene can be obtained quite readily. Values in the range of 100 GPa are obtained for polymers that have an extended chain structure, e.g. polyethylene and polyethylene terephthalate, and values of 20–40 GPa for polymers showing helical structures, e.g. polypropylene. Generally these values are not observed in practice, but special processing methods have been devised to produce fibres with very high stiffness, approaching the theoretical estimates in some instances, such as polyethylene and polypropylene.

It is much more difficult to calculate the other elastic constants because the intermolecular deformations depend critically on fine details of the structure and on intermolecular forces that are more difficult to estimate accurately. Nevertheless there has been considerable success in pursuing such calculations and full sets of elastic constants for the crystalline regions of common polymers such as polyethylene, polyethylene terephthalate and polypropylene have been derived. The most straightforward method is based on determining the potential energy of a crystalline unit cell by adding together all the internal energy contributions, both intramolecular (bond stretching, bond bending and internal rotation) and intermolecular (e.g. Lennard-Jones 6–12 potentials). The elastic constants are then calculated by imposing the appropriate deformation to the unit cell, e.g. tensile deformation along the chain axis, and reminimizing the energy with respect to intermolecular and intramolecular degrees of freedom, allowing the components of the structure to deform in a non-affine manner (i.e. the deformation of elements of structure do not mimic the macroscopic deformation). This procedure defines the relationship between the internal energy and the particular strain considered, hence determining the corresponding elastic constant.

The results of such calculations for semi-crystalline polyethylene have been reviewed elsewhere [37]. A rather wide range of predicted values is obtained, due to the choice of force constants and also to sensitivity to detailed assumptions on the unit cell structure. In spite of these limitations the principal predictions for the elastic anisotropy are clear. These include the anticipated high values for c_{33} and the very low values for the shear stiffnesses c_{44}, c_{55} and c_{66}, which reflect the major differences between bond stretching and bond bending forces that control c_{33} and the intermolecular dispersion forces that determine the shear stiffnesses. It is therefore of value to compare such theoretical results with those obtained experimentally. Table 7.3 shows results for polyethylene where data for the orthorhombic unit cell at 300 K are used to calculate these constants for an equivalent fibre (Voigt averaging procedure; see Section 7.5.2 below) compared with ultrasonic data for a solid sheet made by hot compaction. It can be seen that

Table 7.3 Comparison of hot compacted sheet stiffness constants (in GPa) with theoretical predictions for polyethylene

	c_{33}	c_{11}	c_{13}	c_{12}	c_{44}
Theoretical values for uniaxially oriented sheet	290	9.15	5.15	3.95	2.86
Hot compacted sheet (Tenfor)	62.3	7.16	5.09	4.15	1.63

the predicted and experimental patterns of anisotropy are similar, with $c_{33} \gg c_{11} > c_{44}$. It is not surprising that the experimental values for c_{33} are lower than the theoretical values because these materials have not reached full axial orientation, and other production methods such as hot drawing of gel-spun fibres give values of c_{33} at low temperatures of about 300 GPa. The experimental results for the other elastic constants are very much in line with the theoretical estimates, giving some credence to the latter, in spite of the difficulties associated with defining the force-field parameters and the precise structure.

The determination of the elastic constants for the crystalline regions of polymers is of importance for providing a baseline against which practical achievements can be judged. Comprehensive reviews have been published that deal not only with theoretical calculations but with moduli determined from the changes in the X-ray diffraction pattern on stressing, by Raman spectroscopy and from the inelastic scattering of neutrons [38].

7.4.2 Orientation and morphology

The degree of mechanical anisotropy is usually much less than considerations of the molecular chain would imply; and, in particular, the very high intrinsic modulus along the axial direction is not achieved. This is demonstrated in Table 7.4, where measured values of elastic constants for ultrahigh-modulus polyethylene are compared with theoretical values. Crystalline polymers are essentially composite materials with alternating crystalline and amorphous regions. The former regions can be highly aligned during processing, but the latter are less well oriented. Even when the overall orientation of the chain segments appears to be quite high, measured for instance by birefringence, there are still very few chains

Table 7.4 Elastic constants of ultrahigh-modulus polyethylene

	20 °C	−196 °C	Theoretical (Tadokoro)
Axial modulus (GPa)	70	160	316
Transverse modulus (GPa)	1.3	–	8–10
Shear modulus (GPa)	1.3	1.95	1.6–3.6
Poisson's ratio	0.4	–	0.5

where long lengths of a molecule are axially aligned. It is such molecules that are critical in increasing the stiffness, because there is such a large difference between the stresses involved in bond bending and stretching and in other modes of deformation. Peterlin [39] has proposed that the Young's modulus of an oriented filament is essentially determined by the proportion of extended chain tie molecules that produce links in the axial direction between crystalline blocks. An alternative proposal of crystalline bridges linking the crystalline blocks is discussed in Chapter 8.

Polymers that do not crystallize, such as polymethyl methacrylate, show a good correlation between the (low) degree of mechanical anisotropy and molecular orientation determined from birefringence. There is so much disorder that it seems unlikely that a significant proportion of the chains can achieve the high alignment of a crystalline polymer such as polyethylene. Other polymers such as polyethylene terephthalate, which have a comparatively low overall crystallinity ($\sim$30 per cent is typical), may occupy an intermediate position. The mechanical anisotropy produced by drawing correlates well with overall molecular orientation, but this result may arise because tie molecules play a vital role, and their number increases with overall molecular orientation.

In conclusion, it must be emphasized that although it is convenient for the purpose of constructing models to assume a composite that comprises distinct crystalline and non-crystalline components, on the molecular level a gradual transition must occur, extending over a number of monomer units, between the well-orientated and ordered crystallites and the bulk of the remaining material.

7.5 Experimental studies of anisotropic mechanical behaviour and their interpretation

In general it is to be expected that mechanical anisotropy will depend both on the crystalline morphology and the molecular orientation. Two extreme models form the basis for present understanding: a single-phase aggregate model proposed by Ward in which anisotropy arises through the orientation of anisotropic units of structure; and a microscopic model, proposed by Takayanagi, in which the crystalline and amorphous regions are considered as two distinct phases. It is appropriate to discuss the mechanical anisotropy of polymers in terms of one or other of these models. In the present chapter those polymers whose behaviour best approximates to the Ward aggregate model will be discussed. In the following chapter those polymers that are best considered as composite solids will be discussed, following an introductory presentation to the theoretical understanding of the mechanical behaviour of composite materials.

7.5.1   The aggregate model and mechanical anisotropy

In this model it is proposed that the polymer can be considered as an aggregate of identical units, which in the unstretched state are oriented randomly. As orientation develops, the units rotate and become completely aligned at the maximum achievable orientation. The elastic properties of the units are those of the mostly highly aligned structure, which could be a fibre with transverse isotropy or a film with orthorhombic symmetry [40]. In this chapter the theory will be developed for a fibre, where both the fibre and the structural units possess transverse isotropy.

It is instructive to test the appropriateness of the model in two stages:

1. Can the elastic constants of the isotropic polymer be deduced from measurements on the most highly oriented sample?

2. Does the mechanical anisotropy develop with orientation in the predicted manner?

We shall consider each stage in turn.

7.5.2   Correlation between the elastic constants of a highly oriented and an isotropic polymer

Table 7.5 gives an example of the development of mechanical properties as the level of molecular orientation is increased; we wish to model this effect. The average elastic constants for the isotropic aggregate can be obtained in two ways, either by assuming that the constituent units are in series or in parallel (Figure 7.14). The former case assumes a summation of strains for each unit subjected to the same stress (which implies a summation of compliance constants), and the latter assumes that each unit undergoes the same strain, with the stresses being summed (which implies a summation of stiffness constants). In general the principal axes of stress and of strain do not coincide in an anisotropic solid, with the result that both of the above approaches involve an approximation: for the

Table 7.5   Physical properties of polyethylene terephthalate fibres at room temperature

Birefringence	X-ray crystallinity	Extensional modulus (GPa)	Torsional modulus (GPa)
0	0	2.0	0.77
0	33	2.2	0.89
0.142	31	9.8	0.81
0.159	30	11.4	0.62
0.190	29	15.7	0.79

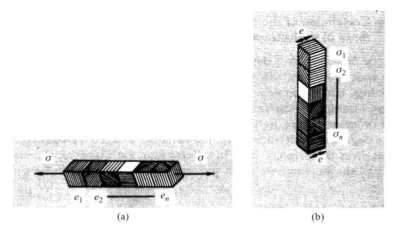

Figure 7.14   The aggregate model for uniform stress (a) and uniform strain (b)

assumption of uniform stress the strains throughout the aggregate are not uniform; for the alternative assumption of uniform strain the stresses are non-uniform. Bishop and Hill [41] have shown that for a random aggreagate the correct response lies between the extremes predicted by the alternative schemes.

Consider the polymer to be comprised of N identical, transversely isotropic but randomly oriented cubes. For the case of uniform stress (Figure 7.14(a)) the cubes are arranged end-to-end, forming a series model in which the direction of elastic symmetry is defined by the angle θ that the cube axis makes with the direction of external stress σ. The strain in a single cube e_1 is given by the compliance formula

$$e_1 = [s_{11} \sin^4 \theta + s_{33} \cos^4 \theta + (2s_{13} + s_{44}) \sin^2 \theta \cos^2 \theta]\sigma$$

where s_{11}, etc. are the compliance constants of the cube. We ignore the fact that the cubes will distort under the applied stress and do not satisfy compatibility of strain throughout the aggregate. Then the average strain e is

$$e = \frac{\sum e_1}{N} = [s_{11} \overline{\sin^4 \theta} + s_{33} \overline{\cos^4 \theta} + (2s_{13} + s_{44}) \overline{\sin^2 \theta \cos^2 \theta}]\sigma$$

where $\overline{\sin^4 \theta}$, etc. now define the average values of $\sin^4 \theta$, etc. for the aggregate of units. Evaluating the average values of the trigonometrical functions we obtain for the average extensional compliance

$$\overline{s'_{33}} = \frac{e}{\sigma} = \frac{8}{15} s_{11} + \frac{1}{5} s_{33} + \frac{2}{15}(2s_{13} + s_{44}) \tag{7.4}$$

For the alternative uniform strain model, where the N elementary cubes are stacked in parallel (Figure 7.14(b)), the stress in a single cube is given by

$$\sigma_1 = [c_{11} \sin^4 \theta + c_{33} \cos^4 \theta + 2(c_{13} + 2c_{44}) \sin^2 \theta \cos^2 \theta]e$$

where c_{11}, etc. are the stiffness constants of the cube. Taking average values as before, we obtain for a random aggregate

$$\overline{c'_{33}} = \frac{\sigma}{e} = \frac{8}{15}c_{11} + \frac{1}{5}c_{33} + \frac{4}{15}(c_{13} + 2c_{44}) \tag{7.5}$$

For an isotropic polymer there are two independent elastic constants, and the two alternative schemes predict a value for the isotropic shear compliance $\overline{s'_{44}}$ and the isotropic shear stiffness $\overline{c'_{44}}$ respectively.

$$\overline{s'_{44}} = \tfrac{14}{15}s_{11} - \tfrac{2}{3}s_{12} - \tfrac{8}{15}s_{13} + \tfrac{4}{15}s_{33} + \tfrac{2}{5}s_{44} \tag{7.6}$$

$$\overline{c'_{44}} = \tfrac{7}{30}c_{11} - \tfrac{1}{6}c_{12} - \tfrac{2}{15}c_{13} + \tfrac{1}{15}c_{33} + \tfrac{2}{5}c_{44} \tag{7.7}$$

Averaging the compliance constants defines the elastic properties of the isotropic aggregate in terms of $\overline{s'_{33}}$ and $\overline{s'_{44}}$, giving the 'Reuss average' [42]. Averaging the stiffness constants defines the properties in terms of $\overline{c'_{33}}$ and $\overline{c'_{44}}$, giving the 'Voigt average' [43]. In the latter case the matrix can be inverted to obtain the corresponding $\overline{s'_{33}}$ and $\overline{s'_{44}}$, in order to make a direct comparison between the two averaging procedures.

Table 7.6 compares the measured compliances for isotropic samples of five polymers with the Reuss and Voigt average compliances calculated from measurements on highly oriented specimens. For polyethylene terephthalate and low-density polyethylene the measured isotropic compliances fall between the calculated bounds, suggesting that here molecular orientation could well be the principal factor that determines mechanical anisotropy. For nylon 6:6 the

Table 7.6 Comparison of calculated and measured extensional and torsional compliances (units are 10^{-10} Pa^{-1}) for unoriented fibres

	Extensional compliance $(\overline{s'_{11}} = \overline{s'_{33}})$			Torsional compliance $(\overline{s'_{44}})$		
	Reuss average	Voigt average	Measured	Reuss average	Voigt average	Measured
Low-density polyethylene	139	26	81	416	80	238
High-density polyethylene	10	2.1	17	30	6	26
Polypropylene	7.7	3.8	14	23	11	2.7
Polyethylene terephthalate	10.4	3.0	4.4	25	7.6	11
Nylon	6.6	5.2	4.8	17	13	12

measured compliances fall just outside the bounds, which suggests that both molecular orientation and structural factors are important. In high-density polyethylene and in polypropylene the extensional compliances lie well outside the calculated bounds, suggesting that factors other than orientation play a major role in determining the mechanical anisotropy.

7.5.3   The development of mechanical anisotropy with molecular orientation

The starting point for the extension of aggregate theory to transversely isotropic polymers of intermediate orientation lay in earlier investigations of the manner in which birefringence increased as a function of the draw ratio imposed on initially isotropic specimens. Some results for low-density polyethylene are shown in Figure 7.15(a). Crawford and Kolsky [44] proposed a model of transversely isotropic rod-like units that rotated towards the draw axis. It was assumed that overall deformation occurred uniaxially at constant volume (a reasonable assumption for most polymers, for which the extensional Poisson's ratio $v_{13} \sim 0.5$ (see Section 7.3)) and that the symmetry axes of the anisotropic units rotated in the same manner as lines joining pairs of points in the macroscopic body. This assumption differs from the 'affine' deformation scheme for the optical anisotropy of rubbers (see Section 3.1) in ignoring any change in length of the units on deformation, and has been called 'pseudo-affine' deformation. As the result of drawing, the angle between the rod axis and the direction of draw changes from θ to θ', and it can be shown that for a draw ratio λ

$$\tan \theta' = \frac{\tan \theta}{\lambda^{3/2}}$$

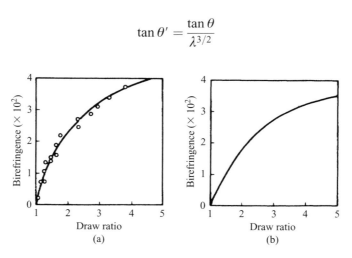

Figure 7.15   Experimental (a) and theoretical (b) curves for the birefringence of low-density polyethylene as a function of draw ratio

which leads to a birefringence Δn, given by

$$\Delta n = \Delta n_{max} \left(1 - \tfrac{3}{2}\overline{\sin^2 \theta}\right)$$

where $\overline{\sin^2 \theta}$ is the average value of $\sin^2 \theta$ for the aggregate of rod-like units and Δn_{max} is the birefringence at full orientation.

The pseudo-affine deformation scheme gives a reasonable approximation to experimental data for low-density polyethylene (Figure 7.15(b)), nylon, polyethylene terephthalate and polypropylene, despite ignoring the distinction between crystalline and disordered regions. It thus provides a basis for extending Ward's aggregate model to predict the compliance constants s'_{ij} and stiffness constants c'_{ij} of the partially oriented polymer in terms of the constants s_{ij} and c_{ij} for the anisotropic elastic unit (in practice, those of the most highly oriented sample obtainable).

$$s'_{11} = \tfrac{1}{8}(3I_2 + 2I_5 + 3)s_{11} + \tfrac{1}{4}(3I_3 + I_4)s_{13} + \tfrac{3}{8}I_1 s_{33} + \tfrac{1}{8}(3I_3 + I_4)s_{44}$$

$$c'_{11} = \tfrac{1}{8}(3I_2 + 2I_5 + 3)c_{11} + \tfrac{1}{4}(3I_3 + I_4)c_{13} + \tfrac{3}{8}I_1 c_{33} + \tfrac{1}{2}(3I_3 + I_4)c_{44}$$

$$s'_{12} = \tfrac{1}{8}(I_2 - 2I_5 + 1)s_{11} + I_5 s_{12} + \tfrac{1}{4}(I_3 + 3I_4)s_{13} + \tfrac{1}{8}I_1 s_{33}$$
$$\qquad + \tfrac{1}{8}(I_3 - I_4)s_{44}$$

$$c'_{12} = \tfrac{1}{8}(I_2 - 2I_5 + 1)c_{11} + I_5 c_{12} + \tfrac{1}{4}(I_3 + 3I_4)c_{13} + \tfrac{1}{8}I_1 c_{33}$$
$$\qquad + \tfrac{1}{2}(I_3 - I_4)c_{44}$$

$$s'_{13} = \tfrac{1}{2}I_3 s_{11} + \tfrac{1}{2}I_4 s_{12} + \tfrac{1}{2}(I_1 + I_2 + I_5)s_{13} + \tfrac{1}{2}I_3 s_{33} - \tfrac{1}{2}I_3 s_{44}$$

$$c'_{13} = \tfrac{1}{2}I_3 c_{11} + \tfrac{1}{2}I_4 c_{12} + \tfrac{1}{2}(I_1 + I_2 + I_5)c_{13} + \tfrac{1}{2}I_3 c_{33} - 2I_3 c_{44}$$

$$s'_{33} = I_1 s_{11} + I_2 s_{33} + I_3(2s_{13} + s_{44})$$

$$c'_{33} = I_1 c_{11} + I_2 c_{33} + 2I_3(c_{13} + 2c_{44})$$

$$s'_{44} = (2I_3 + I_4)s_{11} - I_4 s_{12} - 4I_3 s_{13} + 2I_3 s_{33}$$
$$\qquad + \tfrac{1}{2}(I_1 + I_2 - 2I_3 + I_5)s_{44}$$

$$c'_{44} = \tfrac{1}{4}(2I_3 + I_4)c_{11} - \tfrac{1}{4}I_4 c_{12} - I_3 c_{13} + \tfrac{1}{2}I_3 c_{33}$$
$$\qquad + \tfrac{1}{2}(I_1 + I_2 - 2I_3 + I_5)c_{44}$$

The terms $I_1 \ldots I_5$ are orientation functions that define average values of $\sin^4 \theta(I_1)$, $\cos^4 \theta(I_2)$, $\cos^2 \theta \sin^2 \theta(I_3)$, $\sin^2 \theta(I_4)$ and $\cos^2 \theta(I_5)$ for the aggregate. Only two of these orientation functions are independent parameters (e.g. $I_4 = I_1 + I_3$, $I_5 = I_2 + I_3$, $I_4 + I_5 = 1$).

In Figures 7.9–7.13 the predictions of the aggregate model assuming the pseudo-affine deformation scheme are compared with experimental measurements. For low-density polyethylene the Reuss average predictions show the correct general form, and provide an explanation for the minimum in the extensional modulus. With increasing orientation θ decreases, so that $\overline{\sin^4 \theta}$ decreases monotonically and $\overline{\cos^4 \theta}$ increases with increasing draw ratio, whereas $\overline{\sin^2 \theta \cos^2 \theta}$ shows a maximum at a draw ratio of about 1.2. Thus s'_{33} can pass through a maximum (corresponding to a minimum in the Young's modulus E_0) provided that $(2s_{13} + s_{44})$ is sufficiently large compared with s_{11} and s_{33} (which differ only slightly at low draw ratios). For low-density polyethylene at room temperature s_{44} is much larger than either s_{11} or s_{33}, so fulfilling the condition. At much lower temperatures s_{44} is no longer much greater than the other compliances, so that a more conventional pattern of mechanical anisotropy is observed.

The theoretical curves of Figures 7.9–7.13 differ from those obtained experimentally in two ways. Some features of detail, such as a small minimum in the transverse modulus of low-density polyethylene and a small minimum in the extensional modulus of high-density polyethylene, are not predicted at all. However, it has been shown that such effects may be associated with mechanical twinning [45]. A second deficiency is that the predicted development of mechanical anisotropy with draw ratio is considerably less rapid than is observed. The quantities $\overline{\sin^4 \theta}$, $\overline{\cos^4 \theta}$ and $\overline{\sin^2 \theta \cos^2 \theta}$ can also be determined by X-ray diffraction and nuclear magnetic resonance methods, and for low-density polyethylene a much improved fit was obtained by these means. For this material it appears that the anisotropy relates to the orientation of the crystalline regions, and is predicted to a good approximation by the Reuss averaging scheme. A similar explanation is unlikely for polyethylene terephthalate, where other evidence suggests that the anisotropy resembles that of a deformed network. In this material, moreover, the experimental compliances lie approximately midway between the two bounds, with the median condition applying almost exactly for cold-drawn filaments [46].

In nylon the Voigt average lies closer to the experimental data and, although the quantitative fit is poor, both averaging schemes predict a maximum in the torsional modulus at an intermediate draw ratio. For polypropylene the aggregate model appears appropriate only at low draw ratios, which is consistent with evidence for simultaneous changes in morphology and molecular mobility for highly drawn specimens. The model does not appear appropriate for high-density polyethylene, a highly crystalline material in which a two-phase lamellar texture is often evident (see Section 8.4.1).

In particular polymers either the Reuss or the Voigt averages or a mean of the two lie closest to measured values. It is likely that these differences relate to details of the stress and strain distributions at a molecular level, which should in turn be related to the structure.

Kausch, who has applied the aggregate model to a range of both crystalline and amorphous polymers [47], also noted that for some materials the experimental increase in extensional modulus with stiffness was significantly greater than predicted, and has suggested that the effect may be due to an additional orientation of segments within each unit of the aggregate. Another possibility is that the increase in stiffness of crystalline polymers may be enhanced at the higher draw ratios through the competing process of pulling out more intercrystalline tie molecules. Subsequently [48] Kausch emphasized the value of reformulating the model in terms of a molecular network rather than orienting rods, which would allow the representation of high strain properties.

In summary it is evident that despite the highly simplistic nature of the assumptions made the aggregate model provides an appropriate model for a number of important polymers. For such materials details of the crystal structure can play no more than a subsidiary role in the development of mechanical anisotropy, and the deformation is essentially that expected for a single-phase texture or a distorted network.

7.5.4   The anisotropy of amorphous polymers

Relatively few measurements have been performed on amorphous polymers, but some typical data are summarized in Table 7.7 [49, 50] and Figure 7.16 [51].

Table 7.7   Elastic compliances of oriented amorphous polymers (units of compliance are 10^{-9} Pa^{-1})

Material	Draw ratio	s_{33}	s_{11}	s_{44}
Polyvinyl chloride	1	0.313	0.313	0.820
	1.5	0.276	0.319	0.794
	2.0	0.255	0.328	0.781
	2.5	0.243	0.337	0.769
	2.8	0.238	0.341	0.763
	∞	0.204	0.379	0.730
Polymethyl methacrylate	1	0.214	0.214	0.532
	1.5	0.208	0.215	0.524
	2.0	0.204	0.215	0.518
	2.5	0.200	0.216	0.510
	3.0	0.196	0.217	0.505
Polystyrene	1	0.303	0.303	0.769
	2.0	0.296	0.304	0.769
	3.0	0.289	0.305	0.769
Polycarbonate	1	0.376	0.376	1.05
	1.3	0.314	0.408	0.980
	1.6	0.268	0.431	0.926

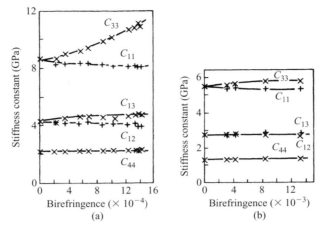

Figure 7.16 Stiffness constants of uniaxially drawn amorphous polymers, measured at room temperature, as a function of birefringence: (a) polymethyl methacrylate; (b) polystyrene. (Reproduced with permission from Wright, *et al.*, *J. Phys. D*, **4**, 2002 (1971))

It is to be anticipated that the aggregate model will be appropriate for considering the development of anisotropy in amorphous polymers, where there are no complications due to crystallinity and changes in morphology. This expectation is borne out by results for polymethylmethacrylate [52], where the experimental stiffnesses as a function of birefringence lie between the predicted Reuss and Voigt bounds. Kausch [48], Rawson and Rider [53] and Hennig [49] have also found the aggregate model appropriate for discussing the mechanical anisotropy of non-crystalline polymers, and the development of anisotropy with draw ratio often can be described in terms of pseudo-affine deformation.

7.5.5   Later applications of the aggregate model

All nine independent elastic constants have been measured for polyethylene terephthalate sheet drawn uniaxially at constant width to provide a material with single crystal texture [54]. The high degree of mechanical anisotropy (Table 7.8) is related to the high chain axis orientation and the preferential orientation of the terephthalate residues. The extensional compliance s_{33} is low, presumably because the deformation involves bond bending and stretching in extended chain molecules. By contrast the much higher transverse compliances s_{11} and s_{22} relate principally to dispersion forces. It follows that, if the material is stretched perpendicular to the draw direction, the principal contraction occurs in the second perpendicular direction rather than in the draw direction. The shear compliances s_{44} and s_{66} are large compared with s_{55}, reflecting easy shear in the 23 and 12 planes, respectively, presumably as the result of planar terephthalate chains sliding over one another, constrained only by weak dispersion forces. The shear com-

Table 7.8 Full set of compliances for an oriented polyethylene terephthalate sheet with orthorhombic symmetry

Compliance	Value ($\times 10^{-10}$ Pa^{-1})
s_{11}	3.61 ± 0.12
s_{22}	9.0 ± 1.6
s_{33}	0.66 ± 0.01
s_{12}	-3.8 ± 0.4
s_{13}	-0.18 ± 0.01
s_{23}	-0.37 ± 0.05
s_{44}	97 ± 3
s_{55}	5.64 ± 0.25
s_{66}	141 ± 8

pliance s_{55} involves distortion of the plane of the poyester molecules, and is of the same order as s_{11} and s_{22}.

It is interesting to apply the aggregate model to these data, calculating bounds for the elastic constants of an 'equivalent fibre' by averaging the sheet constants in the plane normal to the sheet draw direction. This requires an extension of the mathematical treatment of Section 7.5.3 to deal with the case of a transversely isotropic aggregate of orthorhombic units. The basic equations are given in detail elsewhere so only the key results will be summarized here. If the orthorhombic unit constants are $s_{11}, s_{13} \ldots, s_{66}$, the Reuss average fibre constants $s'_{33}, s'_{13}, \ldots, s'_{44}$ obtained by averaging in the 12 plane are given by

$$s'_{33} = s_{33}$$

$$s'_{11} = \tfrac{3}{8}s_{11} + \tfrac{1}{4}s_{12} + \tfrac{3}{8}s_{22} + \tfrac{1}{6}s_{66}$$

$$s'_{12} = \tfrac{1}{8}s_{11} + \tfrac{3}{4}s_{12} + \tfrac{1}{8}s_{22} - \tfrac{1}{8}s_{66}$$

$$s'_{13} = \tfrac{1}{2}(s_{13} + s_{23})$$

$$s'_{44} = \tfrac{1}{2}(s_{44} + s_{55})$$

and the Voigt average fibre constants in similar terms are $c'_{33}, c'_{13}, \ldots, c'_{44}$, where

$$c'_{33} = c_{33}$$

$$c'_{11} = \tfrac{3}{8}c_{11} + \tfrac{1}{4}c_{12} + \tfrac{3}{8}c_{22} + \tfrac{1}{2}c_{66}$$

$$c'_{12} = \tfrac{1}{8}c_{11} + \tfrac{3}{4}c_{12} + \tfrac{1}{8}c_{22} - \tfrac{1}{2}c_{66}$$

$$c'_{13} = \tfrac{1}{2}(c_{13} + c_{23})$$

$$c'_{44} = \tfrac{1}{2}(c_{44} + c_{55})$$

The results of this calculation are shown in Table 7.9 together with the experimental value obtained for a highly oriented fibre monofilament. Although the experimental values do not always lie exactly within the predicted bounds, they are always in the correct range. In Table 7.9 a comparison is given between the calculated and measured compliance constants for isotropic polyethylene terephthalate based on the sheet data. Again the measured values lie between the Reuss and Voigt bounds. Taking into account the very large degree of anisotropy and the very simplistic nature of these calculations, these results afford support for the contention that to a first approximation the mechanical anisotropy can be considered in terms of the single-phase aggregate model.

Table 7.9 Comparison of calculated and measured compliance constants ($\times 10^{-10}$ Pa^{-1}) for polyethylene terephthalate fibres based on the sheet compliances

Compliance constant	Calculated bounds		Experimental value
	Reuss	Voigt	
Highly oriented fibres			
s_{11}	21	7.3	16.1
s_{12}	−19	−5.5	−5.8
s_{13}	−0.28	−0.25	−0.31
s_{33}	0.66	0.66	0.71
s_{44}	51	10.7	13.6
Isotropic fibres			
s_{33}	18	2.4	4.4
s_{44}	53	6.4	11

7.6 The aggregate model for chain-extended polyethylene and liquid crystalline polymers

Annealing polyethylene at high temperatures ($\sim$230–240 °C) and high pressures ($\sim$200 MPa) produces a chain-extended structure consisting of small domains where the crystal thicknesses are $\sim$2 μm, i.e. an aggregate of small, highly aligned units. This material can be aligned by hydrostatic extrusion, where a billet is extruded in the solid phase through a conical die to give a product with moderately high orientation and Young's moduli of $\sim$ 40 GPa in the alignment direction. Not surprisingly, the aggregate model can be applied to the development of orientation, which follows the pseudo-affine deformation scheme, leading to an understanding of the mechanical anisotropy [55].

On the compliance averaging scheme, the tensile modulus of the oriented polymer E_3 is given by

$$\frac{1}{E_3} = s'_{33} = s_{11}\overline{\sin^4\theta} + s_{33}\overline{\cos^4\theta} + (2s_{13} + s_{44})\overline{\sin^2\theta\cos^2\theta} \qquad (7.8)$$

As before, θ represents the angle between the unit of the aggregate and the axis representing total alignment, and $\overline{\sin^4\theta}$, etc. represent average values.

For the very high degree of molecular orientation found in these polymers $\overline{\sin^4\theta} \ll 1$ and $\overline{\cos^4\theta} \equiv \cos^2\theta = 1$. We can then rewrite Equation (7.8) as

$$\frac{1}{E_3} = s'_{33} = \frac{1}{E_c} + s_{44}\overline{\sin^2\theta} \qquad (7.9)$$

where E_c is the tensile modulus of the aggregate unit. To a similar degree of approximation it can be shown that

$$s'_{44} = s_{44} = \frac{1}{G} \qquad (7.10)$$

where G is the shear modulus of the polymer.

Combining Equations (7.9) and (7.10)

$$\frac{1}{E_3} = \frac{1}{E_c} + \frac{\overline{\sin^2\theta}}{G} \qquad (7.11)$$

Figure 7.17 show results for a plot of $1/E_3$ versus $1/G$ for hydrostatic extrusion of pressure-annealed polyethylene taken to a series of extrusion ratios (equivalent to draw ratios). It can be seen that the gradients of the fitted lines reduce with increasing extrusion ratio as the orientation parameter $\overline{\sin^2\theta}$ reduces, and that the lines converge to a value of $E_c \sim 250$ GPa, in the correct range for the chain modulus of polyethylene.

The aggregate model also has been used with success to describe the mechanical anisotropy of several liquid crystalline polymers. Ward and co-workers [56] examined the dynamic mechanical behaviour of several thermotropic polyesters in tension and shear over a wide temperature range, and used the single-phase aggregate model to relate quantitatively the fall in tensile modulus with temperature to the corresponding fall in shear modulus.

Figure 7.18 shows data for the temperature dependence for the tensile modulus and the shear modulus for a highly oriented thermotropic copolyester, on a plot of $1/E_3$ versus $1/G$. The results lie on a reasonable straight line extrapolating to a

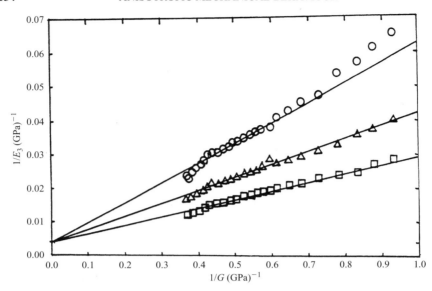

Figure 7.17 A plot of $1/E_3$ against $1/G$ showing the validity of the aggregate model analysis for a range of R006-60 extrudates previously pressure annealed at 23°C for 15 min at 450 MPa pressure: (□)10:1; (△) 7:1; (○) 5:1. (Reproduced with permission from *J. Mater. Sci.*, **25** 3990 (1990).)

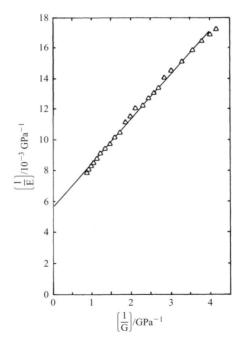

Figure 7.18 Plot of $1/E_3$ versus $1/G$ for a highly oriented thermotropic copolyester

value of 173 GPa for the tensile modulus of a unit of aggregate. This value compares favourably with theoretical estimates based on bond stretching and bond bending modes of deformation.

In a second version of the aggregate model Ward and co-workers assumed, on the basis of observation of the X-ray diffraction pattern, that the aggregate unit averages the deformation over a length of 8–10 monomer units. The chain modulus in this case can be determined experimentally by measuring the change in the X-ray diffraction pattern under stress, and a temperature-dependent E_c was observed. By rearrangement of Equation (7.11) a further plot of

$$\left(\frac{1}{E_3} - \frac{1}{E_c}\right) \text{ versus } \frac{1}{G}$$

was obtained (Figure 7.19) and this gave a good straight line through the origin. The value of $\overline{\sin^2 \theta}$ obtained from the slope of this line is less than that found from the fitting procedure of Figure 7.18, which reflects the misalignment along the length of the liquid crystalline polymer chain. Because of the sinuous nature of the polymer chains the authors have called the effect the 'sinuosity'.

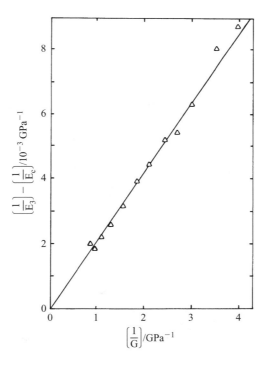

Figure 7.19 Plot of $(1/E_3 - 1/E_c)$ versus $1/G$ for a highly oriented thermotropic copolyester

Northolt and Van Aartsen [57] used the aggregate model to interpret the rapid increase in Young's modulus with increasing crystalline orientation for polyparaphenylene terephthalamide (PPTA) fibres. The PPTA is a lytropic liquid crystalline polyester whose X-ray diffraction pattern suggests a crystallite size of about 70 mm in the axial direction and 5 mm transverse.

The sonic modulus E_{sonic} and the X-ray data for $\overline{\sin^2 \theta}$ were fitted on the assumption that PPTA fibres can be considered an aggregate of small crystals (similar to chain-extended polyethylene discussed above). We have

$$\frac{1}{E_{sonic}} = \frac{1}{E_c} + A \overline{\sin^2 \theta} \qquad (7.12)$$

It was shown that the measured values for E_c and A agreed well with those calculated theoretically. For high orientation $A = 1/G$, so Equation (7.12) is exactly equivalent to Equation (7.11) above.

In more recent research, Northolt and co-workers [58] have proposed explicit theoretical models for the elastic modulus of a fibre consisting of fibrils containing rigid-rod chains. These models take as their starting point a planar zigzag chain. Deformation can occur by extensional strain along the axes of the rigid chain segments (e.g. PQ in Figure 7.20) and by shear. The extensional compliance of the fibre S_{33} is given by

$$S_{33} = \frac{1}{E_c} + \frac{<\sin^2 \theta>}{2G}$$

where E_c is the axial Young's modulus of each segment, G is the shear modulus of the fibre and

$$<\sin^2 \theta> = \frac{\int_0^{\pi/2} R'(\theta)\sin^2 \theta \cos \theta d\theta}{\int_0^{\pi/2} R(\theta_0) \cos \theta_0 d\theta_o}$$

Figure 7.20   Deformation of a rigid chain on the Northolt/Van Aartson model

Note that this equation differs from the classical aggregate model equation in two respects:

1. The denominator of the second term as the left-hand side is $2G$, and not G.

2. The definition of $<\sin^2 \theta>$ differs from that for $\overline{\sin^2\theta}$ in that the average is carried out in the plane of the chains and is not a three-dimensional average; $R(\theta_0)$ refers to the initial chain distribution and $R'(\theta)$ the final distribution under stress.

7.7 Auxetic materials: negative Poisson's ratio

Although the possibility of materials with a negative Poisson's ratio had been recognized theoretically it is only comparatively recently, since the mid 1980s, that examples of such materials have become available. These materials expand laterally when stretched and contract laterally when compressed, and are called auxetic from the classical Greek word *auxetos*, meaning to increase.

A simple example of a material with a negative Poisson's ratio is a cellular foam based on a re-entrant honeycomb, shown in Figure 7.21. On stretching, the axial fibrils straighten to produce lateral expansion.

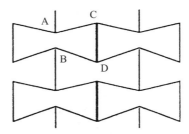

Figure 7.21   Re-entrant honeycomb with negative Poisson's ratio made of bendable ligaments. A similar structure can be made with rigid ligaments if a spring is placed between points of type A and B. (Reproduced with permission from Lakes, *J. Mater. Sci.*, **26**, 2287 (1991).)

Evans and collaborators [59] have shown how an anisotropic microstructure consisting of nodules and fibrils can be produced in polytetrafluoroethylene (PTFE) that gives rise to a very large negative Poisson's ratio. Figure 7.22 is a schematic representation of the deformation of microporous PTFE.

Initially, there is expansion in both the axial and lateral directions as the fibrils

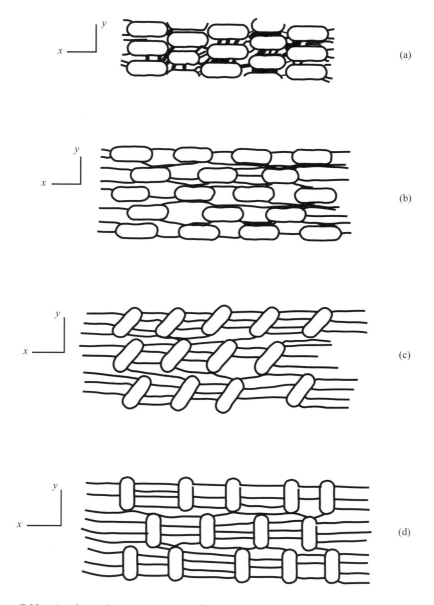

Figure 7.22 A schematic representation of the structural changes observed in microporous polytetrafluoroethylene undergoing tensile loading in the x direction: (a) initial dense microstructure; (b) tension in fibrils causing transverse displacement of anisotropic nodal particles with lateral expansion; (c) rotation of nodes producing further lateral expansion; (d) fully expanded structure prior to further, plastic deformation due to node break-up. (Reproduced with permission from Evans and Caddock, *J. Phys. D: Appl. Phys.*, **22**, 1883 (1989)).

are stretched, similar to the deformation of the re-entrant structure of Figure 7.21. This is followed by rotation of the nodules with a particular handedness, as shown in the figure, to give further lateral expansion. Evans and Caddock [60] showed how two theoretical models – a translational model for stages (a) and (b) of Figure 7.22 and a rotational model for stages (b), (c) and (d) – could provide good confirmation of the observed changed in Poisson's ratio with strain. Their results are illustrated in Figure 7.23, from which it can be seen that experimental values of Poisson's ratio approaching −12 can be observed.

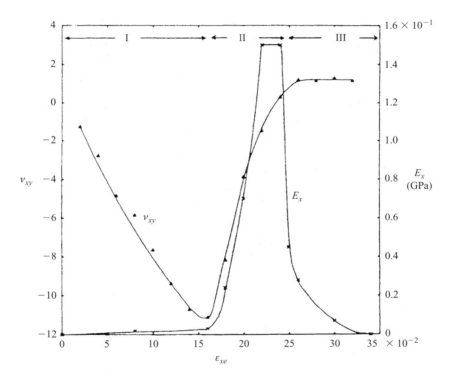

Figure 7.23   Plot of Poisson's ratio v_{xy} and Young's modulus E_x against engineering strain ε_{xe} showing the three regions of behaviour (I, II and III) described in the text. (Reproduced with permission from Caddock and Evans, *J. Phys. D: Appl. Phys.*, **22**, 1877 (1989).)

In a further study, Evans and Alderson [61−63] showed that in both PTFE and ultrahigh molecular weight polyethylene (UHMPE) an isotropic microstructure of nodules and fibrils, shown schematically in Figure 7.24 can give rise to a negative Poisson's ratio. Essentially this arises because when the materials are stretched the extension of the fibrils causes the nodules to move apart.

Figure 7.24 Schematic diagram of an isotropic microstructure consisting of nodules and fibrils that causes a negative Poisson's ratio when pulled in any direction. (Reproduced with permission from Evans and Alderson, *J. Mater. Sci. Lett.*, **11**, 1721 (1992).)

References

1. Biot, M. A., in *International Union of Theoretical and Applied Mechanics Colloquium (Madrid)*, Springer-Verlag, Berlin, 1955, p. 251.
2. Nye, J. F., *Physical Properties of Crystals*, Clarendon Press, Oxford, 1957, p. 138.
3. Ward, I. M., *Mechanical Properties of Solid Polymers*, Wiley, Chichester, 1983.
4. Horgan, C. O., *J. Elast.*, **2**, 169, 335 (1972); *Int. J. Solids Structure*, **10**, 837 (1974).
5. Folkes, M. J. and Arridge, R. G. C., *J. Phys. D*, **8**, 1053 (1975).
6. Wilson, I., Cunningham, A., Duckett, R. A. and Ward, I. M., *J. Mater. Sci.*, **11**, 2189 (1976).
7. Ladizesky, N. H. and Ward, I. M., *J. Macromol. Sci. B*, **5**, 661 (1971).
8. Clayton, D., Darlington, M. W. and Hall, M. M., *J. Phys. E.*, **6**, 218 (1973); Darlington, M. W. and Saunders, D. W., in *Structure and Properties of Oriented Polymers* (ed. Ward, I. M.), Applied Science Publishers, London, 1975, Ch. 10.
9. Wilson, I., Cunningham, A. and Ward, I. M., *J. Mater. Sci.*, **11**, 2181 (1976).
10. Richardson, I. D. and Ward, I. M., *J. Polym. Sci., Polym. Phys.*, **16**, 667 (1978).

11. Lekhnitskii, S. G., *Theory of Elasticity of an Anisotropic Elastic Body*, Mir Publishers, Moscow, 1981.
12. Ladizesky, N. H. and Ward, I. M., *J. Macromol. Sci. B*, **5**, 759, (1971); **9**, 565 (1974); Lewis, E. L. V. and Ward, I. M., *J. Mater. Sci.*, **15**, 2354 (1980).
13. Lewis, E. L. V., Richardson, I. D. and Ward, I. M., *J. Phys. E.*, **12**, 189 (1979).
14. Hadley, D. W., Ward, I. M. and Ward, J., *Proc. R. Soc. A*, **285**, 275 (1965).
15. Pinnock, P. R., Ward, I. M. and Wolfe, J. M., *Proc. Roy. Soc. A*, **291**, 267 (1966).
16. Abdul Jaward, S. and Ward, I. M., *J. Mater. Sci.*, **13**, 1381 (1978).
17. Timoshenko, S. P. and Goodier, J. N., *Theory of Elasticity*, McGraw-Hill, International Editions, New York, 1970.
18. Raumann, G. and Saunders, G. W., *Proc. Phys. Soc.*, **77**, 1028 (1961).
19. Gupta, V. B. and Ward, I. M., *J. Macromol. Sci. B*, **1**, 373 (1967).
20. Hadley, D. W., Pinnock, P. R. and Ward, I. M., *J. Mater. Sci.*, **4**, 152 (1969).
21. Kawabata, S., *J. Text. Inst.*, **81**, 432 (1990).
22. Kotani, T., Sweeney, J. and Ward, I. M., *J. Mater. Sci.*, **29**, 5551 (1994).
23. Hine, P. J. and Ward, I. M., *J. Mater. Sci.*, **31**, 371 (1996).
24. Wilczynski, A. P., Ward, I. M. and Hine, P. J., *J. Mater. Sci.*, **30**, 5879 (1995).
25. Brew, B., Hine, P. J. and Ward, I. M., *Comp. Sci. and Tech.*, **59**, 1109 (1999).
26. Krüger, J. K., Marx, A., Peetz, L., Roberts, R. and Unruh, H.-G., *Colloid Polym. Sci.*, **264**, 403 (1986).
27. Krüger, J. K., Grammes, C., Stockem, K., Zeitz, R. and Dettenmaier, R., *Colloid Polym. Sci.*, **269**, 764 (1992).
28. Wang, C. H., Liu, Q.-L. and Li, B. Y., *Polym. Phys.*, **25**, 485 (1987).
29. Cavanagh, D. B. and Wang, C .H., *J. Polym. Sci. Polym. Phys. Edn.*, **20**, 1647 (1982).
30. Kumar, S. R., Rennesch, D. P. and Grimsditch, M., *Macromolecules*, **33**, 1819 (2000).
31. Chan, O. K., Chen, F. C., Choy, C. L. and Ward, I. M., *J. Phys D: Appl. Phys.*, **11**, 617 (1978).
32. Leung, W. P., Chen, F. C., Clay, C. L., Richardson, A. and Ward, I. M., *Polymer*, **25**, 447 (1984).
33. Choy, C. L., Leung, W. P., Ong, E. L. and Wang, T. *J. Polym. Sci. B, Polym. Phys. Ed.*, **26**, 1569 (1988).
34. Choy, C. L., Leung, W. P. and Huang, C. E., *Polym. Eng. Sci.*, **23**, 910 (1983).
35. Choy, C. L. and Leung, W. P., *J. Polym. Sci. Polym. Phys. Ed.*, **23**, 1759 (1985).
36. Leung, W. P. and Choy, C. L., *J. Polym. Sci. Polym. Phys. Ed.*, **21**, 725 (1983).
37. Hadley, D. W. and Ward, I. M., in *Structure and Properties of Oriented Polymers* (2nd edn.) (ed. I. M. Ward), Chapman and Hall, London, 1997, p. 279.
38. Sakurada. I., Ito, T. and Nakamac, K., *J. Polym. Sci. C*, **15**, 75 (1966); Shaufele, R. F. and Shimanouchi, T., *J. Chem. Phys.*, **47**, 3605 (1967): Feldkamp, L. A., Venkateraman, G. and King, J. S., in *Neutron Inelastic Scattering*, Vol. 2, IAEA, Vienna, 1968, p. 159.
39. Peterlin, A., in *Ultra-High Modulus Polymers*, (eds A. Ciferri and I. M. Ward), Applied Science Publishers, London, 1979, Ch. 10.
40. Ward, I. M., *Proc. Phys. Soc.*, **80**, 1176 (1962).
41. Bishop, J. and Hill, G., *Philos. Mag.*, **42**, 414, 1248 (1951).
42. Reuss, A., *Z. Angew. Math. Mech.*, **9**, 49 (1929).
43. Voigt, W., *Lehrbuch des Krystailphvsik*, Teubner, Leipzig, 1928. p. 410.
44. Crawford, S. M. and Kolsky, H., *Proc. Phys. Soc. B.*, **64**, 119 (1951).

45. Frank, F. C., Gupta V. B. and Ward, I. M., *Philos. Mag.*, **21**, 1127 (1970).
46. Allison, S. W. and Ward, I. M., *Br. J. Appl. Phys.*, **18**, 1151 (1967).
47. Kausch, H. H., *J. Appl. Phys.*, **38**, 4213 (1967); *Kolloidzeitschrift*, **234**, 1148 (1969); **237**, 251 (1970); *Polymer Fracture*, Springer, Berlin, 1978. p. 33.
48. Kausch, H. H., *J. Macromol. Sci. B.*, **5**, 269 (1971).
49. Hennig, J., *Koll. Z.*, **200**, 46 (1964).
50. Robertson, R. E. and Buenker, R. J., *J. Polym. Sci.*, **42**, 4889 (1964).
51. Wright, H., Faraday, C. S. N., White, E. F. T. *et al.*, *J. Phys. D*, **4**, 2002 (1971).
52. Kashiwagi, M., Folkes, M. J. and Ward, I. M., *Polymer*, **12**, 697 (1971).
53. Rawson, F. F. and Rider, G., *J. Phys. D*, **7**, 41 (1974).
54. Lewis, E. L. V. and Ward, I. M., *J. Mater. Sci.*, **15**, 2354 (1980).
55. Maxwell, A. S., Unwin, A. P. and Ward, I. M., *Polymer*, **37**, 3283 (1996).
56. Davies, G. R. and Ward, I. M., in *High Modulus Polymers* (eds A. E. Zachariades and R. S. Porter), Marcel Dekker, New York, 1988, Ch. 2; Troughton, M. J., Davies, G. R. and Ward, I. M., *Polymer*, **30**, 58 (1989); Green, D. I., Unwin, A. P., Davies, G. R. and Ward, I. M., *Polymer*, **31**, 579 (1990).
57. Northolt, M. G. and Van Aarlsen, J. J., *J. Polym. Sci., Polym. Symp.*, **58**, 283 (1997).
58. Northolt, M. G. and van der Hout, R., *Polymer*, **26**, 310 (1985).
59. Caddock, B. D. and Evans, K. E., *J. Phys. D. Appl. Phys.*, **22**, 1872 (1989).
60. Evans, K. E. and Caddock, B. D., *J. Phys. D. Appl. Phys.*, **22**, 1883 (1989).
61. Alderson, K. L. and Evans, K. E., *Polymer*, **33**, 4435 (1992).
62. Evans, K. E. and Alderson, K. L., *J. Mater. Sci. Lett.*, **11**, 1721 (1992).
63. Neale, P. J., Pickles, A. P., Alderson, K. L. *et al.*, *J. Mater. Sci.*, **30**, 4087 (1995).

8

Polymer Composites: Macroscale and Microscale

In this chapter we first give a brief survey of the advantages to be gained by using a composite material whose components often have contrasting but complementary properties, e.g. ductile fibres reinforcing a brittle matrix. We then discuss two distinct applications of these general principles: macroscopic composites, composed of a polymeric matrix in which a second component is embedded; and microscale composites, used to model the morphology of partially crystalline polymers.

8.1 Composites: a general introduction

Many useful engineering materials have a heterogeneous composition. Metals, for instance, are often used in the form of alloys. The addition of a small percentage of another metal, such as copper, magnesium or manganese, is necessary to prevent plastic deformation occurring in aluminium at very low stresses. An increase in carbon content from 0.1 per cent to 3 per cent is a primary determinant in whether a ferrous alloy becomes a mild steel or a cast iron. Concrete, which, like cast iron has good compressive but poor tensile properties, consists of a hard aggregate embedded in a metal silicate network.

Both animal and vegetable life are dependent on natural composites. Bones must be stiff and yet able to absorb significant amounts of energy without fracturing; they also provide anchor points for muscles, which are composite. The skeletal material of plants, and in particular wood, provides a splendid example of the desirable properties of a composite. As a gross simplification, its structure can be considered in terms of an array of relatively stiff fibres embedded in a more compliant matrix. The matrix permits stresses to be redistributed among the fibres, so retarding the onset of fracture at stress concentrations. Wood fails in compression when its fibres buckle. The fracture stress is higher in tension, as a large amount of work must be done in pulling the fibres out of the matrix.

Reinforced concrete has practically a century of use as a building material. Continuous steel rods, prestressed under tension, pass completely through each

An Introduction to the Mechanical Properties of Solid Polymers I. M. Ward and J. Sweeney
© 2004 John Wiley & Sons, Ltd   ISBN: 0471 49625 1 (HB); 0471 49626 X (PB)

structural element and enable the material to withstand tensile as well as compressive stresses, to give a combination that combines the desirable features of each component.

A further form of composite is one where the second component acts as a filler. Carbon black in vehicle tyres is an example of a filler needed to provide the required properties. Each carbon particle provides an anchorage for many rubber molecules, and so assists in the redistribution of stress; and the carbon is also essential to obtain the desired hysteresis behaviour and abrasion resistance. A much simpler application of a filler is the use of sawdust or other cheap powder in mouldings made from a thermosetting plastic. Although the mechanical properties of the base material are degraded (except possibly for impact resistance), they are still adequate for the proposed application, and the product cost is reduced. We shall not consider fillers in the discussion that follows.

Another desirable property of a composite that will not be considered further is the protection that a compliant matrix affords to a brittle reinforcing fibre. Glass and other brittle materials fracture in tension due to the deepening of pre-existing cracks. Because of the absence of plastic flow (unlike the situation for polymers discussed in Chapter 11), blunting of the crack tip cannot occur, and so the stress rapidly approaches that required for fracture. If glass fibres are encapsulated in a soft plastic matrix, the possibility of surface scratches is reduced and the fracture stress is thereby increased. Good adhesion between fibre and matrix will assist in reducing stress concentrations, and transverse cracks will grow only with difficulty across a fibrous composite.

8.2 Mechanical anisotropy of polymer composites

8.2.1 Mechanical anisotropy of lamellar structures

It is instructive to start the discussion of polymer composites by modelling an idealized lamellar composite that consists of a high modulus layer and a more compliant matrix layer. Provided that the bonding between the layers remains intact, the volume fraction of each component and not the thickness of the individual layers is the important factor. As with the aggregate model discussed in Section 7.5, different values of overall stiffness are obtained depending on whether the components are in parallel or in series, yielding the Voigt or Reuss average modulus, respectively.

The maximum stiffness is obtained when a uniaxial stress is applied parallel with the layers, as indicated in Figure 8.1. It is assumed that the strain is the same in all the composite layers, a form of loading known as the isostrain (or homogeneous strain) condition.

The force acting on the composite (F_c) is equal to the sum of the forces acting on the fibre and matrix layers

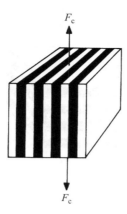

Figure 8.1   Isostrain condition for layered composite

$$F_c = F_f + F_m \tag{8.1}$$

Force is equal to stress multiplied by area. Hence

$$\sigma_c A_c = \sigma_f A_f + \sigma_m A_m$$

where A_f and A_m represent the areas of the end faces occupied by each component. As both components are of length l, areas can be represented by volumes, or rather by volume fractions V_f and V_m. The volume fraction of the composite (V_c) is unity. Hence

$$\sigma_c V_c = \sigma_f V_f + \sigma_m V_m \tag{8.2}$$

Under isostrain conditions this expression can be rewritten in terms of Young's modulus (E) as

$$E_c = E_f V_f + E_m V_m \tag{8.3}$$

which is a Voigt average modulus (see Section 7.5.1).

The modulus is, however, much lower in the direction transverse to the layered structure (Figure 8.2). In this case each layer is subjected to the same force, and hence to the same stress, because the area remains constant through the stack. Loading of this form is known as the isostress (or homogeneous stress) condition.

The total deformation δl_c is equal to the sum of the deformations in each component:

$$\delta l_c = \delta l_f + \delta l_m$$

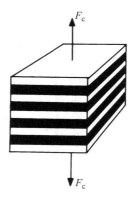

Figure 8.2   Isostress condition for layered composite

Length changes can be converted to strains using $\varepsilon = \delta l / l$:

$$\varepsilon_c l_c = \varepsilon_f l_f + \varepsilon_m l_m \tag{8.4}$$

Substituting modulus E as the ratio of (uniform) stress to strain, we obtain

$$\frac{\sigma l_c}{E_c} = \frac{\sigma l_f}{E_f} + \frac{\sigma l_m}{E_m}$$

As the cross-section of the composite is assumed to be uniform, the length of a component is proportional to its volume fraction. Again, take V_c as unity to give

$$\frac{1}{E_c} = \frac{V_f}{E_f} + \frac{V_m}{E_m} \tag{8.5}$$

This expression can be rewritten as

$$E_c = \frac{E_f E_m}{E_m V_f + E_f V_m} \tag{8.6}$$

and is a Reuss average modulus (see Section 7.5.1).

8.2.2   Elastic constants of highly aligned fibre composites

Although the lamellar composites are of particular interest with regard to semicrystalline polymers, fibre composites, where a polymer matrix phase is reinforced with stiff and strong fibres, are of very great commercial importance and therefore have been the subject of extensive theoretical modelling. Because there is an analogy between aligned fibre composites and highly oriented crystalline polymers, it is valuable to precede discussion of the latter with an introductory account of the mechanical properties of fibre composites.

We take up the discussion following Section 8.2.1 above. In a conventional fibre composite a matrix of moderate stiffness ($\sim$1 GPa) is reinforced with a stiff and strong fibre (of modulus $\sim$100 GPa). Most usually this is glass or carbon fibre, but high-strength fibres such as aramid or polyethylene fibres are also used.

It is instructive to consider the calculation of the elastic constants for an undirectional fibre composite consisting of perfectly aligned fibres of infinite length. It is assumed that there is excellent adhesion between the fibres and the matrix.

The simplest approach is to extend the assumptions of Equations (8.3) and (8.5). Choosing the fibre direction as the 3 axis, the assumptions of homogeneous strain in the 3 direction and homogeneous stress in the 1 direction imply, in terms of the nomenclature proposed in Section 7.1.1, that the five independent elastic constants for the composite (E_1^c, E_3^c, ν_{13}^c, ν_{12}^c and G_4^c) are given by

$$\frac{1}{E_1^c} = \frac{V_f}{E_1^f} + \frac{V_m}{E^m} \tag{8.7a}$$

$$E_3^c = V_f E_3^f + V_m E^m \tag{8.7b}$$

$$\nu_{12}^c = \frac{V_f \nu_{12}^f E^m + V_m \nu^m E_1^f}{V_f E^m + V_m E_1^f} \tag{8.7c}$$

$$\nu_{13}^c = V_f \nu_{13}^f + V_m \nu^m \tag{8.7d}$$

and

$$\frac{1}{G_4^c} = \frac{V_f}{G_4^f} + \frac{V_m}{G^m} \tag{8.7e}$$

where V_f and V_m are the fibre and matrix volume fractions respectively, E_1^f, E_3^f, ν_{13}^f, ν_{12}^f and G_4^f are the fibre elastic constants and E^m, G^m and ν^m are the elastic constants of the isotropic matrix. Equations (8.7b) and (8.7d) give good predictions for the axial Young's modulus E_3^c and the axial Poisson's ratio ν_{13}^c, but the simple isostress equations for the lateral elastic constants E_1^c and ν_{12}^c and the longitudinal shear modulus are not satisfactory because they do not take into account any constraints due to the very high axial stiffness of the fibres. This can be appreciated readily by undertaking the calculation of E_1^c on the assumption that, if the fibre Young's modulus E_3^f is much greater than that of the matrix E_m, when a stress is applied perpendicular to the fibre direction there is no strain in the fibre direction.

Consider a perfectly aligned fibre composite reinforced with isotropic fibres of infinite stiffness in an isotropic matrix of modulus E_m and Poisson's ratio ν_m. In the matrix we have

$$e_1 = V_m \left\{ \frac{\sigma_1}{E_m} - \frac{\nu_m \sigma_2}{E_m} - \frac{\nu_m \sigma_3}{E_m} \right\}$$

$$e_2 = V_m \left\{ -\frac{\nu_m \sigma_1}{E_m} + \frac{\sigma_2}{E_m} - \frac{\nu_m \sigma_3}{E_m} \right\}$$

$$e_3 = V_m \left\{ -\frac{\nu_m \sigma_1}{E_m} - \frac{\nu_m \sigma_2}{E_m} + \frac{\sigma_3}{E_m} \right\}$$

In this situation of plane strain

$$e_3 = 0 \text{ and } \sigma_3 = \nu_m(\sigma_1 + \sigma_2)$$

For stress σ_1 applied in the 1 direction, with $\sigma_2 = 0$, $\sigma_3 = \nu_m \sigma_1$ and

$$e_1 = V_m \left\{ \frac{\sigma_1}{E_m} - \frac{\nu_m^2 \sigma_1}{E_m} \right\}$$

i.e Young's modulus in the 1 direction, E_1^c is given by

$$E_1^c = \frac{E_m}{V_m(1 - \nu_m^2)} \tag{8.8}$$

This result differs significantly from Equation (8.7a) above but still shows that the modulus in the transverse direction is very much less than that in the fibre direction.

Exact analytical expressions of all the elastic constants of a fibre composite with perfectly aligned fibres of infinite length have been obtained recently by Wilczynski [1] and confirmed by finite element calculations [2]. It is, however, common practice to use the equations proposed by Halpin and Kardos [3] based on a generalized self-consistant model developed by Hermans [4] for a composite with continuous aligned fibres. The Halpin-Tsai equations, as they are called, for E_3^c and ν_{13}^c are identical to Equations (8.7b) and (8.7d) above but differ significantly for the transverse modulus E_1^c and the longitudinal shear modulus G_4^c. These are given by

$$\frac{E_1^c}{E_m} = \frac{1 + 2\eta_1 V_f}{1 - \eta_1 V_f}$$

$$\frac{G_4^c}{G_m} = \frac{1 + \eta_2 V_f}{1 - \eta_2 V_f}$$

where

$$\eta_1 = \frac{E_1^f/E_m - 1}{E_1^f/E_m + 1}$$

$$\eta_2 = \frac{G_4^f/G_m - 1}{G_4^f/G_m + 1}$$

For glass fibres where E_1^f and G_4^f are made much greater than E_m and G_m, these equations reduce to

$$\frac{E_1^c}{E_m} = \frac{1 + 2V_f}{1 - V_f} \tag{8.9a}$$

$$\frac{G_4^c}{G_m} = \frac{1 + V_f}{1 - V_f} \tag{8.9b}$$

It is important to note that Equations (8.9a) and (8.9b) show that the values of E_1^c and G_4^c are greater than those for a calculation based on a simplistic homogeneous stress approach, which would give

$$E_1^c = E_m/(1 - V_f)$$

$$G_4^c = G_m/(1 - V_f)$$

assuming that $E_1^f \gg E_m$ and $G_4^f \gg E_m$.

There is an extensive literature on composite materials stemming from the seminal papers of Eshelby [5], who considered the elastic field in and around an elliptic inclusion in an infinite matrix. His theory assumed a single particle in an infinite matrix and therefore was valid only for low-volume fractions (~ 1 per cent). The extension to more concentrated systems was undertaken by Mori and Tanaka [6], whose method was used by Tandon and Weng [7] to derive the elastic constants of an aligned fibre composite. Although these more sophisticated methods are satisfying to theorists, the simpler robust Halpin–Tsai equations give excellent predictions to a good approximation. The reader is referred to a recent paper by Tucker and Liang [8] for a comprehensive review.

8.2.3 Mechanical anisotropy and strength of uniaxially aligned fibre composites

The theoretical considerations outlined above show that a uniaxially aligned fibre composite will show a high degree of mechanical anisotropy, i.e. $E_3^c > E_1^c \sim G_4^c$. Consider, for example, a glass-fibre/resin composite containing a 0.6 volume fraction of glass filaments with a tensile modulus of 70 GPa. For a matrix of modulus 5 GPa and a Poisson's ratio of 0.35, Equation (8.7a) gives the predicted axial Young's modulus E_3^c of 44 GPa and Equation (8.8) gives a transverse Young's modulus E_1^c of 14.2 GPa. The composite has a highly anisotropic stiffness, which decreases very rapidly at quite small angles to the direction of the reinforcing fibres. For this reason, parts made from preimpregnated sheets of uniaxially oriented fibres (called 'prepregs') are generally produced by 0/90 cross-ply laminates or even more sophisticated lay-ups.

The importance of these calculations in this textbook lies in the analogy between a highly aligned fibre composite and a highly oriented polymer where the highly oriented molecules with a very high chain modulus ($\sim$280 GPa in the case of polyethylene) can act in the role of the glass or carbon fibres, and the analogy will be pursued in later sections.

8.3 Short fibre composites

Although continuous filament composites are of considerable commercial importance, their fabrication is a rather complex process and a cheaper though mechanically inferior product is obtained by mixing short lengths of fibre with a thermoplastic polymer.

A prime requirement is good adhesion between fibre and matrix, a condition that will be dependent on features such as chemical bonding and surface cleanliness, as well as mechanical factors. The ratio of surface area to fibre volume should be as high as possible. Considering cylinders of length l and radius r:

$$\text{Area } A = 2\pi r^2 + 2\pi r l \quad \text{and} \quad \text{volume } V = \pi r^2 l$$

$$\therefore \quad \frac{A}{V} = \frac{2}{l} + \frac{2}{r} \tag{8.10}$$

Expressed in terms of the aspect ratio ($a = \frac{l}{2}r$) the above expression becomes

$$\frac{A}{V} + \sqrt[3]{\frac{2\pi}{V}}(a^{-2/3} + a^{1/3}) \tag{8.11}$$

It can be seen, therefore, that for optimum adhesion the aspect ratio should be either very small, where $a^{-2/3}$ becomes very large, corresponding to flat platelets (minerals such as talc and mica have been used), or very high, where $a^{1/3}$ is very large, corresponding to fibres. The latter case will now be examined in more detail.

8.3.1   The influence of fibre length: shear lag theory

Consider a short length of fibre aligned with the tensile stress direction. The stiff fibre will tend to restrain the deformation of the matrix, and so a shear stress will be set up in the matrix at its interface with the fibre, which will be a maximum at the ends of the fibre and a minimum in the middle (Figure 8.3(a)). This shear stress then transmits a tensile stress to the fibre, but as the fibre–matrix bond ceases at the fibre ends there can be no load transmitted from the matrix at each fibre extremity. The tensile stress is thus zero at each end of the fibre and rises to an intermediate maximum or plateau over a critical length $l_0/2$ (Figure 8.3(b)). For effective reinforcement the fibre length must be greater than the critical value l_0, otherwise the stress will be less than the maximum possible.

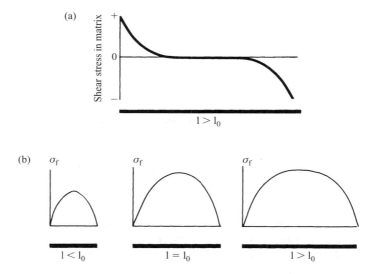

Figure 8.3   Interfacial shear stress (a) and fibre tensile stress (b) as a function of fibre length (schematic)

The reduction in tensile stress towards the ends of each fibre inevitably leads to a decrease in the tensile modulus compared with the continuous filament case. Consider a plane drawn perpendicular to the stress direction in an aligned discontinuous fibre composite (Figure 8.4), which must intercept individual filaments at random positions along their length. Hence the stress carried by the composite must be lower than that for the continuous filament case, and is dependent on the length of each fibre. Cox [9] predicted a correction factor η_l for the tensile modulus in the axial direction that takes into account the finite length of the fibres so that Equation (8.3) is modified to

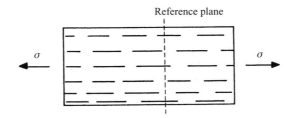

Figure 8.4 Schematic section through a discontinuous fibre composite (fibre fraction shown very low for clarity)

$$E = \eta_l \mathrm{E_f} V_f + E_m V_m \tag{8.12}$$

where

$$\eta_l = 1 - \left[\frac{\tanh ax}{ax}\right] \tag{8.13}$$

where a is the aspect ratio $l/2r$ and x is a dimensionless factor

$$x = \left[\frac{2G_m}{E_f \ln(R/r)}\right]^{1/2} \tag{8.14}$$

where G_m is the shear modulus of the matrix and R is half the separation between the nearest fibres. The basis of this expression is the assumption that the deformation of the complete composite can be modelled by considering a fibre to be embedded in a cylinder of matrix of radius R.

The factor x (Equation 8.14) depends on two key ratios: G_m/E_f, which is typically 0.01–0.02; and R/r, which is not much greater than unity. Figure 8.5 indicates that the length correction factor becomes significant for values of ax less than 10. In practice the corresponding aspect ratio for effective reinforcement is usually greater than 100.

It has been pointed out by Tucker and Liang [8] that to give a result consistent with the self-consistent models Equation (8.12) should be modified to

$$E_c = \eta_l E_f v_f + (1 - \eta_l v_f) E_m$$

For a further discussion the reader is referred to the texts by Kelly [10] and Hull and Clyne [11].

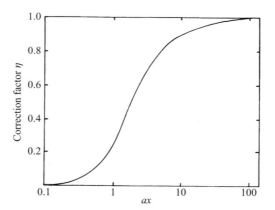

Figure 8.5 Correction factor for the tensile modulus of short fibres as a function of aspect ratio

8.3.2   Debonding and Pull-out

A composite frequently fails as a result of debonding between fibre and matrix. New interfaces are created, which involve the expenditure of energy, as is discussed in Chapter 11. The basic process can be modelled by considering a single fibre embedded in a block of matrix to a depth x, and it can be shown that the debonding energy is a maximum when x is equal to half the critical length. If the embedded length is less than $l_0/2$ the fibre will be pulled out of the matrix rather than fracturing, so involving the expenditure of further energy.

A stress–strain plot can be derived from a tensile load–extension experiment, as depicted schematically in Figure 8.6. The energy of debonding is obtained from the area OAB, and the usually larger pull-out energy is associated with the area

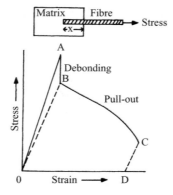

Figure 8.6 Pull-out test and the resulting stress–strain curve showing the difference in magnitude of the energies of debonding (area OAB) and pull-out (area OBCD). (From Anderson *et al.*, *Materials Science*, (4th edn) Chapman and Hall, London, 1990, Ch. 11)

OBCD. A strong fibre, which has not fractured after some debonding has occurred, will bridge the newly formed surfaces in the wake of a propagating crack, and thereby hinder crack opening. This toughening process has a microscopic analogy in the role of extended-chain bridging filaments in semicrystalline polymers.

8.3.3   Partially oriented fibre composites

The fibres in short fibre composites are not usually fully aligned. An early attempt to deal with such systems was made by Brody and Ward [12], who applied the aggregate model of Section 7.5 above assuming that the elastic constants of a representative unit of structure could be predicted by Equations (8.7a)–(8.7e). It was found that this simple theory fitted the results for composites reinforced with short fibres of carbon or glass reasonably well, with the moduli lying close to the lower Reuss bounds. Recent studies by Ward and co-workers [13] have shown that this approach is viable provided that the elastic constants of the representative unit are calculated more exactly.

Again, the Halpin–Tsai equations give a set of elastic constants to a good approximation, provided that the axial Young's modulus E_3^c is modified to

$$E_3^c = \frac{E_m(1 + \xi\eta\nu_f)}{1 - \eta\nu_f}$$

where

$$\eta = \frac{E_f/E_m - 1}{E_f/E_m + \xi}$$

$$\xi = l/r$$

Values for E_3^c obtained in this way are very similar to those calculated as the basis of shear lag theory. Recent work by Wire [14] suggests that the other elastic constants may be calculated better by the methodologies developed by Wilczynski and Tandon and Weng.

Quantitative studies of the elastic anisotropy of short fibre composites have been undertaken by combining quantitative measurements of fibre orientation (using optical image analysis methods) with aggregate models. Ward and co-workers have been very active in this area [12, 14], as have Tucker and co-workers [15].

8.4   Takayanagi models for semicrystalline polymers

It was realized by Takayanagi [17] that oriented highly crystalline polymers with a clear lamellar texture might be modelled in terms of a two-component composite in which the alternating layers corresponded to the crystalline and amorphous phases [18]. The model was later extended to include a parallel component in addition to that in series, and was applied first to describe the relaxation behaviour of amorphous polymers with two distinct phases, and later to crystalline polymers in which the parallel component represented either interlamellar crystalline bridges or amorphous tie molecules threading through the amorphous phase.

8.4.1 The simple Takayanagi model

High-density polyethylene that has been drawn uniaxially (i.e. with fibre symmetry) and then annealed has a distinct lamellar texture, and we will show that the orientation of the lamellae, as distinct from the molecular orientation, plays the dominant role in determining the mechanical anisotropy. The temperature variation of the in-phase (E_1) and out-of-phase (E_2) components of the dynamic modulus, both parallel with ($\parallel$) and perpendicular to ($\perp$) the draw direction, is shown in Figure 8.7, with the parallel modulus ($E_0 = \parallel$) crossing the perpendicular modulus ($E_{90} = \perp$) at high temperatures; at low temperatures $E_0 > E_{90}$, but at high temperatures $E_{90} > E_0$. The principal features can be explained in terms of a simple model of amorphous (A) and crystalline (C) components, which are in series in the draw direction but in parallel in the transverse direction (Figure 8.8). In the orientation direction each component is subject to the same stress, so that compliances are added as in the Reuss averaging scheme. The stiffness above the

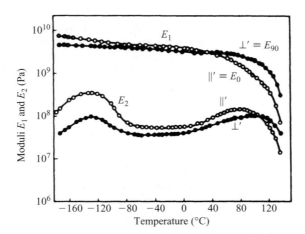

Figure 8.7 Temperature dependence of E_1 and E_2, the components of the dynamic modulus, in directions parallel ($\parallel'$) and perpendicular ($\perp'$) to the initial draw direction for annealed samples of high-density polyethylene. (Reproduced with permission from Takayanagi, Imada and Kajiyama, *J. Polym. Sci. C*, **15**, 263 (1966).

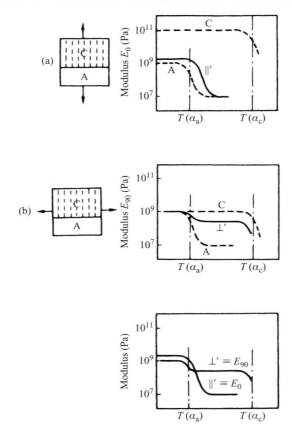

Figure 8.8 Schematic representations of change in modulus E with temperature on the Takayanagi model for (a) the $\|'$ and (b) the $\perp'$ situations corresponding to E_0 and E_{90}, respectively. Calculations assume amorphous relaxation at temperature $T(\alpha_a)$ and crystalline relaxation at temperature $T(\alpha_c)$ and (c) shows combined results. C, crystalline phase; A, amorphous phase. (Reproduced with permission from Takayanagi, Imada and Kajiyama, *J. Polym. Sci. C*, **15**, 263 (1966).

relaxation transition is primarily determined by the compliant amorphous regions (Equation 8.6), so giving a large fall in modulus as the temperature is increased. In the perpendicular direction the parallel components are each subject to the same strain, and stiffnesses are added as in the Voigt scheme. The crystalline regions support the applied stress at temperatures above the relaxation transition, and so a comparatively high stiffness is maintained. Takayanagi and colleagues considered that appropriate values of the modulus might be $E_c(\|) = 100$ GPa, $E_c(\perp) = 1$ GPa, E_A (low T) = 1 GPa, E_A (high T) = 0.01 GPa.

8.4.2   Takayanagi models for dispersed phases

Takayanagi [17] devised series-parallel and parallel-series models as an aid to understanding the viscoelastic behaviour of a blend of two isotropic amorphous polymers in terms of the properties of the individual components. For an A phase dispersed in a B phase there are two extreme possibilities for the stress transfer. For efficient stress transfer perpendicular to the direction of tensile stress we have the series-parallel model (Figure 8.9(a)) in which the overall modulus is given by the contribution for the two lower components in parallel (as in Equation (8.3)) in series with the contribution for the upper component (as in Equation (8.5)):

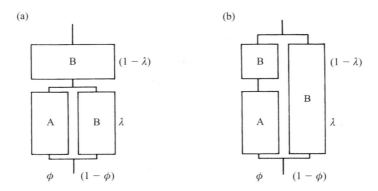

Figure 8.9   Takayanagi models for polymer blends: (a) the series-parallel model; (b) the parallel-series model

$$\frac{1}{E^*} = \frac{\phi}{\lambda E_A^* + (1 - \lambda)E_B^*} + \frac{1 - \phi}{E_B^*} \tag{8.15}$$

where E^*, etc. represent the complex moduli associated with dynamic experiments. If the stress transfer across planes containing the tensile stress is weak, a parallel-series model (Figure 8.9(b)) is appropriate, for which the two left-hand components combine in series (as in Equation (8.5)) before combining in parallel with the right-hand component (Equation (8.3)), giving a modulus

$$E^* = \lambda \left(\frac{\phi}{E_A^*} + \frac{1 - \phi}{E_B^*} \right)^{-1} + (1 - \lambda)E_B^* \tag{8.16}$$

Note that $\lambda\phi$ corresponds to the volume fraction V_f in the earlier notation.

The predictions of both models were then compared with measurements of the temperature variation of storage and loss moduli for a film made from a blend of polyvinyl chloride and nitrile butadiene rubber (Figure 8.10). It is seen that the

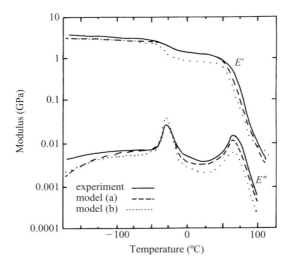

Figure 8.10 Temperature dependence of storage and loss moduli for a polyvinyl chloride–nitrile bidadiere rubber film bonded in parallel to a polyvinyl chloride film. Takayanagi model type (a) gives better fit to experiment. (Reproduced with permission from Takayanagi, *Mem. Fac. Eng. Kyushu Univ.*, **23**, 41 (1963))

series-parallel model (a) gives the better fit. The performance of polymer blends was well represented by a series-parallel model in which the relative values of λ and ϕ were related to the shape of the dispersed phase: $\lambda = \phi$ for homogeneous dispersions, and $\lambda \geqslant \phi$ for dispersions in the form of elongated aggregates. For semicrystalline polymers in general, however, with A and B representing the crystalline and amorphous components, experimental dispersions were usually broader than the predictions, suggesting that at least some of the unordered material was not identical to that in a completely amorphous state.

Gray and McCrum [19] have criticized the Takayanagi model as applied to partially crystalline polymers, and refute the assertion that if mechanical relaxation occurs in the amorphous phase the peak value of the out-of-phase modulus is proportional to the volume fraction of the amorphous phase. They assert that as the model represents a Voigt average solution it can give only upper bounds to moduli. Stress and strain fields must differ between the crystalline and amorphous components, so a Reuss-type average is equally inadmissible, and a correct solution must lie between the two limits. An empirical logarithmic mixing hypothesis is advanced as an acceptable law of mixing

$$\log G^* = V_A \log G_A^* + V_c \log G_c^* \tag{8.17}$$

where G^* represents the complex shear modulus and V represents a volume fraction. The logarithmic decrement of the polymer (Λ) is then given as a weighted combination of the logarithmic decrement of the two phases:

$$\Lambda = V_A \Lambda_A + V_c \Lambda_c \tag{8.18}$$

which is a mathematical statement of the assumptions of some earlier workers.

8.4.3 Modelling polymers with a single-crystal texture

The Takayanagi models were remarkably successful in providing a simple interpretation of the dynamic mechanical behaviour of crystalline polymers and polymer blends. The theoretical basis is contained in Equations (8.1) to (8.6) of Section 8.2, and is deficient in two respects. First, only tensile deformations are considered and shear deformations are ignored. Secondly, as emphasized in Chapter 7, Voigt and Reuss schemes (i.e. parallel and series) only provide bounds to the true behaviour.

These deficiencies became very apparent when Ward and co-workers [20] studied the mechanical behaviour of well-defined lamellar textures. Rolling and annealing processes established by Hay and Keller [21] enable sheets of low-density polyethylene to be produced with well-defined crystallographic and lamellar orientations (Figure 8.11) and it was possible to study the behaviour of three types of special structure sheet illustrated in Figure 8.12: '*bc* sheet', in which the *c* axes of the crystallites lie along the initial draw direction, the *b* axes lie in the plane of the sheet and the *a* axes are normal to the plane of the sheet; '*ab* sheet', in which the *a* axes lie along the draw direction, the *b* axes again lie in the plane of the sheet and the *c* axes are normal to the sheet; and 'parallel lamellae sheet', where the lamellar plane normals lie along the initial draw direction and the *c* axes make an angle of about 45° with this direction. For specimens of *bc* and *ab* sheet a four-point small-angle X-ray diffraction pattern is shown, which is interpreted as indicating the presence of lamellae inclined at about 45° to the direction of the *c* axes. This type of morphology is represented schematically for the *bc* sheet by the model shown in Figure 8.11, in which the solid blocks represent crystalline lamellae, and the intermediate spaces are occupied by disordered material and interlamellar tie molecules, which are relaxed as a result of the annealing treatment. In contrast, the parallel lamellae sheet has a twinned structure with regard to the crystallographic orientation but a single texture structure (only a two-point, low-angle diffraction pattern) as far as lamellar orientation is concerned.

In contrast with the Takayanagi model, which considers only extensional strains, a major deformation process involves *shear* in the amorphous regions. Rigid lamellae move relative to each other by a shear process in a deformable matrix. The process is activated by the resolved shear stress $\sigma \sin \gamma \cos \gamma$ on the lamellar surfaces, where γ is the angle between the applied tensile stress σ and the lamellar plane normals, which reaches a maximum value for $\gamma = 45°$ (see Chapter 11 for discussion of resolved shear stress in plastic deformation processes).

Gupta and Ward found cross-over points in the extensional moduli for *bc* and *ac*

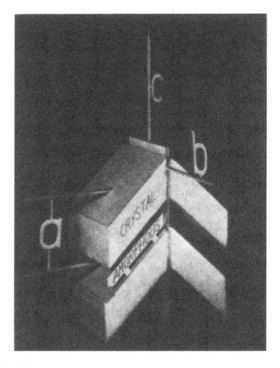

Figure 8.11 Model of morphology of oriented and annealed sheets of low-density polyethylene. This photograph shows the structure of the *bc* sheet; *a*, *b* and *c* axes indicate the crystallographic directions in the crystalline regions. (Reproduced with permission from Stachurski and Ward, *J. Polym. Sci. A2*, **6**, 1817 (1968))

sheets similar to those found by Takayanagi in high-density polyethylene, but at lower temperatures corresponding to the β relaxation (see Chapter 9 for a discussion of relaxation processes). The fall in modulus in the *c* direction in *bc* sheet and the *a* direction in *ab* sheet can be attributed to an interlamellar shear process. As the lamellar planes are approximately parallel to the *b* axis, a tensile stress in this direction will not favour interlamellar shear, so that at temperatures above the relaxation transition $E_b > E_a \sim E_c$ (Figure 8.12). Dynamic mechanical loss spectra show significant anisotropy, both for the lower temperature β relaxation, which corresponds to the cross-over in tensile moduli previously discussed, and for the higher temperature α relaxation, where the results are consistent with the proposal that the α process involves shear in the *c*-axis direction on planes containing the *c* axis of the crystallites (the *c*-shear process). In the *bc* sheet $\tan \delta_{45}$ is larger than $\tan \delta_0$ and $\tan \delta_{90}$ (angles being measured from the original draw direction) because it represents the situation in which there is maximum resolved shear stress parallel to the *c*-axis directions. Similarly, for the parallel lamellae sheets, the greatest losses for the α process occur when the stress is applied parallel to the initial draw direction. Finally, in *ab* sheet the α relaxation is very small, as there are no planes containing the c axis that will shear in the c direction when a tensile stress is applied in the plane of the sheet.

Other applications of the Takayanagi model to oriented polymers have included linear polyethylene that was cross-linked and then crystallized by slow cooling from the melt under a high tensile strain [22], and sheets of nylon with ortho-rhombic elastic symmetry [23]. A fuller discussion is given in the more advanced text by Ward [24].

The studies of the viscoelastic behaviour of the specially oriented sheets are valuable in emphasizing the composite nature of these materials and defining where the lamellar texture determines the mechanical anisotropy, thus carrying our understanding one stage further than the simple series/parallel Takayanagi models. It is, however, also possible to use the composite laminate model to gain a more quantitative understanding of the mechanical anisotropy, and carry the discussion further than the evaluation of Reuss and Voigt bounds for the Young's moduli.

In recent research, Al-Hussein *et al.* [25] prepared an oriented low-density polyethylene with a parallel lamellar stack morphology where the c axes of the crystalline lamellae were parallel to the lamellar plane normals. For this structure, explicit equations can be obtained for the elastic constants in terms of the crystalline volume fraction and the elastic constants of the crystalline lamellae (c_{11}^c, c_{33}^c, c_{44}^c, etc.) and the amorphous layer (c_{11}^a, c_{33}^a, c_{44}^a, etc.).

For example, on the assumption that the lateral dimensions are very large, the elastic constants of the composite (c_{11}^u, c_{33}^u, c_{44}^u, etc.) are given by

$$c_{11}^u = Xc_{11}^c + (1-X)c_{11}^a - \frac{(c_{13}^c - c_{12}^a)^2}{[(c_{33}^c/X) + c_{33}^a/(1-X)]} \tag{8.19a}$$

$$\frac{1}{c_{33}^u} = \frac{X}{c_{33}^c} + \frac{1-X}{c_{33}^a} \tag{8.19b}$$

$$\frac{1}{c_{44}^u} = \frac{X}{c_{44}^c} + \frac{(1-X)}{c_{44}^a} \tag{8.19c}$$

where X is the crystalline volume fraction.

The composite model leading to the rather simple Equations (8.19) above assumes that there is uniform strain in the lateral direction. The amorphous phase is soft and therefore extends more than the crystalline phase in response to an axial stress. The lateral contraction necessary to maintain constant volume for the rubbery elastic phase would therefore be considerably greater than that of the stiffer crystalline phase. But it is assumed that the amorphous layers and the crystalline lamellar layers are tightly bonded together. This means that the lateral contraction of the crystalline phase will be much greater in the composite than that predicted theoretically for an isolated crystal. Hussein *et al.* used wide-angle X-ray diffraction to measure the crystalline compliances S_{13}^c and S_{23}^c for their parallel

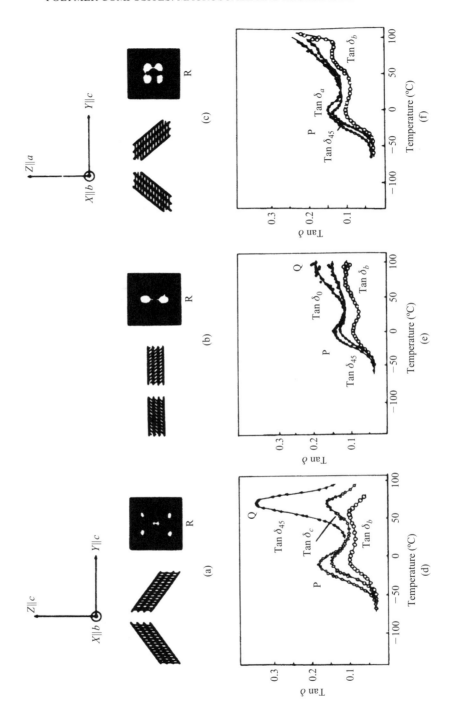

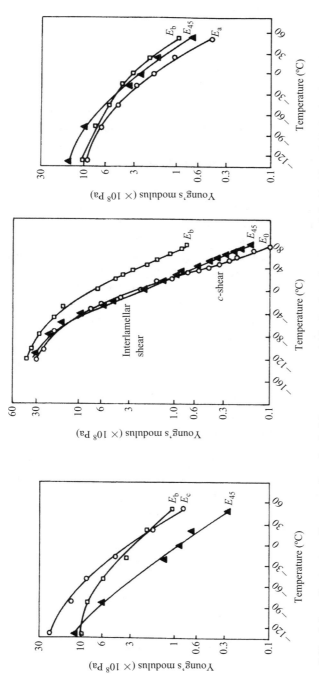

Figure 8.12 Schematic structure diagrams of mechanical loss spectra and 10 s isochronal creep moduli: (a), (d) and (g) for *bc* sheet; (b), (e) and (h) for parallel lamellae sheet; (c), (f) and (i) for *ab* sheet. P, interlamellar shear process; Q, *c*-shear process (note absence of *c*-shear process in (f)); R, small-angle X-ray diagram, beam along X

lamellar structure and showed that the values were indeed much greater than those for the theoretical perfect crystal.

8.5 Ultra high-modulus polyethylene

Conventional drawing processes, which usually involve the polymer being uni-axially extended between two sets of rollers rotating at different speeds, rarely permit a draw ratio greater than 10 × (see Chapter 11). Such materials show an extensional modulus that is only a comparatively small fraction ($\sim$ 10 per cent) of the chain modulus, due to the dominant effect of the relatively compliant unordered component. In the case of polyethylene, however, it is possible to produce an oriented polymer whose Young's modulus at low temperatures approaches the theoretical value of the crystal chain modulus of about 300 GPa (by comparison, the Young's modulus of ordinary steel is about 210 GPa). Several production methods have been used, including solution spinning techniques [26] and a two-stage draw process [27], in which an initial stage of drawing (draw ratio 8 ×) is followed by a continuing stage of extension so that the already drawn material thins down to achieve a final draw ratio of 30 × or more. These production processes are somewhat slower to operate than conventional methods, but nevertheless high-stiffness polyethylene is produced commercially for specific end uses. Young's modulus as a function of draw ratio is shown in Figure 8.13 for a range of initially isotropic polyethylenes drawn at 75 °C. It is seen that the modulus, which even at room temperature can reach an appreciable fraction of the crystal modulus, depends only on the final draw ratio and is independent of the relative molecular mass and the initial morphology, so that an appropriate model appears to be one that depends on the structure produced during deformation rather than on the starting material.

We shall outline two different models used to interpret the elastic behaviour of highly oriented linear polyethylene. Both models take macroscopic composite theory as their starting point but diverge in the way in which they account for the evidence from morphological studies for crystalline regions that can extend for more than 100 nm in the draw direction.

8.5.1 The crystalline fibril model

This model, proposed by a group working at Bristol University [28], is a development of a larger scale model that was used to account for the high mechanical anisotropy of certain copolymers [29]. Electron microscopy demonstrated that, when a three-block polystyrene–polybutadiene–polystyrene copolymer was extruded into a mould, long and completely aligned filaments of glassy polystyrene with a diameter about 15 nm were arranged with hexagonal symmetry in a rubber

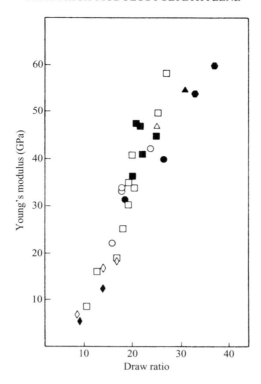

Figure 8.13 A 10 s isochronal creep modulus, measured at room temperature, as a function of draw ratio for a range of quenched (open symbols) and slowly cooled (closed symbols) samples of linear polyethylene drawn at 75 °C. (●), Rigidex 140–60; (△, ▲), Rigidex 25; (□, ■), Rigidex 50; (○, ●), P40; (◇, ◆), H020–54P. (Reproduced with permission from Capaccio, Crompton and Ward, *J. Polym. Sci., Polym. Phys. Ed.*, **14**, 1641 (1976))

matrix. Despite the macroscopic anisotropy, for which the ratio of the longitudinal to the transverse Young's modulus was almost 100:1, both phases were comprised of randomly oriented molecular chains.

The model for highly oriented polyethylene similarly assumed that fibrils of high aspect ratio were arranged with hexagonal symmetry in a compliant matrix, but the discontinuous nature of the fibrils was now the determining factor for the extensional modulus. Fibrils observed in thin sections of the oriented polymer after staining with chlorosulphonic acid and uranyl acetate were considered to represent a needle-like crystal phase with the theoretical stiffness of the polyethylene chain ($E_c \sim 300$ GPa). These crystals, whose concentration (V_f) was estimated to be 0.75, were embedded in a partially oriented matrix containing both amorphous and crystalline components, with a shear modulus $G_m \sim 1$ GPa (Figure 8.14).

Using the Cox model for a fibre composite, already discussed [9], and

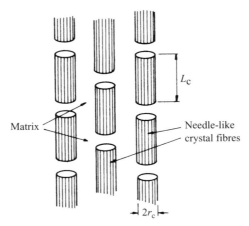

Matrix

Needle-like
crystal fibres

L_c

$2r_c$

Figure 8.14 Schematic diagram of the Barham and Arridge model for ultrahigh-modulus polyethylene

neglecting the very small contribution $E_m V_m$ arising from the tensile modulus of the compliant matrix, the extensional modulus E of the highly oriented polymer becomes

$$E = V_f E_c \left(1 - \frac{\tanh ax}{ax} \right) \qquad (8.20)$$

where a is the fibre aspect ratio

$$\left(\frac{l_c}{2r_c} \right)$$

and

$$x = 2 \left(\frac{G_m}{E_c \ln 2\pi / \sqrt{3 V_f}} \right)^{1/2}$$

which is a restatement of Equation (8.14).

The increased stiffness that results from post-neck drawing is postulated to be a direct consequence of the increased aspect ratio of the crystalline fibrils (from slightly less than 2 to greater than 12 in extreme cases), which thus become more effective reinforcing elements. On the assumption that post-neck drawing is homogeneous on a structural level, so that the fibrils deform affinely, the initial aspect ratio $(l_0/2r_0)$ transforms to $l_c/2r_c = (l_0/2r_0)t^{3/2}$, where t is the draw ratio in the post-neck region. Barham and Arridge [28] show that the observed change in modulus with draw ratio implies that x in Equation (8.20) should depend on $t^{3/2}$.

The good agreement (Figure 8.15) is advanced as a strong argument in favour of the model.

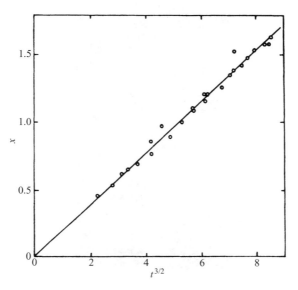

Figure 8.15 Parameter x in Equation (8.19) as a function of the taper draw ratio t to the 3/2 power. (Reproduced with permission from Barham and Arridge, *J. Polym. Sci., Poly. Phys. Ed.*, **15**, 1177 (1977))

8.5.2 The crystalline bridge model

An alternative approach, due to Gibson, Davies and Ward [30], is based on a Takayanagi model, which is then modified to include an efficiency ('shear-lag') factor that takes into account the discontinuous nature of the crystalline reinforcing component. The model was derived by comparing microstructural studies of conventionally drawn polyethylene and ultrahigh-modulus material. At a draw ratio $\sim 10 \times$, wide-angle X-ray diffraction indicates that the crystalline component is highly oriented and, together with a clear two-point small-angle X-ray diffraction pattern, and, suggests a regular stacking of crystal blocks whose length, to accommodate the non-crystalline regions, is less than the long period L, as shown in Figure 8.16. At increasing draw ratio the small-angle pattern retains the same periodicity but diminishes in intensity, and a variety of techniques indicate an increase in the orientation of non-crystalline material. The average crystal length increases to about 50 nm, compared with the constant long period of about 20 nm. The concentration of crystals > 100 nm is low, in contrast with the implied lengths of 100–1000 nm for the crystalline fibrils in the model previously discussed (Section 8.5.1). Reasons for this discrepancy have not been examined in

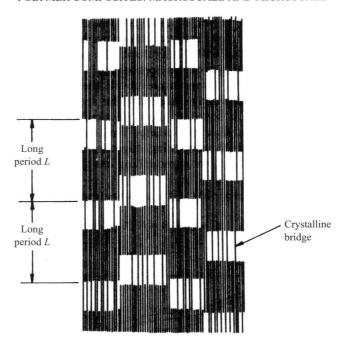

Figure 8.16 Schematic representation of the structure of the crystalline phase in ultrahigh-modulus polyethylene (constructed for $p = 0.4$). (Reproduced with permission from Gibson, Davies and Ward, *Polymer*, **19**, 683 (1978). Copyright IPC Business Press Ltd)

detail and the differences may be a consequence of the specific method used to produce the material, despite the inference from Figure 8.13 that the final drawing process dominates over differences in the initial morphology.

The large increase in stiffness is considered to be a consequence of the linking of adjacent crystal sequences by crystalline bridges (Figure 8.16). In this model the crystalline bridges play a similar role to the taut tie molecules suggested earlier by Peterlin [31], and are equivalent to the continuous phase of a Takayanagi model. The increase in modulus with increasing draw ratio is considered here to arise primarily from an increase in the *proportion* of fibre phase material, and not from the changing aspect ratio of a constant proportion of the fibre phase.

In the absence of information regarding the arrangement of the crystalline bridges, it is assumed that they are randomly placed so that the probability of a crystalline sequence traversing the disordered regions to link adjacent crystalline blocks is given in terms of a single parameter p, defined as

$$p = \frac{\bar{L} - L}{\bar{L} + L}$$

where $\bar{L}$ is the average crystal length determined from wide-angle X-ray diffraction and L is the long period obtained from small-angle X-ray scattering. It can be shown that the volume fraction of continuous phase V_f is given by

$$V_f = Xp(2 - p)$$

where X is the crystallinity expressed as a fraction.

As a first stage, the contribution of the crystalline bridges can be considered as one element of a Takayanagi model (in Figure 8.9(b) this is the continuous phase) that is in parallel with the series combination of the remaining lamellar material and the amorphous component. Young's modulus would then be

$$E = E_c Xp(2 - p) + E_a \frac{\{1 - X + X(1 - p)^2\}^2}{1 - X + X(1 - p)^2 E_a / E_c} \tag{8.21}$$

The first term in this expression, which corresponds to the crystalline bridge sequences, is next treated as an array of short fibres, so introducing the shear lag (efficiency) factor Φ, which is a function of the finite aspect ratio of the crystalline bridges. The analogous equation to Equation (8.20) is

$$E = E_c Xp(2 - p)\Phi' + E_a \frac{\{1 - X + X(1 - p)^2\}^2}{1 - X + X(1 - p)^2 E_a / E_c} \tag{8.22}$$

where Φ' is an average shear lag factor for all materials in the fibre phase. The advantage gained by converting from the one-dimensional Takayanagi model to the Cox short fibre model is that measurable tensile properties are able to yield information about the shear stress development in the matrix (see Section 8.3.1). In particular it can be shown for the case of sinusoidally varying strain that the out-of-phase component of the tensile modulus E_c' is related to the out-of-phase component of the shear modulus of the matrix G_m' by an expression that involves the volume fraction of fibres (i.e. crystalline bridges) and the fibre aspect ratio, together with the ratio G_m'/E_f', where G_m' is the in-phase component of the shear modulus of the matrix. The geometric factors are constant for a given structure, but the modulus ratio varies with temperature because of the temperature dependence of G_m'.

The aspect ratio of the fibre component, which is a measure of the width of the crystalline bridges, is not directly accessible but can be deduced from the value of the shear lag factor Φ' required to give the best match between the predicted and observed patterns of mechanical behaviour as a function of temperature. This exercise yields a radius of 1.5 nm for the crystalline bridge sequences, which suggests that each bridge is comprised of several extended polymer chains. Detailed considerations of the way in which the modulus increases at temperatures below $-50\,°C$ suggest that the modulus of the matrix increases with increasing draw ratio due to an increase in E_a in Equation (8.20), which corresponds to an increase in the modulus of the non-crystalline material.

8.6 Conclusions

A major application of ultrahigh-modulus polymers such as those discussed above is as an inert reinforcing fibre in composite materials [32]. A detailed analysis of the overall anisotropy will thus require appropriate theories on both microscopic and macroscopic scales. There are, however, important and highly significant differences between the two scales. In the macroscopic composite the fibre and the matrix are distinct entities, bonded only by weak secondary forces, whereas the crystalline and amorphous phases of an oriented semicrystalline homopolymer must blend gradually into one another: chain folds and chain ends are associated with crystallites, some of the less regular material is significantly oriented and bridging molecules (either crystalline bridges or tie molecules) that link crystallites traverse amorphous material in the process.

Even for block copolymers, in which the phase separation can be distinguished in electron micrographs, there are problems in matching parameters such as Poisson's ratios of the two components: nevertheless the simple Takayanagi models, particularly when extended by a treatment to account for the finite length of the reinforcing component, can describe numerous features of static and dynamic elastic behaviour.

References

1. Wilczynski, A. P., *Comp. Sci. Tech.*, **38**, 327 (1990).
2. Hine, P. J., Lusti, H. R. and Gusev, A. A., *Comp. Sci. Tech.*, **62**, 1445 (2000).
3. Halpin, J. C. and Kardos, J. L., *Polym. Eng. Sci.*, **16**, 344 (1976).
4. Hermans, J. J., *Proc. Kon. Ned. Akad. Wetensch. B*, **65**, 1 (1967).
5. Eshelby, J. D., *Proc. R. Soc. A*, **241**, 376 (1957).
6. Mori, T. and Tanaka, K., *Acta Metall.*, **21**, 571 (1973).
7. Tandon, G. P. and Weng, G. J., *Polym. Comp.*, **5**, 327 (1984).
8. Tucker, C. L. and Liang, E., *Comp. Sci. Tech.*, **59**, 655 (1999).
9. Cox, H. L., *Br. J. Appl. Phys.*, **3**, 72 (1952).
10. Kelly, A., *Strong Solids*, Clarendon Press, Oxford, 1966.
11. Hull, D. and Clyne, T. W., *An Introduction to Composite Materials* (2nd edn), Cambridge University Press, Cambridge, 1996.
12. Brody, H. and Ward, I. M., *Polym. Eng. Sci.*, **11**, 139 (1971).
13. Hine, P. J., Duckett, R. A. and Ward, I. M., *Comp. Sci. Tech.*, **49**, 13 (1993).
14. Wire, S. L., *PhD Thesis*, University of Leeds, 1998.
15. Gusev, A. A., Hine, P. J. and Ward, I. M., *Comp. Sci. Tech.*, **60**, 535 (2000).
16. Advani, S. G. and Tucker, C. L., *J. Rheol.*, **31**, 751 (1987).
17. Takayanagi, M., *Mem. Fac. Eng. Kyushu Univ.*, **23**, 41 (1963); Takayanagi, M., Imada, I. and Kajiyama, T., *J. Polym. Sci. C*, **15**, 263 (1966).
18. Wu, C. T. D., McCullough, R. C., in *Developments in Composite Materials* (ed. G. S. Holister), Applied Science Publishers, London, 1977, pp. 119–187.
19. Gray, R. W. and McCrum, N. G., *J. Polym. Sci. A2*, **7**, 1329 (1969).

20. Gupta, V. B. and Ward, I. M., *J. Macromol. Sci. B*, **2**, 89 (1968); Stachurki, Z. H. and Ward, I. M., *J. Polym. Sci. A2*, **6**, 1083 (1969); *J. Macromol. Sci. B*, **3**, 427, 445 (1969); Davies, G. R., Owen, A. J., Ward, I. M., *et al.*, *J. Macromol. Sci. B*, **6**, 215 (1972); Davies, G. R. and Ward, I. M., *J. Polym. Sci. B*, **6**, 215 (1972).
21. Hay, I. L. and Keller, A., *J. Mater. Sci.*, **1**, 41 (1966).
22. Kapuscinski, M., Ward, I. M. and Scanlan, J., *J. Macromol. Sci. B*, **11**, 475 (1975).
23. Lewis, E. L. V. and Ward, I. M., *J. Macromol. Sci. B*, **18**, 1 (1980); **19**, 75 (1981).
24. Ward, I. M., *Mechanical Properties of Solid Polymers*, Wiley, Chichester, 1983, Ch. 10.
25. Al-Hussein, M., Davies, G. R. and Ward, I. M., *J. Polym. Sci. B, Polym. Phys.*, **38**, 755 (2000).
26. Zwijnenburg, A. and Pennings, A. J., *J. Polym. Sci., Polym. Lett.*, **14**, 339 (1976); Smith, P. and Lemstra, P. J., *J. Mater. Sci.*, **15**, 505 (1980).
27. Capaccio, G. and Ward, I. M., *Nature Phys. Sci.*, **243**, 143 (1973); *Polymer*, **15**, 223 (1974).
28. Arridge, R. G. C., Barham, P. J. and Keller, A., *J. Polym. Sci., Polym. Phys.*, **15**, 389 (1977); Barham, P. J. and Arridge, R. G. C., *J. Polym. Sci., Polymer Phys.*, **15**, 1177 (1977).
29. Arridge, R. G. C. and Folkes, M. J., *J. Phys. D.*, **5**, 344 (1972).
30. Gibson, A. G., Davies, G. R. and Ward, I. M., *Polymer*, **19**, 683 (1978).
31. Peterlin, A., in *Ultra-High Modulus Polymers* (eds A. Ciferri and I. M. Ward), Applied Science Publishers, London, 1979, Ch. 10.
32. Ladizesky, N. H. and Ward, I. M., *Pure Appl. Chem.*, **57** (1985).

Problems for Chapters 7 and 8

1. Explain how the generalized Hooke's law defines the compliance matrix S_{ij} for an anisotropic material.

   The non-zero components of this matrix for an oriented polymer are given in units of 10^{-10} m^2 Pa^{-1} by the following:

$$S_{11} = S_{22} = 10 \qquad S_{33} = 1$$

$$S_{12} = S_{21} = -4$$

$$S_{13} = S_{31} = S_{23} = S_{32} = -0.45$$

$$S_{44} = S_{55} = 15$$

$$S_{66} = \tfrac{1}{2}(S_{11} - S_{12}) = 7$$

Calculate the value of Young's modulus for the following directions in the polymer:

(a) the Ox_3 and Ox_1 directions;

(b) any direction making an angle of $45°$ with Ox_3.

2. The elastic properties of a [0, 90] cross-ply laminate constructed using pre-impregnated tapes are given by

$$\sigma_{11} = 45\varepsilon_{11} + 7\varepsilon_{22}$$

$$\sigma_{22} = 7\varepsilon_{11} + 12\varepsilon_{22}$$

Calculate the Young's modulus of the laminate in the direction of one set of fibres, explaining clearly your assumptions.

3. A composite of A and B, with $\lambda\phi$ being the volume fraction of A, is modelled as indicated below

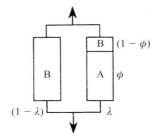

Show that the modulus of the composite is given by

$$E_c = \lambda \left(\frac{\phi}{E_a} + \frac{1 - \phi}{E_b} \right)^{-1} + (1 - \lambda) E_b$$

where E_a, E_b represent the moduli of the components. If

$$E_a = 10 \text{ GPa}; \qquad E_b = 1 \text{ GPa}$$

What is the modulus of the composite if $\phi = 1$; $\phi = 0.8$? In each case $\lambda = 0.5$.

9

Relaxation Transitions: Experimental Behaviour and Molecular Interpretation

We shall discuss the assignment of viscoelastic relaxations in a molecular sense to different chemical groups in the molecule, and in a physical sense to features such as the motion of molecules in crystalline or amorphous regions. Because amorphous polymers exhibit fewer structure-dependent features than those that are semicrystalline, we shall use these simpler materials to illustrate some general characteristics of relaxation behaviour.

9.1 Amorphous polymers: an introduction

It is customary to label relaxation transitions in polymers as α, β, γ, δ, etc. in alphabetical order with decreasing temperature. Three of the four transitions in polymethyl methacrylate (PMMA):

$$\left[\begin{array}{c} H_3C-O-C=O \\ | \\ -C-CH_2- \\ | \\ CH_3 \end{array} \right]_n$$

are shown in Figure 9.1, which summarizes data obtained using a low-frequency torsion pendulum. The highest temperature relaxation, the α relaxation, is the glass transition and is associated with a large change in modulus. Comparative studies on similar polymers, together with nuclear magnetic resonance (NMR) and dielectric measurements [1–5], have shown the β relaxation to be associated with side-chain motions of the ester group. The γ and δ relaxations involve motion of the methyl groups attached to the main chain and the side chain, respectively.

Many other amorphous polymers display a high-temperature transition, the glass transition, associated with the onset of main-chain segmental motion, and secondary transitions that have been assigned to either motion of side groups or

An Introduction to the Mechanical Properties of Solid Polymers I. M. Ward and J. Sweeney
© 2004 John Wiley & Sons, Ltd   ISBN: 0471 49625 1 (HB); 0471 49626 X (PB)

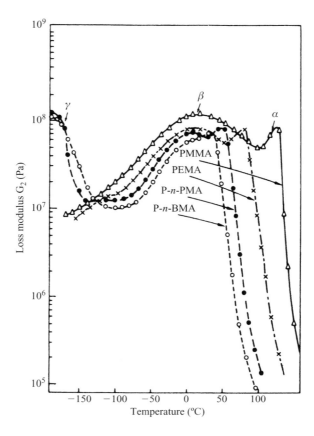

Figure 9.1 Temperature dependence of loss modulus G_2 for polymethyl methacrylate (PMMA), polyethyl methacrylate (PEMA), poly-*n*-propyl methacrylate (P-*n*-PMA) and poly-*n*-butyl methacrylate (P-*n*-BMA). (Reproduced with permission from Heijboer, in *The Physics of Non-crystalline Solids*, North-Holland, Amsterdam, 1965, p. 231)

restricted motion of the main chain of end-group motions. It must be emphasized that in many amorphous polymers the assignment of a relaxation is by no means as straightforward as in PMMA and some related polymers. The case of amorphous polyethylene terephthalate (PET), where there are no side groups, is considered in Section 9.3.2, in conjunction with semicrystalline forms of this polymer.

9.2 Factors affecting the glass transition in amorphous polymers

Two distinct models have been used for interpreting the influence of features such as chemical structure, molecular mass, cross-linking and plasticizers on the glass transition in amorphous polymers. The first approach considers changes in molecular flexibility, which modify the ease with which conformational changes can take place. The alternative approach relates all these effects to the amount of free volume, which is assumed to attain a critical value at the glass transition.

9.2.1   Effect of chemical structure

Although these factors have been intensively studied, because of their importance in selecting polymers for commercial exploitation, much of our knowledge is empirical in nature, due primarily to the difficulty in distinguishing between intra- and intermolecular effects. Some general features are, however, evident.

Main-chain structure

Flexible groups such as an ether link will enhance main-chain flexibility and reduce the glass transition temperature, with the opposite effect being shown by the introduction of an inflexible group, such as a terephthalate residue.

Side groups [6]

Bulky, inflexible side groups increase the temperature of the glass transition, as is illustrated in Table 9.1 for a series of substituted poly-α-olefins:

$$\left[-CH_2-\underset{R}{CH}- \right]_n$$

Table 9.1   Glass transition of some vinyl polymers. [Reproduced with permission from Vincent, in *The Physics of Plastics* (ed. P. D. Ritchie), Iliffe, London, 1965.]

Polymer	R	Transition temperature in °C at ~1 Hz
Polypropylene	CH_3	0
Polystyrene	C_6H_5	116
Poly-*N*-vinylcarbazole		211

A difference between the effect of rigid and flexible side groups is shown in Table 9.2 for a series of polyvinyl butyl ethers:

$$\left[\begin{array}{c} -CH_2-CH- \\ | \\ OR_1 \end{array} \right]_n$$

Table 9.2 Glass transition of some isomeric polyvinyl butyl ethers. [Reproduced with permission from Vincent, in *The Physics of Plastics* (ed. P. D. Ritchie) Iliffe, London, 1965.]

Polymer	R_1	Transition temperature in °C at ~1 Hz
Polyvinyl *n*-butyl ether	$CH_2CH_2CH_2CH_3$	−32
Polyvinyl isobutyl ether	$CH_2CH(CH_3)_2$	−1
Polyvinyl *t*-butyl ether	$C(CH_3)_3$	+83

All these polymers contain the same atoms in the side group OR_1 (where R_1 represents the butyl isomeric form), but more compact arrangements reduce the flexibility of the molecule and give a marked increase in the transition temperature.

Increasing the length of flexible side groups reduces the temperature of the glass transition, as is evident from Table 9.3 for a series of polyvinyl ethers:

$$\left[\begin{array}{c} -CH_2-CH- \\ | \\ OR_2 \end{array} \right]_n$$

where R_2 represents the *n*-alkyl group. Here the increase in length is associated with an increase in free volume at a given temperature.

Table 9.3 Glass transition of some polyvinyl *n*-alkyl ethers. [Reproduced with permission from Vincent, in *The Physics of Plastics* (ed. P. D. Ritchie), Ilffe, London, 1965.]

Polymer	R_2	Transition temperature in °C at ~1 Hz
Polyvinyl methyl ether	CH_3	−10
Polyvinyl ethyl ether	CH_2CH_3	−17
Polyvinyl *n*-propyl ether	$CH_2CH_2CH_3$	−27
Polyvinyl *n*-butyl ether	$CH_2CH_2CH_2CH_3$	−32

Main-chain polarity

In a series of polymers of similar main-chain composition, the temperature of the glass transition may be significantly depressed as the number of successive —CH_2 or —CH_3 groups in the side groups is increased (Figure 9.2). It is evident that the temperature of the glass transition increases with main-chain polarity, and it is assumed that the associated reduction in main-chain mobility is due to the increase in intermolecular forces. In particular it is suggested that the higher values for the polychloracrylic esters arise from the increased valence forces associated with the chlorine molecules.

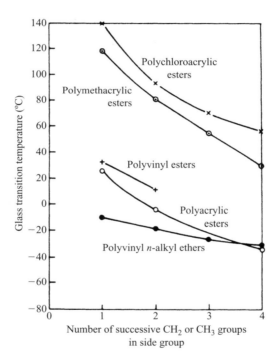

Figure 9.2 The effect of polarity on the position of the glass transition temperature for five polymer series. [Reproduced with permission from Vincent, in *The Physics of Plastics* (ed. P. D. Ritchie), Iliffe, London, 1965]

9.2.2 Effect of molecular mass and cross-linking

The length of the main chain does not affect the dynamic mechanical properties of polymers in the glassy state, where molecular motions are of restricted extent, but the glass transition temperature is depressed at very low relative molecular masses as a consequence of the additional free volume introduced by the increased proportion of chain ends [7].

As already discussed (Section 6.1.1 above) molecular mass has a large effect in the glass transition range, where viscous flow transforms to a plateau range of rubber-like behaviour due to entanglements between the longer molecular chains.

Chemical cross-links reduce the free volume by bringing adjacent chains close together and so raise the temperature of the glass transition, as is shown in Figure 9.3 for phenol–formaldehyde resin cross-linked withhexamethylene tetra-mine to different extents. The transition region is greatly broadened [8], so that in very highly cross-linked materials, where motions of extensive segments of the main chain are not possible, there is no glass transition.

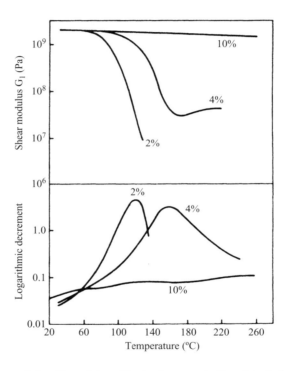

Figure 9.3 Shear modulus G_1 and logarithmic decrement of a phenol–formaldehyde resin cross-linked with hexamethylene tetramine at stated concentrations. (Reproduced with permission from Drumm, Dodge and Nielsen, *Ind. Eng. Chem.*, **48**, 76 (1956))

9.2.3   Blends, grafts and copolymers

The mechanical properties of blends and graft polymers are determined primarily by the mutual solubility of the two homopolymers. For complete solubility the properties of the mixture are close to those of a random copolymer of the same composition, as is shown in Figure 9.4, which compares a 50:50 mixture of polyvinyl acetate and polymethyl acrylate with the equivalent copolymer [8]. Note that the damping peak occurs at 30 °C, compared with 15 °C for the polymethyl acrylate and 45 °C for polyvinyl acetate.

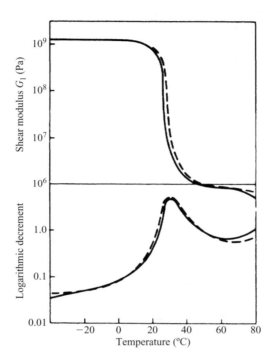

Figure 9.4   Shear modulus G_1 and logarithmic decrement for a miscible blend of polyvinyl acetate and polymethyl acrylate ($- - -$) and a copolymer of vinyl acetate and methyl acrylate (____). (Reproduced with permission from Neilsen, *Mechanical Properties of Polyms*, Van Nostrand-Reinhold, New York, 1962)

A theoretical interpretation of the glass transition temperature of a copolymer is based on the assumption that the transition occurs at a constant fraction of free volume. Gordon and Taylor [9] assume that in an ideal copolymer the partial specific volumes of the two components are constant and equal to the specific volumes of the two homopolymers. The specific volume–temperature coefficients for the two components in the rubbery and glassy states are assumed to remain the same in the copolymer as in the homopolymers, and to be independent of temperature. The glass transition temperature T_g for the copolymer is then given by [10]

$$\frac{1}{T_g} = \frac{1}{(w_1 + Bw_2)} \left[\frac{w_1}{T_{g1}} + \frac{Bw_2}{T_{g2}} \right]$$

where w_1 and w_2 are the mass fractions of the two monomers whose homopolymers have transition temperatures T_{g1} and T_{g2}, respectively, and B is a constant close to unity.

Where the two polymers in a mixture are insoluble they exist as separate phases, so that two glass transitions are observed, as shown in Figure 9.5 for a polyblend of polystyrene and styrene–butadiene rubber [8]. The two loss peaks are very close to those for pure polystyrene and pure styrene–butadiene rubber.

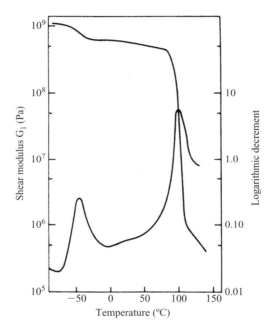

Figure 9.5 Shear modulus G_1 and logarithmic decrement for an immiscible polyblend of polystyrene and a styrene–butadiene copolymer (Reproduced with permission from Nielsen, *Mechanical Properties of Solids*, Van Nostrand-Reinhold, New York, 1962)

9.2.4 Effects of plasticizers

Plasticizers, which are relatively low molecular mass organic materials added to soften rigid polymers, must be soluble in the polymer and usually they dissolve it completely at high temperature. Figure 9.6 shows the change in the loss peak associated with the glass transition of polyvinyl chloride (PVC) when plasticized with varying concentrations of di(ethylhexyl)phthalate [11]. In this polymer plasticization is of major commercial importance: rigid PVC is used in applica-

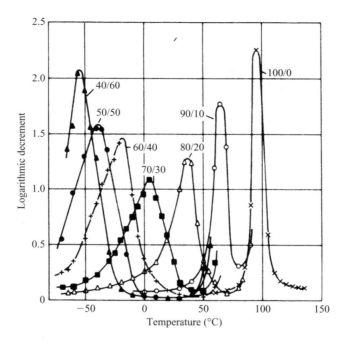

Figure 9.6   The logarithmic decrement of polyvinyl chloride plasticized with various amounts of di(ethylhexyl) phthalate. (Reproduced with permission from Wolf, *Kunststoffe*, **41**, 89 (1951))

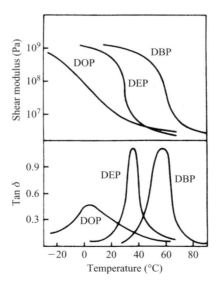

Figure 9.7   Shear modulus and loss factor tan δ for PVC plasticized with diethyl phthalate (DEP), dibutyl phthalate (DBP) and *n*-dioctyl phthalate (DOP). (Reproduced with permission from Neilsen, Buchdahl and Levreault, *J. Appl. Phys.*, **21** 607 (1950))

tions such as replacement window frames, and the plasticized material supplies flexible sheeting and inexpensive footwear.

Plasticizers make it easier for changes in molecular conformation to occur, and so lower the temperature of the glass transition. They also broaden the loss peak, with the extent of broadening depending on the nature of the interactions between the polymer and the plasticizer. A broad damping peak is found where the plasticizer has a limited solubility in the polymer or tends to associate in its presence. The increased width of the damping peak as the plasticizer becomes a poorer solvent is shown in Figure 9.7 for plasticized PVC [8]. Diethyl phthalate is a relatively good solvent, dibutyl phthalate is intermediate and n-dioctyl phthalate is a very poor solvent.

9.3 Relaxation transitions in crystalline polymers

9.3.1 General introduction

Semicrystalline polymers are less sensitive to wide variations of stiffness with temperature than those that are totally amorphous, but even so stiffnesses may vary by an order of magnitude over the useful working range of a given material. Oriented crystalline polymers may additionally show contrasts between extensional and shear deformations, and also angular-dependent changes in relaxation strength.

Some polymers, notably low-density polyethylene, show clearly resolved α, β and γ processes. The high-temperature α relaxation is frequently related to the proportion of crystalline material present, the β process is related to a greatly broadened glass–rubber relaxation and the γ relaxation has been associated, at least in part, with the amorphous phase. Other materials – an example to be discussed shortly being that of PET – show only two relaxation processes. In these cases the α relaxation is akin to the β process in polymers where all three relaxations are evident.

In earlier editions of this work it was noted that the interpretation of viscoelastic relaxations in crystalline polymers was at a very speculative stage but, as a working hypothesis, it was assumed that the tangent of the phase lag angle ($\tan \delta$), or its equivalent the logarithmic decrement (Λ), was an appropriate measure of the relaxation strength. Boyd, in two important review articles [12, 13], has demonstrated that the situation is more complex: for instance the apparent trend of relaxation strength with changing crystallinity can depend on whether $\tan \delta$ or the real (in-phase) or imaginary (out-of-phase) components of modulus (G_1 and G_2, respectively) or compliance (J_1, J_2) are used to record the relaxations; and the interpretation of the data depends on the composite model used to determine the interaction between crystalline and amorphous phases. However, it is still the case that $\tan \delta_{max}$ (but not necessarily G''_{max} or J''_{max}) usually may be expected to correlate directly with phase origin, although a plot versus crystallinity may not be linear.

We shall begin with a brief and simplified discussion of the main features of experimental observations and proceed to consider the interpretation of these features. Three polymers are selected as paradigms: PET, which can exist in the wholly amorphous state but also as a partially crystalline polymer; polyethylene, which is a high crystalline polymer; and a liquid crystalline polymer, the thermotropic copolyester whose mechanical anisotropy was discussed in Section 7.5.4 above.

9.3.2   Relaxation in low crystallinity polymers

The temperature variation of the complex modulus of PET as a function of crystallinity has been studied by Takayanagi [14] in extension at 138 Hz (Figure 9.8) and by Illers and Breuer [15] in shear at $\sim$1 Hz. At the lowest levels of crystallinity there is a sudden and severe drop in stiffness associated with the α process that is characteristic of amorphous polymers. With increasing crystallinity the α peak broadens as the change in stiffness is greatly reduced. This behaviour is consistent with that of a composite for which only one phase softens, with the broadening of the peak resulting from restriction of long-range segmented motions

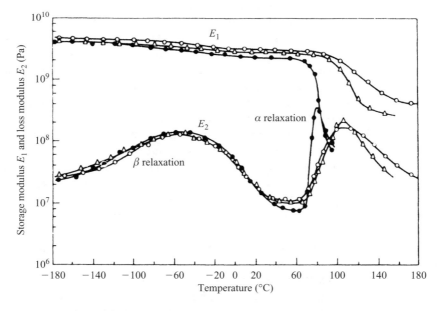

Figure 9.8   Storage modulus E_1 and loss modulus E_2 as a function of temperature at 138 Hz for PET samples of differing degrees of crystallinity ($\bullet$, 5 per cent; $\triangle$, 34 per cent; $\circ$, 50 per cent). (Reproduced with permission from Takayanagi, *Mem. Fac. Eng., Kyushu Univ.*, **23**, 1 (1963))

in the amorphous phase by the remaining crystals. Illers and Breuer noted also that the temperature at which the loss peak (G_2) was a maximum increased up to 30% crystallinity and then decreased slightly at high crystallinities. Studies involving small-angle X-ray scattering indicate that high-crystallinity specimens have both thicker crystal layers and thicker amorphous layers than those of low crystallinity [16]. This latter feature will reduce the constraints imposed by crystal surfaces.

In contrast to the α relaxation both the shape and location of the subglass β process are insensitive to the degree of crystallization. Dielectric studies [17] yield the same conclusion. The process is therefore consistent with localized molecular motions, in contrast with the restrained long-range segmental motions involved in the glass–rubber α relaxation.

Boyd [13] has analysed the dynamic mechanical behaviour of PET in terms of a composite model of crystalline and amorphous phases. The relaxation strengths for the α and β processes, determined from the shear modulus results of Illers and Breuer, were shown to be related to the crystallinity, indicating that both relaxation processes related to the amorphous regions. For the β processes, the shear modulus for the amorphous phase lies between the upper (Voigt) and lower (Reuss) bounds in both the relaxed and unrelaxed states. The α process, however, appeared to fit best near the lower bound behaviour.

Although the phenomenological approach of Boyd gives some insight into the dynamic mechanical behaviour, it does not provide any significant understanding of the molecular origins of the relaxation processes. In recent publications, Ward and co-workers [18, 19] have shown how it is necessary to bring together dynamic mechanical, NMR and dielectric measurements to achieve such an understanding, and NMR spectroscopy is probably the most powerful tool in this respect. The main advantage of NMR is that it allows the mobility in different parts of the molecule to be examined. There are several investigations of the molecular relaxations in PET using NMR. Using selectively deuterated samples to distinguish between molecular motions in the aliphatic and aromatic parts of the PET molecule, English [20] identified significant motion of the ethylene glycol units at the glass transition temperature, which have been assumed to be due to a unique *trans–gauche* movement of these units (refer to Figure 1.7). It is important to note that the ethylene glycol units do not contribute to the relaxation processes present in the β peak. This result is consistent with much earlier NMR studies on selectively deuterated PET samples by Ward [21], which showed that the α relaxation in PET involved significant motions of both the aliphatic and aromatic moieties.

Dielectric and dynamic measurements by Maxwell *et al.* [19] showed that the β peak consists of two different relaxation processes, one on the high temperature side and one on the low temperature side. High-resolution carbon-13 chemical shift and deuterium NMR experiments showed that both the phenyl rings and the carbonyl groups undergo small-angle oscillations at temperatures below the glass transition, and that the phenyl groups also undergo rapid 180° flips. It was confirmed that the ethylene glycol group does not contribute to the β relaxation and it was concluded that the high temperature side of the relaxation is due to the

180° flips and the low temperature side to motion of the carbonyl groups, which has a significantly lower activation energy and activation enthalpy (see Section 6.3.1).

9.3.3   Relaxation processes in polyethylene

Polyethylene is the obvious choice for investigating relaxations in the more highly crystalline polymers. Its structure has been studied in great detail, and the material is readily obtainable in two forms. Low-density polyethylene (LDPE) typically contains about three short side branches per 100 carbon atoms, together with about one longer branch per molecule. High-density polyethylene (HDPE) is much closer to the pure $(CH_2)n$ polymer, and the proportion of side branches is often less than five per 1000 carbon atoms. The main features of the temperature dependence of $\tan \delta$ for each material are indicated schematically in Figure 9.9. The LDPE shows clearly resolved α, β and γ loss peaks. In HDPE the low-temperature γ peak is very similar to that in LDPE, but the β relaxation is hardly resolved and the α relaxation is often considerably modified, appearing to consist of at least two processes (α and α') with different activation energies. The high-temperature behaviour is also dependent on whether loss angle or loss modulus is the quantity being measured. At one time some investigators questioned the existence of the β relaxation in HDPE, but further work involving a range of

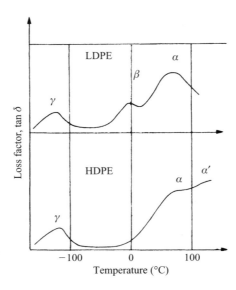

Figure 9.9   Schematic diagram showing α, α', β and γ relaxation processes in LDPE and HDPE

polymers intermediate between the extremes indicated in Figure 9.9 has established its presence.

We must conclude that α, β and γ relaxations occur in all forms of polyethylene. The first stage in the process of analysis is to determine whether a given relaxation is related to the crystalline or the amorphous component, or to an interaction involving both phases. Next, one deduces a process that is able to account for each relaxation. Finally, an attempt must be made to model a molecular mechanism that can cause the process to occur.

In two review articles Boyd [12, 13] presents evidence for two major conclusions regarding the origins of the α, β and γ relaxations in polyethylene. First, the mechanical strengths of all three relaxations relate to the amorphous fraction. Secondly, both mechanical and dielectric measurements show that the location of the α relaxation depends on the crystal lamellar thickness. These conclusions might appear to be in conflict regarding the α relaxation, but we will see that recent research by Ward and co-workers [22, 23] resolves this issue and also the relationship between the relaxations and LDPE and HDPE, as to whether they are of similar molecular origin. Two extra ingredients were required to answer this question: dynamic mechanical measurements on specially produced oriented samples; and measurements over a wide range of frequencies to determine the activation energies for the processes, as discussed in Section 6.3.1 above.

Key results for the mechanical anisotropy of LDPE sheets have already been discussed in Section 8.4.3 above. Oriented and annealed sheets can be considered as composite solids, where the β relaxation is an interlamellar shear process, consistent with its assignment by Boyd to an amorphous process. Figure 9.10 shows results for cold-drawn and annealed HDPE, where no β relaxation is observed [24]. In these sheets the crystal lamellae make an acute angle of about 40° with the initial draw direction [25]. Applying the stress along the initial draw direction then gives the maximum resolved shear stress parallel to lamellar planes. We see from Figure 9.10 that the maximum loss is tan δ_0, confirming that the α relaxation in HDPE is primarily an interlamellar shear process from a macroscopic mechanical viewpoint.

The situation is apparently made more confusing by comparison of the results of Stachurski and Ward [26] for cold-drawn (a) and annealed (b) sheets of LDPE illustrated in Figure 9.11. For the cold-drawn sheet the maximum loss in the 0°C region occurs at 45° to the draw direction. Only one loss process is observed and this shows the anisotropy appropriate for a relaxation that involves shear parallel to the draw direction in a plane containing the draw direction. This material does not show a clear lamellar texture, so it is reasonable to associate this process with the α relaxation. In the cold-drawn and annealed sheet the process has moved to about 70 °C. These results, taken together with the measurements on specially oriented sheets, suggest that the draw or c-axis orientation of the crystalline regions is the governing factor. Because the relaxation involves shear in the c-axis direction in planes containing the c axis it has been termed the 'c-shear relaxation'. The annealed LDPE sheet also shows a β relaxation below 0 °C with

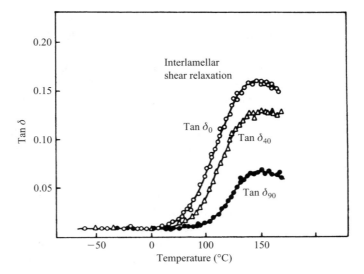

Figure 9.10 Temperature dependence of tan δ in a cold-drawn and annealed HDPE sheet in different directions at 50 Hz. (Reproduced with permission from Stachurski and Ward, *J. Macromol. Sci. B*, **3**, 445 (1969))

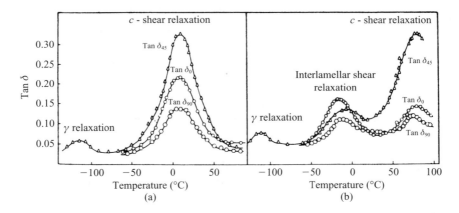

Figure 9.11 Temperature dependence of tan δ in three directions in cold-drawn (a) and cold-drawn and annealed (b) LDPE sheets at approximately 500 Hz. (Reproduced with permission from Stachurski and Ward, *J. Polymer Sci. A-2*, **6**, 1817 (1986))

an anisotropy consistent with interlamellar shear, as discussed in Section 8.4.3 above.

The identification of the α process as a c-shear relaxation and the β process as interlamellar shear in a drawn and annealed LDPE sheet was nicely confirmed by measurements of the anisotropy of dielectric relaxation [27]. Pure polyethylene shows no dielectric response, so experiments were made on specimens that had been lightly decorated with dipoles by means of oxidation, to such a small extent that the overall relaxation behaviour was not significantly affected. The dielectric relaxation data showed marked anisotropy for the relaxation, consistent with its assignment to the c-shear relaxation, but the β relaxation showed no anisotropy, confirming that the mechanical anisotropy observed related to the lamellar texture and not to anisotropy at a molecular level.

We are still left with an apparent paradox that the α relaxation in HDPE relates to interlamellar shear whereas in LDPE it is the c-shear relaxation. Recent measurements of the activation energies for these materials by Matthews *et al.* (Table 9.4) [23] show, however, that the α relaxation in both polymers has a comparatively low activation energy consistent with the relaxation being the c-shear mechanism at a molecular level, i.e. the α relaxation in both HDPE and LDPE is associated with identical molecular mechanisms. It appears that inter-lamellar shear in the mechanical α process in HDPE requires coupled motions of the chains that run through the lamellae (c-shear) together with chains on the lamellar surface. In contrast, c-shear and interlamellar shear in LDPE are two distinct mechanical relaxations, and interlamellar shear has a much higher activa-tion energy, akin to a glass transition.

Table 9.4   Activation energies for α and β relaxations in HDPE and LDPE

Sample	β Relaxation activation energy ($kJ\,mol^{-1}$)	α Relaxation activation energy ($kJ\,mol^{-1}$)
Isotropic LDPE	430	120
Oriented LDPE	500	110
Isotropic HDPE	Not present	120
Oriented HDPE	Not present	80–90

Mansfield and Boyd [28] have proposed that the dielectric α process can be represented by the torsional movement of a segment of chain about $12CH_2$ units in length. This motion, which will cause the short twisted mismatch region to move through the crystal one carbon atom at a time, is consistent with the dependence of activation energy on crystal thickness (Figure 9.12).

In the mechanical situation the translational component of the crystal process can lead to reorganization of the crystal surface and hence modify the connections of amorphous chains to the crystal surface. An example is shown in Figure 9.13, where translational motion of a decorated dipole (indicated by a horizontal arrow) permits lengthening of the tight tie chain in (a), and so enables further deformation of the amorphous fraction.

The β relaxation is very broad compared with that in completely amorphous polymers due to the immobilizing effect of the crystals on the amorphous fraction.

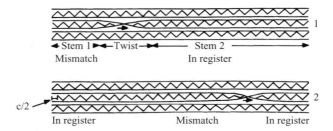

Figure 9.12 Propagation of a localized smooth twist along the chain. As the twist starts (1) it leaves behind it a translational mismatch. As the twist proceeds (2) the mismatch becomes attenuated at large distances from the twist by elastic distortion of the stem valence angles and bond length. (Reproduced with permission from Mansfield and Boyd, *J. Polymer Sci., Phys. Ed.*, **13**, 1407 (1975))

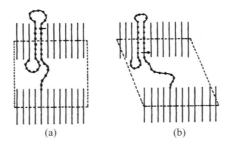

Figure 9.13 Further relaxation of the amorphous fraction resulting from translational mobility in the crystal (the latter acquired in the α process). Illustrated in this case is reorganization of the interface in (a) through shortening of two loops that in (b) permits lengthening of a tight tie chain, which in turn permits more deformation of the amorphous fraction. Also shown is a decorating dipole in the crystal that moves through a number of translational, rotational steps. One such step suffices for dielectric activity. (Reproduced with permission from Boyd, *Polymer*, **26**, 1123 (1985))

Boyd speculates that the shortest relaxation times may be associated with motions of very loose folds and relatively non-extended tie chains; conversely tight folds are unable to relax. The relative prominence of the β relaxation in LDPE compared with HDPE is enhanced by the lower value of the relaxed β process modulus in LDPE, which will increase the relative intensity of the β and decrease that of the α. On a molecular basis the branching of LDPE gives a more loosely organized amorphous component, capable of relaxing to a lower limiting rubbery modulus.

As it occurs below the glass transition temperature the γ relaxation will involve simple conformational motions that are relatively short range in character. Such motions must leave the molecular stems adjacent to the bonds undergoing transition relatively undisturbed, they must require only a modest activation energy, and the swept-out volume during the relaxation should be small. Following Willbourn's [29] suggestion that the γ relaxation in many amorphous and semicrystalline polymers can be attributed to a restricted motion of the main chain that involves at least four successive —CH_2 groups, both Shatzki [30] and Boyer [31] have proposed that subglass relaxations can be modelled in terms of a so-called 'crankshaft' mechanism (Figure 9.14). Shatzki's five-bond mechanism involves the simultaneous rotation about bonds 1 and 7 such that the intervening carbon bonds move as a crankshaft. Boyer's proposal involves only three inter-mediate carbon bonds. The mechanisms are, as Boyd [12] has pointed out, the simplest allowed moves of a tetrahedrally bonded chain on a diamond lattice that leave the adjacent stem bonds in place. The internal energetics of the five-bond model are modest, but the swept-out volume is large in the context of a motion in a glassy matrix. For the three-bond transition a double energy barrier system with an intermediate energy minimum is implied, and the motion associated with one of these barriers requires a significant free volume, and so is inhibited by the matrix. Despite these drawbacks a crankshaft mechanism has been proposed as being relevant to the γ relaxation in polyethylene.

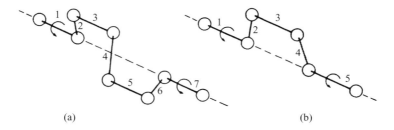

(a)   (b)

Figure 9.14   The crankshaft mechanisms of Shatzki (a) and Boyer (b). (Reproduced with permission from McCrum, Read and Williams, *Anelastic and Dielectric Effects in Polymer Solids*, Wiley, London, 1967)

Boyd [12] discusses a motion related to the three-bond mechanism that can accomplish the appropriate shape change without encountering problems asso-ciated with free volume. It involves the conformational sequence GTG' occurring in an otherwise all-*trans* chain (G and G' represent alternative *gauche* transforma-tions). From Figure 9.15 it can be seen that this sequence, known as a kink [32], has the effect of displacing the separated *trans* components of a planar zigzag yet leaving them parallel to one another. Interchanging the senses of the *gauche* bonds

$$\dots \text{TTGTG}'\text{TT} \dots \; \rightarrow \; \dots \text{TTG}'\text{TGTT} \dots$$

converts the kink into a mirror image of itself. The kink inversion process, shown in Figure 9.16, involves only a small swept-out volume and requires a modest activation energy. The stem displacement causes a localized shape change that can

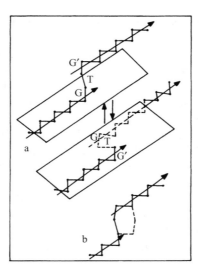

Figure 9.15   Kinks and kink inversion. (a) The conformational sequence ... TTTGTG'TTT ... has parallel offset planar zigzag stems (indicated by arrows) on either side of the GTG' portion. The transition TGTG' → TG'TGT (called here kink inversion) creats a mirror image of the kink about the displaced stems. (b) A three-bond crankshaft move is shown at a kink site (as dashed line). This move advances the kink along the chain by 2CH$_2$ units. (Reproduced with permission from Boyd, *Polymer*, **26**, 1123 (1985))

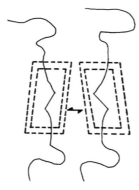

Figure 9.16   Strain fields set up by stem displacement accompanying kink inversion. (Reproduced with permission from Boyd, *Polymer*, **26**, 1123 (1985))

propagate through the specimen as a shear strain. The kink inversion process is therefore a possible candidate on which to base a molecular model of the γ relaxation, but it must be emphasized that there is no direct evidence to demonstrate that it is appropriate for modelling the behaviour of polyethylene.

9.3.4 Relaxation processes in liquid crystalline polymers

Liquid crystalline polymers form another class of polymers from a structural viewpoint and can be produced either from a liquid crystalline solution (termed lyotropic) or a liquid crystalline melt (thermotropic). In this section we will consider thermotropic liquid crystalline polymers, of which the simplest chemically are those invented by Calundann [33], which are random copolymers of hydroxybenzoic acid (HBA) and hydroxynaphthoic acid (HNA) (Figure 9.17). The random arrangement of HBA and HNA units along the chain prevents normal crystallization into three-dimensional order (although there is debate about the extent to which there may be some regions of three-dimensional order [34]). Molecular alignment of the chains can be achieved readily by melt spinning. This produces an oriented liquid crystalline structure (termed a mesophase) in which there is axial alignment of the chains, which are close packed on hexagonal or orthorhombic lattices without any regularity within the chains along the axial direction.

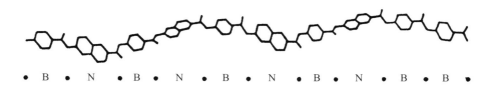

Figure 9.17 Monomers of HBA and HNA and a projection of the random chain. (Reproduced with permission from Davies and Ward, in *High Modulus Polymers* (eds Zachariades and Porter), Marcel Dekker, New York, 1988)

The dynamic mechanical properties of these oriented HBA/HNA copolymers have been studied by Yoon and Jaffe [35], Blundell and Buckingham [36] and Ward and co-workers [37]. It is of particular interest to compare different compositions based on HBA and HNA only, with two other copolymers that incorporate terephthalic acid (TA) dihydroxynaphthalene (DNA) and biphenol (BP) [38]. The compositions of the four copolymers to be discussed are shown in Table 9.5. The most instructive comparisons come from consideration of the dynamic mechanical loss factors in shear (Figure 9.18) and the dielectric loss data (Figure 9.19). It can be seen that there are three relaxation processes, labelled α, β and γ. Because these polymers are essentially single phase, it is possible to seek an understanding entirely in terms of molecular relaxation processes. The relatively high intensity of the β relaxation in CO 30/70 identifies this relaxation with the naphthalene residue because this has the highest concentration in all these polymers. It is clear from comparison with CO 2,6 (where the naphthalene residue is linked by oxygen) that the relaxation does not depend on whether it is linked to carbonyl or ether

Table 9.5 Chemical compositions of thermotropic liquid crystalline polymers

Polymer	Composition (mole fraction in %)				
	HBA	HNA	TA	DHN	BP
CO 73/27	73	27			
CO 30/70	30	70			
CO 2,6	60		20	20	
COTBP	60	5	17.5		17.5

HBA, 4-Hydroxybenzoic acid; HNA, 2-Hydroxy 6-naphthoic acid; TA, terephthalic acid; DHN, 2,6 dihydroxynaphthalene; BP, 4,4′ biphenyldiol.

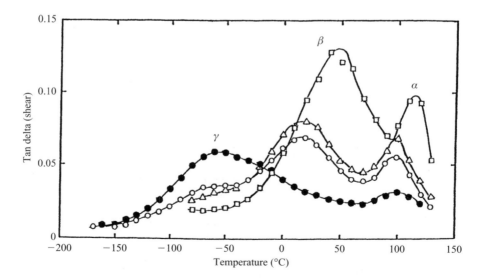

Figure 9.18 Dynamic mechanical loss factor (shear) for CO 30/70 ($\square$), CO 73/27 ($\triangle$), CO 2,6 ($\circ$) and COTBP ($\bullet$). (Reproduced with permission from Green, Ahaj-Mohammed, Abdul Jawad, Davies and Ward, *Polymers for Advanced Technologies*, **1**, 41 (1989))

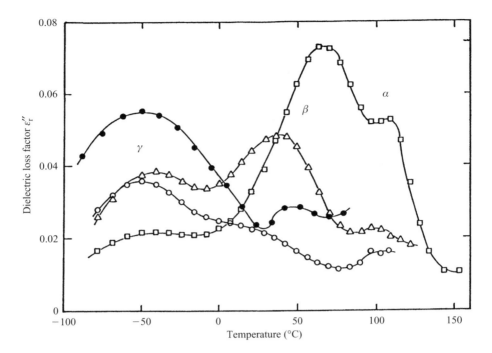

Figure 9.19   Dielectric loss data for CO 73/27 (△), CO 30/70 (□), CO 2,6 (○) and COTBP (●). (Reproduced with permission from Ward, *Macromol. Chem. Macromol. Symp.*, **69**, 75 (1993))

oxygen. This is not true for the dielectric relaxation ; the results in Figure 9.19 show that the β relaxation is not observed in CO 2,6, where the carbonyl groups are not attached to the benzene rings.

The γ relaxation is associated with the motion of phenylene groups, and this is shown most clearly by comparison of the dielectric relaxations for the CO 73/27 and CO 30/70 copolymers. These results suggest that the carbonyl group is strongly coupled to the aromatic ring to which it is attached, so that in this case there is a direct correlation between the mechanical and dielectric relaxations.

Further information is obtained by measurements at different frequencies to determine the activation energies of these relaxations. As shown in Figure 9.20 and Table 9.6, measurements at frequencies from 10^{-2} to 10^4 Hz, combining dynamic mechanical and dielectric measurements, yield activation energies of ≈ 120 kJ mol^{-1} for the β relaxation, similar energies for the γ relaxation but a very high activation energy for the α process. These results are consistent with assigning the α relaxation to the glass transition and the other relaxations to local processes. These conclusions have been confirmed by NMR studies, including measurements on deuterated polymers by Ward and co-workers [39].

There are several conclusions to be made regarding the relaxation behaviour of these thermotropic liquid crystalline polymers that have broad implications. First,

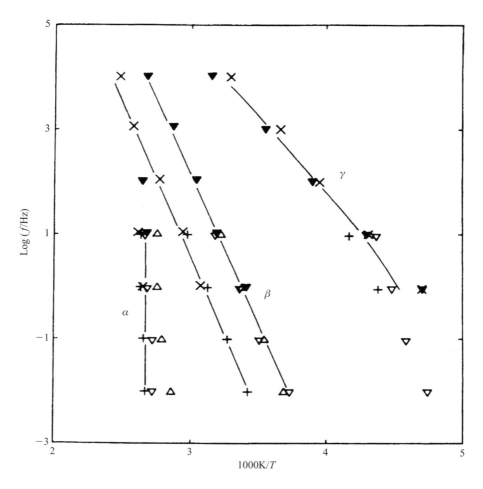

Figure 9.20 Loci of loss maxima. Mechanical $\tan \delta$ for oriented specimens: ($\triangle$) CO 73/27; ($\square$) CO 30/70: Dielectric ε'' for isotropic specimens: ($\blacktriangledown$) CO 73/27; ($\times$) CO 30/70. (Reproduced with permission from Troughton, Davies and Ward, *Polymer*, **30**, 58 (1989))

Table 9.6 Activation energies for mechanical and dielectric data

Sample (U, unannealed; A, annealed)	Activation energy (kJ mol^{-1})			
	Method	α	β	γ
CO 73/27 U	Tensile $\tan \delta$	460	130	
CO 73/27 A	Tensile $\tan \delta$	880	110	160
CO 30/70 A	Tensile $\tan \delta$	1600	130	
Isotropic CO 73/27 and CO 30/70	Dielectric ε''	~700	100	50

there is a very clear identification of the relaxations with the molecular structure. Secondly, the tensile and shear modulus fall very greatly with increasing temperature. This is the downside of incorporating sufficient mobility within the chains to permit melt processing, rather than the solution processing of the stiff chain lyotropic liquid crystalline polymers. Finally, it is to be noted that the shear moduli are low (≈ 1 GPa), which leads to low compressive strengths [40].

9.4 Conclusions

We have seen that there is a general understanding of the main factors that can modify the relaxation behaviour of non-crystalline polymers. With semicrystalline polymers it is frequently possible to attribute each relaxation to either the crystalline component or the amorphous component or to an interaction whereby the crystalline component constrains motions in the less-well-ordered material. For polyethylene considerable progress has been made in unravelling the complex relaxation processes, although it is far from clear whether measurements indicate that two or more mechanisms can operate simultaneously, with the dominant mechanism being dependent on structural features such as the density of branch points. Progress has also been made towards understanding relaxation mechanisms in other polymers, which we have no space to discuss. For these materials a smaller amount of structural information is available than is the case for polyethylene, so we cannot hope for a complete picture of relaxation behaviour. Nevertheless the methods used for elucidation of the relaxations in polyethylene can provide guidelines for future advancement.

References

1. Deutsch, K., Hoff, E. A. and Reddish, W., *J. Polymer Sci.*, **13**, 365 (1954).
2. Powles, J. G., Hunt, B. I. and Sandiford, D. J. H., *Polymer*, **5**, 505 (1964).
3. Sinnot, K. M., *J. Polymer Sci.*, **42**, 3 (1960).
4. Heijboer, J., *Physics of Non-Crystalline Solids*, North-Holland, Amsterdam, 1965, p. 231.
5. McCrum, N. G., Read, B. E. and Williams, G., *Anelastic and Dielectric Effects in Polymic Solids*, Wiley, London, 1967.
6. Vincent, P. I., *Physics of Plastics* (ed. P. D. Ritchie), Iliffe, London, 1965.
7. Fox, T. G. and Flory, P. J., *J. Appl. Phys.*, **21**, 581 (1950); *J. Polymer Sci.*, **14**, 315 (1954).
8. Nielsen, L. E., *Mechanical Properties of Polymers,* Van Nostrand-Reinhold, New York, 1962.
9. Gordon, M. and Taylor, J. S., *J. Appl. Chem.*, **2**, 493 (1952).
10. Mandelkern, L., Martin, G. M. and Quinn, F. A., *J. Res. Natl. Bur. Stand.*, **58**, 137 (1959); Fox, T. G. and Loshaek, S., *J. Polymer Sci.*, **15**, 371 (1955).
11. Wolf, K., *Kunstoffe*, **41**, 89 (1951).

12. Boyd, R. H., *Polymer*, **26**, 1123 (1985).
13. Boyd, R. H., *Polymer*, **26**, 323 (1985).
14. Takayanagi, M., *Mem. Fac. Eng, Kyushu Univ.*, **23**, 1 (1963).
15. Illers, K. H. and Breuer, H., *J. Colloid Sci.*, **18**, 1 (1963).
16. Kilian, H. G., Halboth, H. and Jenckel, E., *Kolloid Z.*, **176**, 166 (1960).
17. Coburn, J. C., PhD Dissertation, University of Utah, 1984.
18. Maxwell, A. S., Ward, I. M., Lauprétre, F., *et al.*, *Polymer*, **39**, 6835 (1998).
19. Maxwell, A. S., Monnerie, L. and Ward, I. M., *Polymer*, **39**, 6851 (1998).
20. English, A. D., *Macromolecules*, **17**, 2182 (1984).
21. Ward, I. M., *Trans. Faraday Soc.*, **56**, 648 (1960).
22. Matthews, R. G., Ward, I. M. and Cappaccio, G., *J. macromol. Sci. Phys.*, **37**, 51 (1999).
23. Matthews, R. G., Unwin, A. P., Ward, I. M. *et al.*, *J. Macromol. Sci. Phys. B*, **38**, 123 (1999).
24. Stachurski, Z. H. and Ward, I. M., *J. Macromol. Sci (Phys)*, **B3**, 445 (1969).
25. Hay, I. L. and Keller, A., *J. Mater. Sci.*, **2**, 538 (1967); Seto, T. and Hara, T., *Rep. Prog. Polymer Phys. (Japan)*, **7**, 63 (1967).
26. Stachurski, Z. H. and Ward, I. M., *J. Polymer Sci. A-2*, **6**, 1817 (1968).
27. Davies, G. R. and Ward, I. M., *J. Polymer Sci.*, **137**, 353 (1969).
28. Mansfield, M. and Boyd, R. H., *J. Polymer Sci., Phys. Ed.*, **16**, 1227 (1978).
29. Willbourn, A. H., *Trans. Faraday Soc.*, **54**, 717 (1958).
30. Shatzki, T. F., *J. Polymer Sci.*, **57**, 496 (1962).
31. Boyer, R. F., *Rubber Rev.*, **34**, 1303 (1963).
32. Pechold, W., Blasenbrey, S. and Woerner, S., *Kolloid-Z, Z. Polymer*, **189**, 14 (1963).
33. Calundann, G. W., *British Patent 1,585,511* (priority 13 May 1976, USA) (US Patent 4 067 852).
34. Donald, A. M. and Windle, A. H., *Liquid Crystalline Polymers*, Cambridge University Press, Cambridge, 1992, p. 162.
35. Yoon, H. N. and Jaffe, M., *Abstr. Pap., 185th Am. Chem. Soc. Natl. Meet.*, Seattle, WA, ANYL-72.
36. Blundell, D. J. and Buckingham, K. A., *Polymer*, **26**, 1623 (1985).
37. Davies, G. R. and Ward, I. M., in *High Modulus Polymers* (eds A. E. Zachariades and R. S. Porter), Dekter, New York, 1988, pp. 37–69.
38. Ward, I. M., *Makromol. Chem. Macromol. Symp.*, **69**, 75 (1993).
39. Allen, R. A. and Ward, I. M., *Polymer*, **32**, 202 (1991).
40. Ward, I. M. and Coates, P. D., in *Solid Phase Processing of Polymers*, (eds I. M. Ward, P. D. Coates and M. M. Dumoulin), Hanser, Munich, 2001, pp. 1–10.

10

Creep, Stress Relaxation and Non-linear Viscoelasticity

In Chapter 4 we introduced linear viscoelasticity. In this scheme, observed creep or stress relaxation behaviour can be viewed as the defining characteristic of the material. The definition of the creep compliance function $J(t)$, which is given as the ratio of creep strain $e(t)$ to the constant stress σ, may be recalled as

$$J(t) = \frac{e(t)}{\sigma} \tag{10.1}$$

The most significant aspect of J is that it is a function of time only. This implies an exact proportionality between the magnitudes of strain and stress at any given time, and it is in this sense that the material is *linear*: a doubling of stress always produces a doubling of strain. A single experiment is sufficient to define the material's creep behaviour in all circumstances. Similar comments apply to the stress relaxation modulus $G(t)$, defined in terms of the stress $\sigma(t)$ at constant strain e:

$$G(t) = \frac{\sigma(t)}{e} \tag{10.2}$$

It can be shown that if a material is linear in the sense of Equation (10.1) then it is also linear in the sense of Equation (10.2), and vice versa; J and G are mathematically related [1]. Thus, a linear material is one for which the creep compliance function or the stress relaxation modulus is a function of time only. When this is not the case, the material is *non-linear*. For example, the following simple forms are characteristic of non-linear viscoelastic materials:

$$J(t, \sigma) = \frac{e(t)}{\sigma} \tag{10.3}$$

$$G(t, e) = \frac{\sigma(t)}{e} \tag{10.4}$$

An Introduction to the Mechanical Properties of Solid Polymers I. M. Ward and J. Sweeney
© 2004 John Wiley & Sons, Ltd   ISBN: 0471 49625 1 (HB); 0471 49626 X (PB)

Once modifications to functions of this kind have been made, the Boltzmann superposition principle can no longer be assumed to apply, and there is no simple replacement for it. This marks a significant change in the level of difficulty when moving from linear to non-linear theory. In the linear case, the material behaviour is defined fully by single-step creep and stress relaxation, and the result of any other stress or strain history then can be calculated using the Boltzmann integral. In the non-linear case we have lost the Boltzmann equation, and it is not even clear what measurements are needed for a full definition of the material.

In this chapter we draw attention to some of the problems in this area and outline a few methods of approach, which are amplified in the advanced text by Ward [2]. A theoretical survey, using sophisticated mathematical methods, is given in the text by Lockett [3], and the contrasting approach of the practical engineer is typified in the work of Turner [4]. Ferry's textbook [5] covers both theoretical and practical aspects of polymer viscoelasticity. The more recent work of Lakes [6] covers some aspects of non-linearity.

10.1 The engineering approach

10.1.1 Isochronous stress–strain curves

The aim here is to predict the behaviour for a proposed application, using the minimum of experimental data. Empirical relations between stress, strain and time are obtained that may have no physical significance, and their use may be restricted to very specific stress or strain programmes. In the general case, creep curves need to cover the complete range of stresses over as long a period of time as feasible. However, Turner [4] has shown that when stress and time dependence are approximately separable it may be possible to interpolate creep curves at intermediate stresses from a knowledge of two creep curves, combined with a knowledge of the stress–strain curve for a fixed time (say 100 s): so-called isochronal stress–strain curves are represented by the vertical lines in Figure 10.1. These data cannot, however, be used to predict the response in multistep loading situations for non-linear viscoelastic materials, because the Boltzmann super-position principle does not then hold. In a number of cases it is claimed that simplifying regularities become evident if data are replotted in terms of *fractional recovery,* defined as the strain recovered as a fraction of the maximum creep strain, and *reduced time,* which is the ratio of recovery time to creep time.

10.2 The rheological approach

There is no general method of dealing with non-linear viscoelastic behaviour, and here we summarize approaches that have been applied with some success in

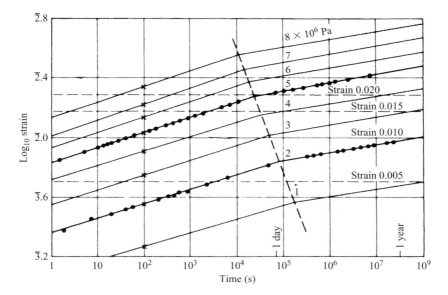

Figure 10.1 Tensile creep of polypropylene at 60 °C. The stress and time dependence are approximately separable and therefore creep curves at intermediate stresses can be interpolated from a knowledge of two creep curves (●) and the isochronous stress–strain relationship (×). (Reproduced with permission from Turner, *Polym. Eng. Sci.*, **6**, 306 (1966))

specific instances. The first approach takes linear viscoelasticity as the starting point; then we consider a functional representation in which the separability of stress, strain and time is preserved, thus leading to a single-integral representation that is consistent with rigorous continuum mechanics; finally we consider a series of developments based on a single-integral methods. In all cases the approach is phenomenological, with constitutive equations being found that describe, on the basis of continuum mechanics, the viscoelastic behaviour of a particular set of specimens. As a consequence the relationships derived cannot be interpreted directly at the molecular level.

10.2.1 Adaptations of linear theory – differential models

It is possible to take the equations of linear spring and dashpot models and adapt them to non-linear conditions. Thus, Smith [7] has described the large-strain behaviour of elastomers by taking as his starting point the (linear) Maxwell element. More recently, in the study of polypropylene by Kitagawa, Mori and Matsutani [8] and of polyethylene by Kitagawa and Takagi [9], the standard linear solid forms the basis of the approach. The differential equation of the standard linear solid is given by Equation (4.13):

$$\sigma + \tau \frac{d\sigma}{dt} = E_a e + (E_a + E_m)\tau \frac{de}{dt}$$

Recall that the first term on the right is equal to the stress after a long time, or equivalently the stress produced by loading at an infinitely slow rate. The equation can be generalized to

$$\sigma + K \frac{d\sigma}{dt} = f(e) + M \frac{de}{dt} \tag{10.5}$$

where $f(e)$ is simply the stress response of the material at infinitely slow strain rate, which we no longer require to be linear. Both M and K are in general functions of stress, strain and their time derivatives, so that Equation (10.5) defines a non-linear material. It can be rearranged as

$$\sigma - f(e) = M \frac{de}{dt} - K \frac{d\sigma}{dt} \tag{10.6}$$

The quantity $\sigma - f(e)$ on the left – the excess of stress over that obtaining at very slow strain rate – is termed the *overstress*; theories formulated in this way are termed *overstress theories*. This approach has been applied to metals by Lui and Krempl [10], who used the term *viscoplastic* to categorize their model. From the form of Equation (10.6) it is clear that the function K could be measured experimentally using stress relaxation tests ($de/dt = 0$), or that M could be measured using creep tests ($d\sigma/dt = 0$). Additional information can be gained from small-strain, high-strain rate experiments, where the response of the model can be assumed to be linear elastic. In these conditions, we may revert to Equation (4.13) and recall that the elastic response of the Maxwell model is characterized by the elastic modulus $E_a + E_m$. Then, inspection of Equation (10.5) reveals that the instantaneous modulus E is given by

$$E = \frac{M}{K} \tag{10.7}$$

for small strains and fast loading. The model is simplified if Equation (10.7) is assumed to apply under all conditions, so that K and M have the same functional form; this approach was adopted by Lui and Krempl [10], Kitagawa *et al.* [8] and Kitagawa and Takagi [9].

Kitagawa *et al.*, in their work on polypropylene, and Kitagawa and Takagi, working on polyethylene, used torsion testing. Their equivalent of Equation (10.6) in terms of shear stress and shear strain takes the form

$$\tau - f(\gamma) = M \frac{d\gamma}{dt} - K \frac{d\tau}{dt} \tag{10.8}$$

and Equation (10.7) becomes

$$G = \frac{M}{K} \tag{10.9}$$

where G is the shear modulus. In both cases, K was found to depend on the overstress and on the strain, and took the form

$$K = K_0 \exp\{-K_1(\gamma)[\tau - f(\gamma)]\} \tag{10.10}$$

where K_0 was found to be constant and

$$K_1(\gamma) = p_0 + \frac{p_1}{p_2 + \gamma} \tag{10.11}$$

with p_0, p_1 and p_2 material constants.

Stress relaxation tests were used to evaluate K. According to Equation (10.10)

$$\ln K = \ln K_0 - K_1(\gamma)[\tau - f(\gamma)] \tag{10.12}$$

suggesting the form required for graphs to show linear relationships at constant shear strain. Such a plot is shown for the polypropylene data [8] in Figure 10.2, taken at a series of strain levels. There is a good approximation to linearity, with

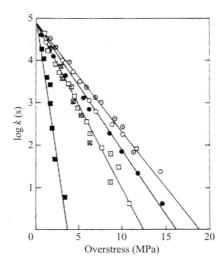

Figure 10.2 The function K of Equations (10.10) and (10.12) for polypropylene obtained from stress relaxation at various strains, with strain applied at varying rates. Symbols correspond to: (■) strain 0.01, strain rate 1.4×10^{-3} s^{-1}; (⊡) 0.045, 1.4×10^{-3}; (⊠) 0.055, 1.4×10^{-3}; (□) 0.053, 1.4×10^{-2}; (●) 0.108, 1.4×10^{-3}; (⊙) 0.253, 1.4×10^{-4}; (○) 0.267, 1.4×10^{-3}. (Reproduced with permission of Kitagawa, Mori and Matsutani, *J. Polym. Sci. B*, **27**, 85 (1989))

the common intercept corresponding to the constancy of the parameter K_0. The efficacy of the model in predicting torsional stress–strain curves is shown in Figure 10.3. Here, strain rates are initially constant but are changed abruptly to different constant rates so that transient stresses are generated. This provides a severe test for the model, which performs well.

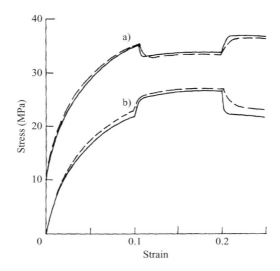

Figure 10.3 Modelling of stress–strain curves with step changes in strain rate: (a) $1.1 \times 10^{-2} \rightarrow 1.4 \times 10^{-3} \rightarrow 1.4 \times 10^{-2}\,\mathrm{s}^{-1}$. (b) $1.0 \times 10^{-3} \rightarrow 1.4 \times 10^{-2} \rightarrow 6 \times 10^{-4}\,\mathrm{s}^{-1}$. Curve (b) is shifted by 10 MPa along the vertical axis. (Reproduced with permission of Kitagawa, Mori and Matsutani, *J. Polym. Sci. B*, **27**, 85 (1989))

Similar work with polyethylene [9] revealed a comparable level of accuracy. The differential approach has been applied also to high-density polyethylene by Zhang and Moore [11, 12].

10.2.2 Adaptations of linear theory – integral models

In the course of extensive studies of the creep and recovery behaviour of textile fibres already referred to, Leaderman [13] became one of the first to appreciate that the simple assumptions of linear viscoelasticity might not hold even at small strains. For nylon and cellulosic fibres he discovered that although the creep and recovery curves may be coincident at a given level of stress – a phenomenon associated with linear viscoelasticity (Section 4.2.1) – the creep compliance plots indicated a softening of the material as stress increased, except at the shortest times (Figure 10.4). Thus, the creep compliance function is non-linear and of the

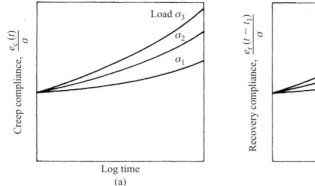
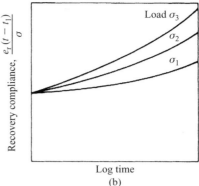

Figure 10.4 Comparison of creep compliance (a) and recovery compliance (b) at three load levels σ_1, σ_2 and σ_3 for a non-linear viscoelastic material obeying Leaderman's modified Boltzmann superposition principle. Note that the creep and recovery curves for a given load level are identical

form of Equation (10.3). For materials such as the polypropylene filament illustrated in Figure 10.5, the non-linearity is even more pronounced: at a given stress the instantaneous recovery is greater than the instantaneous elastic deformation, although the delayed recovery proceeds at a slower rate than the preceding creep.

Leaderman's approach was to modify the basic Boltzmann superposition principle of linear viscoelasticity, so that the strain was given by

$$e(t) = \int_{-\infty}^{t} \frac{\mathrm{d}f(\sigma)}{\mathrm{d}\tau} J(t - \tau)\mathrm{d}\tau \qquad (10.13)$$

where $f(\sigma)$ is an empirical function of stress that depends arbitrarily on the test fibres. This form of empiricism is inadequate to describe the behaviour in loading programmes of greater complexity than creep and recovery, and emphasizes that no general treatment is known to cope with the problems of non-linear viscoelasticity.

Another simple adaptation of the Boltzmann superposition principle is that of Findlay and Lai [14], who worked with step stress histories applied to specimens of poly(vinylchloride). Their theory was reformulated by Pipkin and Rogers [15] for general stress and strain histories. Pipkin and Rogers took a non-linear stress relaxation modulus $R(t, e)$, defined somewhat differently from G in Equation (10.4):

$$R(t, e) = \frac{\partial \sigma(t, e)}{\partial e} \qquad (10.14)$$

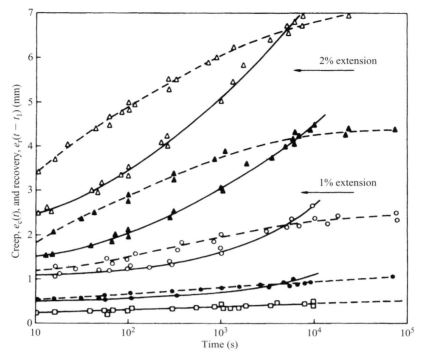

Figure 10.5 Successive creep (—) and recovery (- - -) for an oriented monofilament of polypropylene of total length 302 mm. The load levels are 587 g ($\triangle$), 401.8 g ($\blacktriangle$), 281 g ($\bullet$) and 67.7 g ($\square$). (Reproduced with permission from Ward and Onat, *J. Mech. Phys. Solids*, **11**, 217 (1963))

Note that Equations (10.4) and (10.14) reduce to the same form in the linear case but may differ significantly in the non-linear case. The Pipkin and Rogers integral for the stress in terms of the strain history is

$$\sigma(t) = \int_{-\infty}^{t} \frac{\mathrm{d}e}{\mathrm{d}\tau}(\tau) R[t - \tau, \, e(\tau)] \mathrm{d}\tau \qquad (10.15)$$

Similarly, for the strain in terms of the stress history a creep function C was defined as

$$C(t, \sigma) = \frac{\partial e(t, \sigma)}{\partial \sigma} \qquad (10.16)$$

with a corresponding integral law:

$$e(t) = \int_{-\infty}^{t} \frac{d\sigma}{d\tau}(\tau)C[t - \tau, \sigma(\tau)]d\tau \tag{10.17}$$

Pipkin and Rogers tested their model (Equation (10.17)) using published data on poly(vinylchloride) under multistep creep conditions. This kind of testing programme consists of a constant stress being applied for a predetermined period and then abruptly changing to a different value for a second time interval, and so on (see, for example, Figure 4.6); this testing regime gives a more severe test of the theory than, for instance, constant strain rates. As an illustration, we show their results for a five-step stress history (Figure 10.6).

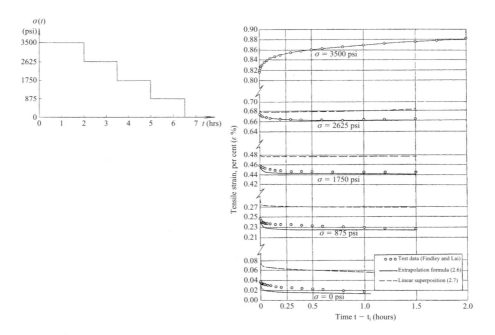

Figure 10.6 (a) Step stress history. (b) Strain resulting from stress history in (a). Successive steps are shown from the top down. (Reproduced with permission of Pipkin and Rogers, *J. Mech. Phys. Solids*, **16**, 59 (1968))

The most significant difference between the Leaderman and the Pipkin and Rogers approach is that in the former the material response is separable into time and stress dependence. Thus, in Equation (10.13) there is a function f of stress multiplied by a function J of time. By contrast, in Equation (10.17) C is explicitly a function of two variables that may or may not be separable in this sense. The Pipkin and Rogers approach is thus more general and we would expect it to be capable of modelling a greater range of material behaviour.

10.2.3   More complicated single-integral representations

Schapery [16, 17] has used the theory of the thermodynamics of irreversible processes to produce a model that may be viewed as a further extension of Leaderman's. Schapery continued Leaderman's technique of replacing the stress by a function of stress $f(\sigma)$ in the superposition integral, but also replaced time by a function of time, the *reduced time* ψ. The material is assumed to be linear viscoelastic at small strains, with a creep compliance function of the form [17]

$$J(t) = \frac{e(t)}{\sigma} = D_0 + \Delta D(t) \tag{10.18}$$

The constant term D_0 corresponds to the instantaneous elastic response (the unrelaxed compliance J_u of Equation (4.19)). In general, the strain in terms of the stress history is given by

$$e(t) = g_0 D_0 \sigma + g_1 \int_0^t \Delta D(\psi - \psi') \frac{dg_2 \sigma}{d\tau} d\tau \tag{10.19}$$

where the stress history is assumed to start at zero time and g_0, g_1 and g_2 are stress-dependent parameters for which $g_0 = g_1 = g_2 = 1$ at sufficiently small stresses. The reduced times are defined by

$$\psi = \psi(t) = \int_0^t \frac{dt'}{a_\sigma[\sigma(t')]}$$

$$\psi' = \psi(\tau) = \int_0^\tau \frac{dt'}{a_\sigma[\sigma(t')]} \tag{10.20}$$

where the stress-dependent factor $a_\sigma = 1$ at sufficiently small stresses. The small-stress values of unity ensure that the linear Boltzmann integral is returned under these conditions. When $g_0 = g_1 = a_\sigma = 1$, and g_2 is allowed to depend on stress, Leaderman's theory results. An entirely analogous system of equations involving the stress relaxation behaviour gives the stress in terms of the strain history [17].

In the theory of Equation (10.19), four functions of stress g_0, g_1, g_2 and a_σ characterize the non-linearity and must be evaluated over the required stress range. Experimental regimes that involve periods of constant stress, during which the functions are constants, have proved useful for this purpose. For a single-step creep test at stress σ applied at time $t = 0$, Equation (10.19) can be evaluated, noting that it contains a Duhamel integral like Equation (4.3), to give the result

$$e(t) = g_0 D_0 \sigma + g_1 g_2 \Delta D \left(\frac{t}{a_\sigma} \right) \sigma \tag{10.21}$$

where Equations (10.20) have been used. Clearly, even if the low-stress linear behaviour defined by D_0 and ΔD is known, the creep test does not allow for the separation of functions g_1, g_2 and a_σ. Schapery [17] showed how the addition of two-step creep experiments, including creep and recovery tests (in which the second step is at zero stress), could be used to generate distinct values for the parameters. This was aided by the use of a power-law approximation for the creep compliance function, such that

$$\Delta D(\psi) = D_1 \psi^n \tag{10.22}$$

He showed how the use of double logarithmic plots of the recovery strain against time, obtained for different stress levels, could be related to one another by shift factors; the shift factors then could be related simply to g_1 and a_σ. The technique of step loading combined with Equation (10.22) has been used also by Crook [18] and Lai and Bakker [19]. Schapery's model has been applied to nitrocellulose, fibre-reinforced phenolic resin and polyisobutylene [17], polycarbonate [18] and high-density polyethylene [19].

The theory of Bernstein, Kearsley and Zapas [20] and developments of it (e.g. Zapas and Craft [21]) – so-called BKZ theories – are aimed in particular at large deformation behaviour. The Gaussian model of rubber elasticity tells us that in uniaxial stretching the true stress σ is in the form

$$\sigma = C(\lambda^2 - 1/\lambda) \tag{10.23}$$

where C is a constant and λ is the extension ratio. This follows from Equation (2.6) or (3.24), which gives the nominal or engineering stress; when nominal stress is replaced by true stress via the use of the incompressibility condition, Equation (10.23) results. The form of Equation (10.23) suggests that a theory in which the quantity $\sigma/(\lambda^2 - 1/\lambda)$ plays a central role might be particularly appropriate for large strains.

Perhaps the most important feature of the BKZ model, which distinguishes it from all the models discussed so far, is the choice of strain measure. Hitherto, all the materials have been assumed to be solids, in that they have an initial undeformed, stress-free state that acts as a reference relative to which all strained states are measured. In BKZ theories there is no such special state, and the material therefore may be classed as a fluid. At any 'present' time t, the state of strain is measured relative to the state at previous times τ. This is done by adopting as the strain measure the quantity $\lambda(t)/\lambda(\tau)$. Reflecting on the remarks in the paragraph above now suggests the importance of the quantity $[\lambda^2(t)/\lambda^2(\tau)] - [\lambda(\tau)/\lambda(t)]$ in a theory in which the stress depends on the strain history. The BKZ form given by Zapas and Craft [21] in uniaxial stretching is

$$\sigma(t) = \int_{-\infty}^{t} \left[\frac{\lambda^2(t)}{\lambda^2(\tau)} - \frac{\lambda(\tau)}{\lambda(t)} \right] h \left[\frac{\lambda(t)}{\lambda(\tau)}, \, t - \tau \right] d\tau \qquad (10.24)$$

Because τ runs through all values previous to t, the stress depends on the strain at time t as measured relative to all previous strain states. For a specimen that is unstrained prior to time $t = 0$, there are contributions to the stress at positive times from the part of the strain history occurring at times less than zero. In this respect, the theory differs from those mentioned so far. Splitting the integral in Equation (10.24) into parts before and after zero time, and setting $\lambda(\tau) = 1$ for $\tau < 0$, we obtain

$$\sigma(t) = \left[\lambda^2(t) - 1/\lambda(t) \right] \int_{-\infty}^{0} h[\lambda(t), \, t - \tau] d\tau + \int_{0}^{t} \left[\frac{\lambda^2(t)}{\lambda^2(\tau)} - \frac{\lambda(\tau)}{\lambda(t)} \right] h \left[\frac{\lambda(t)}{\lambda(\tau)}, \, t - \tau \right] d\tau$$

$$(10.25)$$

In particular, for stress relaxation starting at zero time the second term is zero. We may then write

$$\sigma(t) = \left(\lambda^2 - 1/\lambda \right) H(\lambda, \, t) \qquad (10.26)$$

where

$$H(\lambda, \, t) = -\frac{\partial h}{\partial t}(\lambda, \, t) \qquad (10.27)$$

Here the assumption has been made that the function h is zero at large times. This is appropriate for a fluid and in any case causes no significant loss of generality. It is clear from Equations (10.26) and (10.27) that the function h is entirely determined by single-step stress relaxation measurements. Therefore, the stress for a general strain history can be calculated once stress relaxation data have been gathered over the appropriate range of strain. There is no parallel formulation for the strain in terms of the stress history.

In the original paper [20], the authors reported work on the uniaxial tension of plasticized poly(vinyl chloride), sulphur vulcanizates of butyl rubber and polyisobutylene. Very successful predictions were made at extension ratios of up to ~5. Zapas and Craft [21] applied their formulation to multistep stress relaxation and creep and recovery of both plasticized poly(vinyl chloride) and polyisobutylene. McKenna and Zapas applied a modified form of the model to the torsional deformation of poly(methyl methacrylate) [22]. McKenna and Zapas [23] have used the model in analysis of the tensile behaviour of carbon-black-filled butyl rubbers.

10.2.4   Comparison of single-integral models

The three principal single-integral theories are that of Pipkin and Rogers, Schapery's thermodynamic theory and the BKZ model. The first two concern solid material, with the Pipkin and Rogers approach being the simpler of the two. The Schapery approach is more complex as a result of its basis in thermodynamics, whereas Pipkin and Rogers' theory is purely a continuum model and is essentially devoid of physics. The BKZ fluid theory is of interest at large strains. Smart and Williams [24] compared the performance of the three models when applied to tensile stretching of polypropylene and poly(vinyl chloride) fibres, but only up to moderate strains (~4 per cent). The BKZ model appeared to be of little interest at these strains. The Pipkin and Rogers approach, although having the advantage over the Schapery theory of simplicity, gave a somewhat worse performance.

10.3   Creep and stress relaxation as thermally activated processes

We have shown (Section 4.2.2) that the standard linear solid, a three-component spring and dashpot model, provides a description of linear viscoelastic behaviour to a first approximation. Eyring and colleagues [25] assumed that the deformation of a polymer was a thermally activated rate process involving the motion of segments of chain molecules over potential barriers, and modified the standard linear solid so that the movement of the dashpot was governed by the activated process. The model, which now represents non-linear viscoelastic behaviour, is useful because its parameters include an activation energy and an activation volume that may give an indication of the underlying molecular mechanisms. The activated rate process may also provide a common basis for the discussion of creep and yield behaviour.

10.3.1   The Eyring equation

In the following account, we outline the application of the Eyring process to mechanical behaviour, and leave more specific physical interpretations until the next chapter. Macroscopic deformation is assumed to be the result of basic processes that are either intermolecular (e.g. chain-sliding) or intramolecular (e.g. a change in the conformation of the chain), whose frequency v depends on the ease with which a chain segment can surmount a potential energy barrier of height ΔH. In the absence of stress, dynamic equilibrium exists, so that an equal number of chain segments move in each direction over the potential barrier at a frequency given by

$$v = v_0 \exp\left(-\frac{\Delta H}{RT}\right)$$

The equation above is identical to Equation (6.16) describing the frequency of a molecular event.

An applied stress σ is assumed to produce linear shifts $\beta\sigma$ of the energy barriers in a symmetrical way (Figure 10.7), where β has the dimensions of volume. The flow in the direction of the applied stress is then given by

$$v_1 = v_0 \exp\left[-\frac{(\Delta H - \beta\sigma)}{RT}\right]$$

compared with a smaller flow in the backward direction of

$$v_2 = v_0 \exp\left[-\frac{(\Delta H + \beta\sigma)}{RT}\right]$$

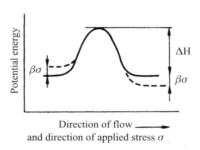

Direction of flow ——————▶
and direction of applied stress σ

Figure 10.7 The Eyring model for creep

The net flow in the forward direction is then

$$v' = v_1 - v_2 = v_0 \exp\left(-\frac{\Delta H}{RT}\right)\left[\exp\left(\frac{\beta\sigma}{RT}\right) - \exp\left(-\frac{\beta\sigma}{RT}\right)\right] \qquad (10.28)$$

The resemblance of the large bracket to the sinh function should be noted.

Assuming that the net flow in the forward direction is directly related to the rate of change of strain, we have

$$\frac{de}{dt} = \dot{e} = \dot{e}_0 \exp\left(-\frac{\Delta H}{RT}\right) \sinh\left(\frac{v\sigma}{RT}\right) \qquad (10.29)$$

where $\dot{e}_0$ is a constant pre-exponential factor and v, which replaces β, is termed the activation volume for the molecular event.

The rate of strain equation (Equation (10.29)) defines an 'activated' viscosity, which is then incorporated in the dashpot of the standard linear solid model and leads to a more complicated relationship between stress and strain than that for the linear model. The activated dashpot model was tested against Leaderman's data for several fibres [13] and, by a suitable choice of model parameters, gave a good fit at a given level of stress over the four decades of time observed.

Subsequently the limitations of simple viscoelastic models have been recognized and it is accepted that exact fitting of data requires a retardation or relaxation time spectrum, so we must consider why the activated dashpot model was so successful. Although creep curves are sigmoidal when plotted on a logarithmic time-scale, over a long intermediate time region they are, to good approximation, linear. The model predicts creep of the form $e = a' + b' \log t$, which is appropriate to this central region.

10.3.2   Applications of the Eyring equation to creep

Sherby and Dorn [26] investigated the creep under constant stress of glassy poly(methyl methacrylate) at different temperatures by applying step temperature changes, and constructed plots of creep rate versus total creep strain at a given stress level (Figure 10.8). The data then were superposed by assuming that the temperature dependence at each stress level followed an activated process, to give a relation between strain rate and strain (Figure 10.9). The temperature shifts were interpreted in terms of an activated process where the activation energy fell in a linear manner with increasing stress, to give a creep rate of the form

$$\dot{e} \exp[(\Delta H - v)/RT] = F(e) \qquad (10.30)$$

which is the high stress approximation of the Eyring equation, where $\sinh x \approx \frac{1}{2} \exp x$ and v is the activation volume.

Mindel and Brown [27] performed a Sherby–Dorn type of analysis on data for the compressive creep of polycarbonate. Superposition was achieved using an equation of the form of Equation (10.30) with an activation volume of 5.7 nm^3, which was very close to the values of the activation volume obtained from measurements of the strain rate dependence of the yield stress (Section 11.5.1).

The results suggest that creep rate can be represented by a general equation of the form

$$\dot{e} = f_1(T)f_2(\sigma/T)f_3(e) \qquad (10.31)$$

where $f_1(T)$, $f_2(\sigma/T)$ and $f_3(e)$ are separate functions of the variables T, σ and e. Although $f_1(T)$ has the exponential form expected for a thermally activated process, the exponential form $f_2(\sigma/T)$ is modified to take into account the hydrostatic component of stress, giving different activation volumes for tensile, shear and pressure measurements. Mindel and Brown also proposed that in the

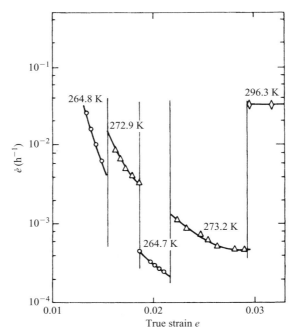

Figure 10.8   Creep rates as a function of total creep strain for poly(methyl methacrylate) at indicated temperatures for a stress level of 56 MPa. (Reproduced with permission from Sherby and Dorn, *J. Mech. Phys. Solids*, **6**, 145 (1958))

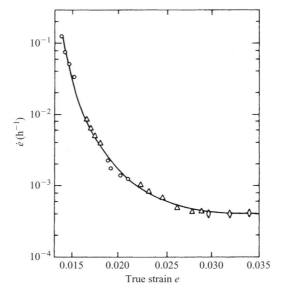

Figure 10.9   Superposition of creep data for poly(methyl methacrylate) at different temperatures at a stress level of 56 MPa. (Reproduced with permission from Sherby and Dorn, *J. Mech. Phys. Solids*, **6**, 145 (1958))

region where the creep rate is falling rapidly with increasing strain $f_3(e)$ has the form

$$f_3(e) = \text{constant} \times \exp(-ce_R)$$

where e_R is the recoverable component of the creep strain and c is a constant. We then have

$$\dot{e} = \dot{e}_0 \exp[-(\Delta H - \tau V + p\Omega)/RT]\exp(-ce_R) \qquad (10.32)$$

where τ and p are the shear and hydrostatic components of stress and V and Ω are the shear and pressure activation volumes. The equation may be rewritten as

$$\dot{e} = \dot{e}_0 \exp\left[-(\Delta H - \{\tau - \tau_i\}V + p\Omega)/RT\right] \qquad (10.33)$$

where $ce_R = \tau_i V/RT$; τ_i, which has the character of an internal stress, increases with strain and is proportional to absolute temperature, as would be expected for the stress in a rubber-like network.

Wilding and Ward [28] have used the Eyring rate process to model the creep of ultrahigh-modulus polyethylene and show that at high strains, which correspond to long creep times, the creep rate reaches a constant value called the plateau (or equilibrium) creep rate (Figure 10.10). For polymers of low relative molecular

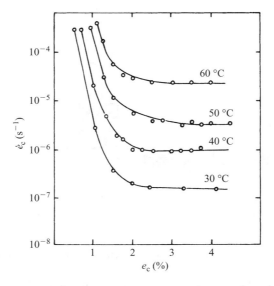

Figure 10.10 Sherby–Dorn plots of creep of ultrahigh-modulus polyethylene at different temperatures. (Reproduced with permission from Wilding and Ward, *Plastics and Rubber Processing and Applications*, **1**, 167 (1981))

mass the stress and temperature dependence of the final creep rate can be modelled by a single activated process with an activation volume of 0.08 nm^3. In molecular terms this volume is that swept out by a single molecular chain moving through the lattice by a discrete distance.

For polymers of a higher molecular mass, and for copolymers, the permanent flow process was activated only at high stress levels, which suggested that there are two Eyring processes coupled in parallel (Figure 10.11). This suggestion is akin to the representation proposed to describe the strain rate dependence of the yield stress in polymers [29–31]. Process A has the smaller tensile activation volume ($\sim$0.05 nm^3) and larger pre-exponential factor, and is activated only at high stress levels. Process B has the larger tensile activation volume ($\sim$1 nm^3) and a smaller pre-exponential factor, and is operative at low stress levels. At low stresses there will be little permanent flow because process B carries almost the whole load. Although the overall creep and recovery behaviour can be represented by a model containing two activated dashpots, a spectrum of relaxation times would be required to give an accurate fit to experimental data over the complete time and strain scales.

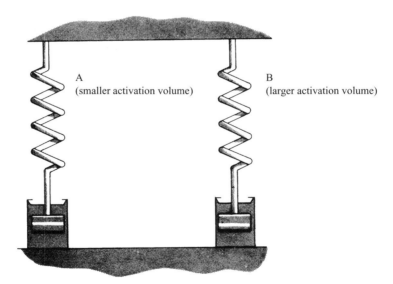

Figure 10.11   The two-process model for permanent flow creep. (Reproduced with permission from Wilding and Ward, *Plastics and Rubber Processing and Applications*, **1**, 167 (1981))

10.3.3   Applications of the Eyring equation to stress relaxation

Guiu and Pratt [32] have shown how a model consisting of an Eyring dashpot in series with an elastic element leads to a simple equation to describe stress relaxation curves in tension. Suppose that there is a total strain e on the system of

Figure 10.12, consisting of an elastic component e_E and an Eyring or viscous component e_V, such that

$$e = e_E + e_V \qquad (10.34)$$

Figure 10.12   Series model spring plus Eyring dashpot

Differentiating with respect to time gives

$$\dot{e} = \dot{e}_E + \dot{e}_V \qquad (10.35)$$

We now replace the viscous strain rate $\dot{e}_V$ with the high stress adaptation of Equation (10.29), which is

$$\dot{e}_V = \dot{e}_0 \exp\left(-\frac{\Delta H}{RT}\right) \frac{1}{2} \exp\left(\frac{v\sigma}{RT}\right) = A \exp(B\sigma) \qquad (10.36)$$

where the constants A and B have been introduced for brevity. Assuming a linear relation for the elastic component, with modulus E, Equation (10.35) now can be rewritten so that only stresses appear on the right-hand side:

$$\dot{e} = \frac{\dot{\sigma}}{E} + A \exp(B\sigma)$$

Under conditions of stress relaxation, the total strain rate is zero and the stress decays in a manner governed by the relation

$$0 = \frac{\dot{\sigma}}{E} + A \exp(B\sigma)$$

(10.37)

This can be solved by separation of variables to give the stress as

$$\sigma_0 - \sigma = \frac{1}{B} \ln\left(1 + \frac{t}{c}\right)$$

(10.38)

where σ_0 is the stress at time $t = 0$ and c is a constant.

Equation (10.38), the Guiu and Pratt expression, has been shown to be remarkably effective in representing stress relaxation curves for polymers. Escaig [33] has discussed its general utility. Sweeney and Ward [34] applied the expression successfully to the stress relaxation behaviour of highly oriented polyethylene fibres at small strains. They also showed that the two-process model, which in some circumstances will generate stress relaxation predictions approximating to Guiu–Pratt curves, gave a more satisfactory model of the overall behaviour.

10.3.4   Applications of the Eyring equation to yield

Suppose that the system of Figure 10.12 is subjected to a constant total strain rate $\dot{e}$, starting at zero load. Initially the stress is low and causes only a small rate of strain in the Eyring dashpot, and the elastic spring is stretched. Continued stretching of the spring increases the stress and thus the rate of strain $\dot{e}_V$ in the Eyring dashpot, until eventually it becomes equal to the total applied strain rate $\dot{e}$. At this point the spring ceases to extend, and so the stress in it, which is equal to the total stress on the system, becomes constant. Thus, behaviour resembling yield is predicted. It is easily shown that this yield stress depends on strain rate. Adopting the notation of the previous section, once the spring reaches its state of constant strain, $\dot{e}_E = 0$. Then, using Equation (10.36)

$$\dot{e} = \dot{e}_V$$

and from Equation (10.36)

$$\dot{e} = A \exp(B\sigma_Y)$$

(10.39)

where σ_Y denotes the yield stress. Rearranging gives the relation

$$\sigma_Y = \frac{1}{B} \ln\left(\frac{\dot{e}}{A}\right)$$

or, regaining the expressions for A and B implicit in Equation (10.36),

$$\sigma_Y = \frac{RT}{\nu} \ln \left[\frac{2\dot{e}}{\dot{e}_0} \exp\left(\frac{\Delta H}{RT}\right) \right]$$ (10.40)

These equations imply a linear relationship between σ_Y/T and $\ln(\dot{e})$. This has been demonstrated for a number of polymers. For example, the work of Bauwens-Crowet et al. [30] on polycarbonate is illustrated in Figure 10.13.

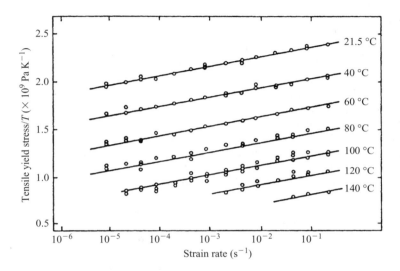

Figure 10.13 Measured ratio of yield stress to temperature as a function of the logarithm of strain rate for polycarbonate. The set of parallel straight lines is calculated from Equation (11.10) (Reproduced with permission from Bauwens-Crowet, Bauwens and Homes, *J. Polym. Sci. A2*, **7**, 735 (1969))

The stress–strain behaviour of models such as that of Figure 10.12 can be explored by solving the associated equations using numerical techniques. In the work of Sweeney et al. on PET fibres [35] a model similar to that of Figure 10.12, but with the Eyring dashpot restrained by a Gaussian network, was solved in this way. The strain at which yield occurs, and the general shape of the stress–strain curve, were predicted and compared with experiment. The predictions were realistic at temperatures both above and below the glass transition.

References

1. Wineman, A. S. and Rajagopal, K. R., *Mechanical Response of Polymers: an Introduction*, Cambridge University Press, Cambridge, 2000, p. 68.
2. Ward, I. M., *Mechanical Properties of Solid Polymers*, Wiley, Chichester, 1983.
3. Lockett, F. J., *Non-Linear Viscoelastic Solids*, Academic Press, London, 1972.

4. Turner, S., *Polym. Eng. Sci.*, **6**, 306 (1966).
5. Ferry, J. D., *Viscoelastic Properties of Solids*, Wiley, New York, 1980.
6. Lakes, R. S., *Viscoelastic Solids*, CRC Press, Boca Raton, FL, 1999.
7. Smith, T. L., *Trans. Soc. Rheol.*, **6**, 61 (1962).
8. Kitagawa, M., Mori, T. and Matsutani, T., *J. Polym. Sci. B*, **27**, 85 (1989).
9. Kitagawa M. and Takagi, H., *J. Polym. Sci. B*, **28**, 1943 (1990).
10. Lui, M. C. M. and Krempl, E., *J. Mech. Phys. Solids*, **27**, 377 (1979).
11. Zhang, C. and Moore, I. D., *Polym. Eng. Sci.*, **37**, 404 (1997).
12. Zhang, C. and Moore, I. D., *Polym. Eng. Sci.*, **37**, 415 (1997).
13. Leaderman, H., *Elastic and Creep Properties of Filamentous Materials and Other High Polymers*, Textile Foundation, Washington, DC, 1943.
14. Findlay, W. N. and Lai, J. S. Y., *Trans. Soc. Rheol.*, **11**, 361 (1967).
15. Pipkin, A. C. and Rogers, T. G., *J. Mech. Phys. Solids*, **16**, 59 (1968).
16. Schapery, R. A., *Int. J. Solids Struct.*, **2**, 407 (1966).
17. Schapery, R. A., *Polym. Eng. Sci.*, **9**, 295 (1969).
18. Crook, R. A., *Polym. Eng. Sci.*, **33**, 56 (1993).
19. Lai, J. and Bakker, A., *Polym. Eng. Sci.*, **35**, 1339 (1995).
20. Bernstein, B., Kearsley, B. A. and Zapas, L. P., *Trans. Soc. Rheol.*, **7**, 391 (1963).
21. Zapas, L. J. and Craft, T., *J. Res. Nat. Bur. Stand. A*, **69**, 541 (1965).
22. McKenna, G.B. and Zapas, L. J., *J. Rheol.*, **23**, 151 (1979).
23. McKenna, G. B. and Zapas, L. J., *Rubber Chem. Technol.*, **54** 718 (1981).
24. Smart, J. and Williams, J. G., *J. Mech. Phys. Solids*, **20**, 313 (1972).
25. Halsey, G., White, H. J. and Eyring, H., *Text Res. J.*, **15**, 295 (1945).
26. Sherby, O. D. and Dorn, J. B., *J. Mech. Phys. Solids*, **6**, 145 (1958).
27. Mindel, M. J. and Brown, N., *J. Mater. Sci.*, **8**, 863 (1973).
28. Wilding, M. A. and Ward, I. M., *Polymer*, **19**, 969 (1978); **22**, 870 (1981).
29. Roetling, J. A., *Polymer*, **6**, 311 (1965).
30. Bauwens-Crowet, C., Bauwens, J. C. and Homes, G., *J. Polym. Sci. A2*, **7**, 735 (1969).
31. Robertson, R. E., *J. Appl. Polym. Sci.*, **7**, 443 (1963).
32. Guiu, F. and Pratt, P. L., *Phys. Status Solidi*, **6**, 111 (1964).
33. Escaig, B., in *Plastic Deformation of Amorphous and Semi-crystalline Materials*, (eds B. Escaig and C. G'Sell), Les Éditions de Physique, Les Ulis, France, 1982, pp. 187–225.
34. Sweeney, J. and Ward, I. M., *J. Mater. Sci.*, **25**, 697 (1990) .
35. Sweeney, J., Shirataki, H., Unwin, A. P., *et al.*, *J. Appl. Polym. Sci.*, **74**, 3331 (1999).

11

Yielding and Instability in Polymers

As we observed at the end of the previous chapter, the non-linear behaviour of polymers, as represented by the Eyring model, gives rise to a phenomenon resembling yield. The observed maximum stress can be treated as a yield stress, although Equation (10.40) shows that this yield stress depends on the rate of strain. Wineman and Waldron [1] have pointed out that there appear to be two approaches to the modelling of yield in polymers: the use of non-linear viscoelasticity; and the direct application of metal plasticity. The use of the Eyring model is one example of the former approach. Relatively simple theories of plasticity, where there is no rate dependence, are available from the metals field. These theories, which embody the classical concepts of plasticity, still may be applied usefully to polymers, for instance in cases where changes in strain rate are small.

Allied to the subject of yielding is that of instability. Necking in a tensile test specimen is an example of instability and is caused by the underlying yield properties of the material. Yielding may lead to a maximum in the applied force, which may then allow the strain in the specimen to increase with no increase in force – the unstable condition. We should note that yielding and instability are qualitatively different phenomena, in that yielding is an intrinsic material characteristic whereas instability is a function of the geometry and loading conditions of the loaded body.

A number of different factors contribute to the present interest in the yield behaviour. First, it has been recognized that the classical concepts of plasticity are relevant to forming, rolling and drawing processes in polymers. Secondly, there has been a number of striking experimental studies of 'slip bands' and 'kink bands' in polymers that suggest that deformation processes in polymers might be similar to those in crystalline materials such as metals and ceramics. Finally, it is evident that distinct yield points are observed and there is much interest in understanding these in the context of other ideas in polymer science.

Our first task in this chapter is to discuss the relevance of classical ideas of plasticity to the yielding of polymers. Although the yield behaviour is temperature and strain rate dependent it will be shown that, provided that the test conditions are chosen suitably, yield stresses can be measured that satisfy conventional yield

An Introduction to the Mechanical Properties of Solid Polymers I. M. Ward and J. Sweeney
© 2004 John Wiley & Sons, Ltd   ISBN: 0471 49625 1 (HB); 0471 49626 X (PB)

criteria. The temperature and time dependence often obscure some generalities of the yield behaviour. For example, it might be concluded that some polymers show necking and cold-drawing, whereas others are brittle and fail catastrophically. Yet another type of polymer (a rubber) extends homogeneously to rupture. A salient point to recognize is that polymers in general can show all these types of behaviour depending on the exact conditions of test (see Figure 12.1), quite irrespective of their chemical nature and physical structure. Thus explanations of yield behaviour which involve, for example, cleavage of crystallites or lamellar slip or amorphous mobility are only relevant to specific cases. As in the case of linear viscoelastic behaviour or rubber elasticity we must first seek an understanding of the relevant phenomenological features, decide on suitable measurable quantities and then provide a molecular interpretation of the subsequent constitutive relations.

11.1 Discussion of the load–elongation curves in tensile testing

The most dramatic consequence of yield is seen in a tensile test when a neck or deformation band occurs, as in Figure 11.1, with the plastic deformation concentrated either entirely or primarily in a small region of the specimen. The precise

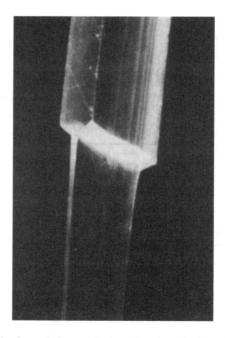

Figure 11.1   Photograph of a neck formed in the redrawing of oriented polyethylene

nature of the plastic deformation depends both on the geometry of the specimen and on the form of the applied stresses, and will be discussed more fully later.

The characteristic necking and cold-drawing behaviour is as follows. On the initial elongation of the specimen, homogeneous deformation occurs and the conventional load–extension curve shows a steady increase in load with increasing elongation (AB in Figure 11.2). At the point B the specimen thins to a smaller cross-section at some point, i.e. a neck is formed. Further elongation brings a fall in load. Continuing extension is achieved by causing the shoulders of the neck to travel along the specimen as it thins from the initial cross-section to the drawn cross-section. The existence of a finite or natural draw ratio is an important aspect of polymer deformation and is discussed in Section 11.6.1 below. Ductile behaviour in polymers does not always give a stabilized neck, so the requirements for necking and cold-drawing must now be considered in some detail.

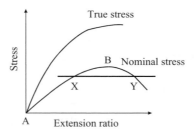

Figure 11.2   Comparison of the nominal stress–elongation curve (load–elongation curve) and the true stress–elongation curve

11.1.1   Necking and the ultimate stress

It is important to distinguish between the nominal stress, which is the load at any time during deformation divided by the initial cross-sectional area, and the true stress, which is the load divided by the actual cross-section at any time. The cross-section of the sample decreases with increasing extension, so the true stress may be increasing when the apparent or conventional stress or load remains constant or even decreasing. This has been discussed very well by Nadai [2] and Orowan [3].

Consider the conventional stress–strain curve or the load–elongation curve for a ductile material (Figure 11.2). The ordinate is equal to the nominal stress obtained by dividing the load P by the original cross-sectional area A_0:

$$\sigma_a = P/A_0$$

This gives a stress–strain curve of the form shown. The load reaches its maximum value at the instant the extension of the sample ceases to be uniform. At this elongation the specimen begins to neck and consequently the load falls, as shown by the last part of the stress–strain curve. Finally, the sample fractures at the narrowest point of the neck.

It is instructive to plot the true tensile stress at any elongation rather than the nominal stress σ_a. This is given by $\sigma = P/A$, where A is the actual cross-section at any time. We now assume, as is usual for plastic deformation, that the deformation takes place at constant volume. Then $Al = A_0 l_0$, and if we put $l/l_0 = \lambda$, where λ is the extension ratio,

$$A = \frac{A_0 l_0}{l} = \frac{A_0}{\lambda}$$

The true stress is given by

$$\sigma = \frac{P}{A} = \frac{\lambda P}{A_0} = \lambda \sigma_a \qquad (11.1)$$

Thus if we know σ_a, the true stress as a function of λ (i.e. the true stress–strain curve) can be computed. The nominal and true stress–strain curves are compared in Figure 11.2.

Consideration of the nominal stress can lead to insight into the mechanical instability associated with necking. For a tensile specimen of initially uniform cross-section, equilibrium of forces ensures that the nominal stress σ_a is the same all along its length. Therefore, when the curve of σ_a against λ possesses a maximum such as that in Figure 11.2, a small strain (point X) can coexist with a large strain (point Y). Point X corresponds to the unnecked region of the specimen, and point Y to a region that is beginning to neck. If the specimen were stretched to the same elongation without necking, the strain would be uniform at a level somewhere between that of X and that of Y. It is clear that such a state of strain corresponds to a nominal stress higher than the line XY, and therefore to a strain energy higher than that in the necked specimen. On this basis we would expect the necked state, corresponding to the lower strain energy, to be preferred.

The argument above is based on the assumption that the stress–strain curve completely defines the material behaviour. In reality, with polymers, stress depends on the strain rate and because necking is associated with a local increase in strain rate the issue is more complex. A strong dependence of stress on strain rate can inhibit necking even when the nominal stress reaches a maximum; the existence of the maximum is a necessary condition for necking, but not a sufficient one. Necking in rate-dependent materials has been discussed recently by Sweeney et al. [4].

Another aspect of the energy/equilibrium argument that at first seems problematic is the question of how the material moves from the unnecked state at X to the necked state at Y without violating equilibrium. Intermediate strain states are clearly (Figure 11.2) at a higher nominal stress than that pertaining at X and Y. However, as the neck develops and material moves from one state to the other, the cross-sectional area is non-uniform and the true stress thus varies along the neck. Equilibrium equations then ensure that shear stresses are acting, and so the material is not in a state of uniaxial stress or strain. Therefore, the stress–strain state of the transitional material cannot be plotted on the curve of Figure 11.2. Its combined state of normal and shear stress is such that it maintains equilibrium. This has been explained by Vincent [5], and corresponding work on metals has revealed the complexity of the stress field in the neck [6].

In mathematical terms, the existence of a stress maximum as the condition necessary for necking is

$$\frac{\mathrm{d}\sigma_a}{\mathrm{d}\lambda} = 0 \qquad (11.2)$$

This can be re-expressed in terms of the true stress σ. From Equation (11.1), Equation (11.2) is equivalent to

$$\frac{\mathrm{d}}{\mathrm{d}\lambda}\left(\frac{\sigma}{\lambda}\right) = 0$$

which becomes

$$\frac{1}{\lambda}\frac{\mathrm{d}\sigma}{\mathrm{d}\lambda} - \frac{\sigma}{\lambda^2} = 0$$

i.e.

$$\frac{\mathrm{d}\sigma}{\mathrm{d}\lambda} = \frac{\sigma}{\lambda} \qquad (11.3)$$

Equation (11.3) defines a geometric condition for the true stress–strain curve, corresponding to the simple construction due to Considère shown in Figure 11.3. The ultimate stress is obtained when the tangent to the true stress–strain curve $\mathrm{d}\sigma/\mathrm{d}\lambda$ is given by the line from the point $\lambda = 0$ on the extension axis. The angle α in Figure 11.3 is defined by

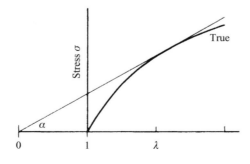

Figure 11.3   The Considère construction

$$\tan \alpha = \frac{\mathrm{d}\sigma}{\mathrm{d}\lambda}$$

The ultimate stress has a much greater significance than is the case for metals because it is a determining factor in deciding whether a polymer will neck and cold-draw.

The significance of the argument at this stage relates to the failure of plastics in the ductile state. Orowan [3] first pointed out that for ductile materials the ultimate stress is entirely determined by the stress–strain curve, i.e. by the plastic behaviour of the material, without any reference to its fracture properties provided that fracture does not occur before the load maximum corresponding to $\mathrm{d}\sigma/\mathrm{d}\lambda = \sigma/\lambda$ is reached. Yield stress is thus an important property in many plastics, and defines the practical limit of behaviour much more than the ultimate fracture, unless the plastic fails by brittle fracture.

11.1.2   Necking and cold-drawing: a phenomenological discussion

Figure 11.4 shows that there are four distinct regions on the nominal stress strain curve of a typical cold-drawing polymer:

1. Initially the stress rises in an approximately linear manner as the applied strain increases.

2. The nominal stress reaches a maximum. During the subsequent fall in stress, the neck shape develops. From this point on, the strain in Figure 11.4 should be understood to be that at the centre of the neck.

3. The nominal stress reaches a minimum. The strain at this point corresponds to the natural draw ratio. In tensile stretching, the strain stays at this approximately constant level for some considerable time as the neck propagates through the specimen and the specimen continues to be elongated – a phenomenon termed 'stable necking'.

4. The specimen is of finite dimensions, so that at some stage the propagating neck occupies the whole of its length – it reaches the grips at both ends of the

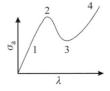

Figure 11.4   Nominal stress–strain behaviour of a necking tensile specimen

tensile specimen. Now, further elongation causes an increase in strain and a corresponding increase in nominal stress as molecular chains are stretched – phenomenologically, this process may be termed 'strain-hardening'.

In some materials, notably metals, there is no minimum region 3 and the stress continues to decrease. Then, there is no stable necking and the neck continues to stretch, becoming continuously thinner until fracture.

The maximum mentioned above as occurring in region 2 corresponds to the condition shown in Equation (11.2). This condition involves the nominal stress only, and the existence or otherwise of a maximum in the true stress is unspecified by the physical arguments employed so far. Observations show that a maximum in true stress may or may not arise, depending on the polymer and test conditions. This is well illustrated in the work of Amoedo and Lee [7]. In Figure 11.5, two sets of true tensile stress–strain curves are shown, one set for polycarbonate and the other for polypropylene. The stress maximum is clearly present in the case of polycarbonate and, equally clearly, absent in the case of polypropylene. It is of significance here that the comparison is between an amorphous polymer (poly-carbonate) and a semicrystalline polymer (polypropylene).

There are two ways in which a neck may be initiated. First, if the cross-section of the sample is not uniform, perhaps as a result of a flaw, the element with the smallest effective cross-section will be subjected to the highest true stress, and so will reach the yield point before any other element in the sample. Secondly, a fluctuation in material properties may cause a localized reduction of the yield stress in a given element so that this element reaches the yield point at a lower applied load. When a particular element has reached its yield point it is easier to continue deformation entirely within this element, because it has a lower flow stress stiffness than the surrounding material. Hence, further deformation of the sample is accompanied by straining in only one region and a neck is formed.

11.1.3   Use of the Considère construction

Based on the above description of the four regions of the nominal stress–strain curve, a parallel description is possible using true stress. The maximum and minimum in regions 2 and 3 are replaced by tangents to the stress–strain curve corresponding to Equation (11.3). In Figure 11.6 two such tangent lines have been drawn to the true stress–strain curve from the point $\lambda = 0$. Inspection of the figure reveals that the slopes of the tangents $\partial\sigma/\partial\lambda$ are as given by Equation (11.3). The regions 1–4 correspond to the same physical stages as those in Figure 11.4. When the curve is such that it is possible to construct the tangent to region 2, the polymer has the potential to neck; the existence of the tangent at region 3 implies a potential for stable necking and cold-drawing. Similar deductions can be made from the presence or otherwise of maxima and minima in the nominal stress–strain relations of Figure 11.4.

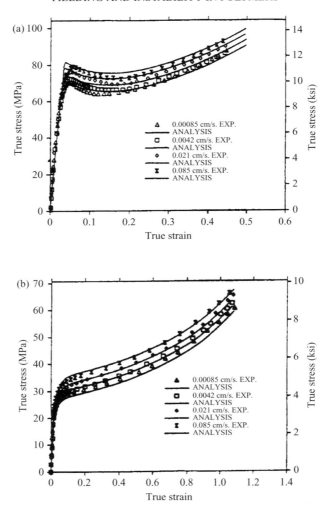

Figure 11.5   True stress–strain curves at room temperature for polycarbonate (a) and poly-propylene (b). (Reproduced with permission from Amoedo and Lee, *Polym. Eng. Sci.*, **32** 1055 (1992)).

An example of a stress–strain relation for which a Considère tangent is not possible is the upwardly curving line shown in Figure 11.7(a). In Figure 11.7(b) one tangent is possible that corresponds to necking, but the second one associated with stable necking is not. The typical polymer behaviour of necking followed by strain hardening and stable necking is only possible for curves resembling those of Figure 11.6.

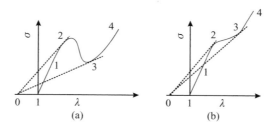

Figure 11.6 Considère tangents. In (a), the true stress has reached a maximum, whereas in (b) there is no maximum

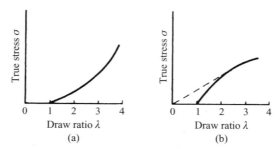

Figure 11.7 Stress–strain curves for which: (a) no Considère tangent is possible; (b) only one tangent, corresponding to the onset of necking, has been drawn (Reproduced with permission from Vincent, Polymer, **1**, 7 (1960) (c) IPC Business Press Ltd)

11.1.4 Definition of yield stress

Yield stress may be regarded most simply as the minimum stress at which permanent strain is produced when the stress is subsequently removed. Although this deformation is satisfactory for metals, where there is a clear distinction between elastic recoverable definition and plastic irrecoverable deformation, in polymers the distinction is not so straightforward. In many cases, such as the tensile tests discussed above, yield coincides with the observation of a maximum load in the load–elongation curve. The yield stress then can be defined as the true stress at the maximum observed load (Figure 11.8(a)). Because this stress is achieved at a comparatively low elongation of the sample, it is often adequate to use the engineering definition of the yield stress as the maximum observed load divided by the initial cross-sectional area.

In some cases there is no observed load drop and another definition of yield stress is required. One approach is to determine the stress where the two tangents to the initial and final parts of the load–elongation curve intersect (Figure 11.8(b)). An alternative is to attempt to define an initial linear slope on the stress–strain curve and then to draw a line parallel to this that is offset by a specified strain, say 2 per cent. The interception of this line with the stress–strain curve then

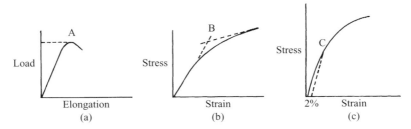

Figure 11.8 (a) The yield stress is defined as the load divided by the cross-sectional area at point A. (b) The yield stress is defined as the stress at point B. (c) The yield stress is defined as the stress at point C

defines the offset or proof stress, which is considered to be the yield stress (Figure 11.8(c)).

11.2 Ideal plastic behaviour

11.2.1 The yield criterion: general considerations

The simplest theories of plasticity exclude time as a variable and ignore any feature of the behaviour that takes place below the yield point. In other words, we assume a rigid-plastic material whose stress–strain relationship in tension is shown in Figure 11.9. For stresses below the yield stress there is no deformation. Yield can be produced by a wide range of stress states and not just by simple tension. In general, therefore, it must be assumed that the yield condition depends on a function of the three-dimensional stress field. In a Cartesian axis set, this is defined by the six components of stress: σ_{xx}, σ_{yy}, σ_{zz}, σ_{xy}, σ_{yz} and σ_{zx}. However, the numerical values of these components depend on the orientation of the axis set, and it is crucial that the yield criterion be independent of the observer's chosen viewpoint. It is therefore more straightforward to make use of the principal stresses, whose directions and values are determined uniquely by the nature of the stress field. If the material itself is such that its tendency to yield is independent of

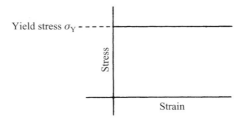

Figure 11.9 Stress–strain relationship for an ideal rigid-plastic material

the direction of the stresses, i.e. if it is isotropic, then the yield criterion is a function of the principal stresses only. Thus, the stress tensor in a general axis set

$$\begin{bmatrix} \sigma_{xx} & \sigma_{xy} & \sigma_{xz} \\ \sigma_{xy} & \sigma_{yy} & \sigma_{yz} \\ \sigma_{xz} & \sigma_{yz} & \sigma_{zz} \end{bmatrix}$$

corresponds to the stress tensor in axes along principal directions

$$\begin{bmatrix} \sigma_1 & 0 & 0 \\ 0 & \sigma_2 & 0 \\ 0 & 0 & \sigma_3 \end{bmatrix}$$

where σ_1, σ_2 and σ_3 are the principal stresses. If we confine ourselves to isotropic material, we can construct a *yield criterion* of the form

$$f(\sigma_1, \sigma_2, \sigma_3) = \text{constant}$$

11.2.2 The Tresca yield criterion

The earliest yield criterion to be suggested for metals was Tresca's proposal that yield occurs when the maximum shear stress reaches a critical value [8], i.e.

$$\sigma_1 - \sigma_3 = \text{constant}$$

with $\sigma_1 > \sigma_2 > \sigma_3$.

Of a similar nature is the Schmid critical resolved shear-stress law for the yield of metal single crystals [9].

11.2.3 The Coulomb yield criterion

The Tresca yield criterion assumes that the critical shear stress is independent of the normal pressure on the plane on which yield is occurring. Although this assumption is valid for metals, it is more appropriate in polymers to consider the possible applicability of the Coulomb yield criterion [10], which states that the critical shear stress τ for yielding to occur in any plane varies linearly with the stress normal to this plane, i.e.

$$\tau = \tau_c - \mu\sigma_N \tag{11.4}$$

The Coulomb criterion was originally conceived for the failure of soils and τ_c was termed the 'cohesion' and μ the coefficient of internal friction. For a compressive stress, σ_N has a negative sign so that the critical shear stress τ for yielding to occur on any plane increases linearly with the pressure applied normal to this plane.

The Coulomb criterion is often written as

$$\tau = \tau_c - \tan \phi \sigma_N \tag{11.5}$$

where μ has been written as $\tan \phi$ for reasons that will now become apparent.

Consider uniaxial compression under a compressive stress σ_1 where yield occurs on a plane whose normal makes an angle θ with the direction of σ_1 (Figure 11.10). The shear stress is $\tau_1 = \sigma \sin \theta \cos \theta$ and the normal stress is $\sigma_N = -\sigma_1 \cos^2 \theta$. Yield occurs when

$$\sigma_1 \sin \theta \cos \theta = \tau_c + \sigma_1 \tan \phi \cos^2 \theta$$

i.e. when

$$\sigma_1 (\cos \theta \sin \theta - \tan \phi \cos^2 \theta) = \tau_c$$

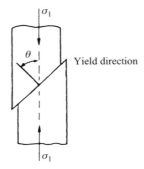

Figure 11.10 The yield direction under a compressive stress σ_1 for a material obeying the Coulomb criterion

For yield to occur at the lowest possible value of σ_1, $(\cos \theta \sin \theta - \tan \phi \cos^2 \theta)$ must be a maximum, which gives

$$\tan \phi \tan 2\theta = -1 \quad \text{or} \quad \theta = \frac{\pi}{4} + \frac{\phi}{2} \tag{11.6}$$

Thus $\tan \phi$ determines the direction of yield and conversely the direction of yielding can be used to define ϕ, where $\tan \phi$ is the coefficient of friction. If the stress σ_1 is tensile, the angle θ is given by

$$\theta = \frac{\pi}{4} - \frac{\phi}{2}$$

We see that the Coulomb yield criterion therefore defines both the stress condition required for yielding to occur and the directions in which the material will deform. Where a deformation band forms, its direction is one that is neither rotated nor distorted by the plastic deformation, because its orientation marks the direction that establishes material continuity between the deformed material in the deformation band and the undistorted material in the rest of the specimen. If volume is conserved, the band direction denotes the direction of shear in a simple shear (by the definition of a shear strain). Thus for a Coulomb yield criterion the band direction is defined by Equation (11.6).

11.2.4   The von Mises yield criterion

The von Mises yield criterion [11] assumes that the yield behaviour is independent of hydrostatic pressure and that the yield stresses in simple tension and compression are equal. It is expressed most simply in terms of the principal components of stress so that

$$(\sigma_1 - \sigma_2)^2 + (\sigma_2 - \sigma_3)^2 + (\sigma_3 - \sigma_1)^2 = \text{constant} \qquad (11.7)$$

The constant term in Equation (11.7) can be expressed easily in terms of the yield stress σ_Y in uniaxial testing. Then we can assign the values $\sigma_1 = \sigma_Y$ and $\sigma_2 = \sigma_3 = 0$, and the constant on the right is found to be $2\sigma_Y^2$.

In rather more sophisticated terms the von Mises yield criterion assumes that the yield criterion depends only on the components of the deviatoric stress tensor obtained by subtracting the hydrostatic components of stress from the total stress tensor. In terms of principal components of stress the deviatoric stress tensor is

$$\begin{bmatrix} \sigma_1' & 0 & 0 \\ 0 & \sigma_2' & 0 \\ 0 & 0 & \sigma_3' \end{bmatrix} = \begin{bmatrix} \sigma_1 + p & 0 & 0 \\ 0 & \sigma_2 + p & 0 \\ 0 & 0 & \sigma_3 + p \end{bmatrix}$$

where $p = -\frac{1}{3}(\sigma_1 + \sigma_2 + \sigma_3)$ is the hydrostatic pressure. The von Mises yield criterion then can be written as

$$\sigma_1'^2 + \sigma_2'^2 + \sigma_3'^2 = \text{constant} \qquad (11.8)$$

The von Mises yield criterion is often written in terms of the so-called octahedral shear stress τ_{oct}, where

$$\tau_{oct} = \frac{1}{3}[(\sigma_1 - \sigma_2)^2 + (\sigma_2 - \sigma_3)^2 + (\sigma_3 - \sigma_1)^2]^{\frac{1}{2}}$$

giving the yield criterion as $\tau_{oct} = \text{constant}$.

We have seen that the Coulomb yield criterion defines both the stresses required for yield and also the directions in which the material deforms. In the case of the von Mises yield criterion we require a further development of the theory to predict the directions in which plastic deformation starts.

It is important to appreciate that plasticity is different in kind from elasticity, where there is a unique relationship between stress and strain defined by a modulus or stiffness constant. Once we achieve the combination of stresses required to produce yield in an idealized rigid plastic material, deformation can proceed without altering stresses and is determined by the movements of the external constraints, e.g. the displacement of the jaws of the tensometer in a tensile test. This means that there is no unique relationship between the stresses and the total plastic deformation. Instead, the relationships that do exist relate the stresses and the incremental plastic deformation, as was first recognized by St Venant, who proposed that for an isotropic material the principal axes of the strain increment are parallel to the principal axes of stress.

If the material is assumed to remain isotropic after yield, then there is no dependence on the deformation or stress history. Furthermore, if we assume that the yield behaviour is independent of the hydrostatic component of stress, then the principal axes of the strain increment are parallel to the principal axes of the deviatoric stress tensor.

Levy [12] and von Mises [11] independently proposed that the principal components of the strain-increment tensor

$$\begin{bmatrix} de_1 & 0 & 0 \\ 0 & de_2 & 0 \\ 0 & 0 & de_3 \end{bmatrix}$$

and the deviatoric stress tensor

$$\begin{bmatrix} \sigma_1' & 0 & 0 \\ 0 & \sigma_2' & 0 \\ 0 & 0 & \sigma_3' \end{bmatrix}$$

are proportional, i.e.

$$\frac{de_1}{\sigma_1'} = \frac{de_2}{\sigma_2'} = \frac{de_3}{\sigma_3'} = d\lambda \qquad (11.9)$$

where $d\lambda$ is not a material constant but is determined by our choice of the extent of deformation of the material, e.g. by the displacement of the jaws of the tensometer.

Rewriting the individual deviatoric stresses

$$\sigma_1' = \sigma_1 + p$$

$$\sigma_2' = \sigma_2 + p$$

$$\sigma_3' = \sigma_3 + p$$

and adding them gives

$$\sigma_1' + \sigma_2' + \sigma_3' = \sigma_1 + \sigma_2 + \sigma_3 - 3 \times \tfrac{1}{3}(\sigma_1 + \sigma_2 + \sigma_3) = 0$$

From Equation (11.9) we have three relations

$$\sigma_1' = \frac{de_1}{d\lambda}, \ \sigma_2' = \frac{de_2}{d\lambda}, \ \sigma_3' = \frac{de_3}{d\lambda}$$

that, when added, give

$$\sigma_1' + \sigma_2' + \sigma_3' = \frac{de_1 + de_2 + de_3}{d\lambda}$$

It follows that $de_1 + de_2 + de_3 = 0$, i.e. that deformation takes place at constant volume.

If the stress–strain relations are referred to other than principal axes we have

$$de_{ij} = \sigma_{ij}' \, d\lambda$$

i.e.

$$\frac{de_{xx}}{d\sigma_{xx}'} = \frac{de_{yy}}{d\sigma_{yy}'} = \frac{de_{zz}}{d\sigma_{zz}'} = \frac{de_{yz}}{d\sigma_{yz}'} = \frac{de_{zx}}{d\sigma_{zx}'} = \frac{de_{xy}}{d\sigma_{xy}'}$$

These equations are called the Levy–Mises equations.

11.2.5 Geometrical representations of the Tresca, von Mises and Coulomb yield criteria

The assumption of material isotropy, which implies that σ_1, σ_2 and σ_3 are interchangeable, means that the Tresca and von Mises yield criteria take very simple analytical forms when expressed in terms of the principal stresses. Thus the yield criteria form surfaces in principal stress space, i.e. that space where the three rectangular Cartesian axes are parallel to the principal stress directions. Points lying closer to the origin than the yield surface represent combinations of stress where yield does not occur; points on or outside the surface represent combinations of stress where yield does occur.

Because the yield criterion is independent of the hydrostatic component of stress, we can replace σ_1, σ_2 and σ_3 by $\sigma_1 + p$, $\sigma_2 + p$ and $\sigma_3 + p$, respectively, without affecting the material's state with regard to yield. Thus, if the point in principal stress space $(\sigma_1, \sigma_2, \sigma_3)$ lies on the yield surface, so does the point $(\sigma_1 + p, \sigma_2 + p, \sigma_3 + p)$. This shows that the yield surface must be parallel to the $\{111\}$ direction, and has the appearance as sketched in Figure 11.11. The material isotropy implies equivalence between σ_1, σ_2 and σ_3 and hence that the section has a threefold symmetry about the $\{111\}$ axis. The assumption that the behaviour is the same in tension and compression implies an equivalence between σ_1 and $-\sigma_1$, etc., and hence we have finally sixfold symmetry about the $\{111\}$ direction. This is most clearly shown by the Tresca yield surface in Figure 11.11.

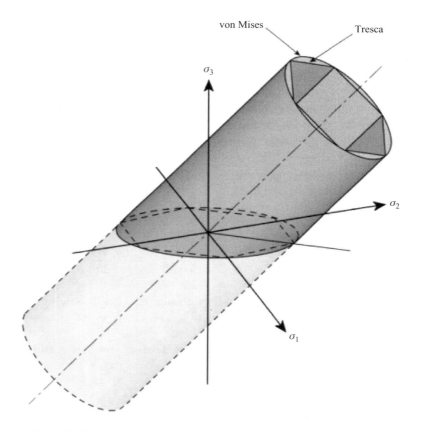

Figure 11.11   Tresca and von Mises yield surfaces in principal stress space

11.2.6   Combined stress states

For the analysis of combined stress in the two-dimensional situation the Mohr circle diagram (see Appendix A1.8) is of value. Normal stresses are represented along the x axis and shear stresses along the y axis, so the Mohr circle thus represents a state of stress, with each point representing the stresses on a particular plane. The direction of the plane normal is given relative to the directions of the principal stresses by the rule that a rotation in real space of θ in a clockwise direction corresponds to a rotation in Mohr circle space of 2θ in an anticlockwise direction. For further details the reader is referred to standard texts [13]. In Figure 11.12(a) two states of stress that produce yield with principal stresses σ_1 and σ_2, σ_3 and σ_4, respectively, are represented by two circles of identical radius tangential to the yield surface. The yield criterion in this case is assumed to be that of Tresca and the yield surface degenerates for the two-dimensional case to two lines parallel to the normal stress axis.

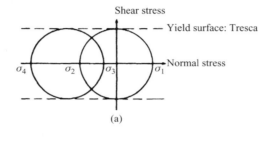

(a)

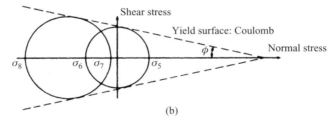

(b)

Figure 11.12 Mohr circle diagram for two states of stress that produce yield in a material satisfying the Tresca yield criterion (a) and the Coulomb yield criterion (b)

In Figure 11.12(b) two states of stress causing yield for a material that satisfies the Coulomb criterion are shown as σ_5 and σ_6, σ_7 and σ_8, respectively. In this case the yield stress depends on the magnitude of the (negative) normal stress, and so the diameters of the Mohr circles will vary with applied stress, increasing as we move to a more compressive stress field. The tangents to the Mohr circles represent the Coulomb yield surface, with the critical shear yield stress for yield decreasing as the normal stress becomes more tensile. It can be shown that these tangents make an angle ϕ with the normal stress axis, where $\tan \phi$ is the coefficient of friction as defined in Section 11.2.3.

11.3 Historical development of understanding of the yield process

We have seen that yield is often associated with a load drop on the load–extension curve and always involves a change in slope on the true stress–strain curve. This load drop sometimes has been attributed either to adiabatic heating of the specimen or to the geometrical reduction in cross-sectional area on the formation of a neck. Present knowledge leads to the conclusion that yielding is an intrinsic property of the material, and that temperature rises and necking are secondary consequences. This is supported by the observations in the previous chapter (Section 10.3.4), to the effect that the Eyring process gives a consistent model of yield in polymers. Localized or geometrical effects can have no relevance to this molecular-based model. However, temperature and geometrical effects are present during yielding and must be understood if the phenomenon is to be interpreted correctly.

11.3.1 Adiabatic heating

Under conventional conditions of cold-drawing, where the specimen is extended at strain rates of the order of $10^{-2}\,\mathrm{s}^{-1}$ or higher, a considerable rise of temperature occurs in the region of the neck. Marshall and Thompson [14], following Müller [15], proposed that cold-drawing involves a local temperature rise and that necking occurs because of strain softening produced by the consequent fall in flow stress with rising temperature. The stability of the drawing process then was attributed to the stability of a localized process of heat transfer through the shoulders of the neck, with extension taking place at constant tension throughout the neck. Hookway [16] later attempted to explain the cold-drawing of nylon 6:6 on somewhat similar grounds, suggesting that there is a possibility of local melting in the neck due to a combination of hydrostatic tension and temperature.

There is no doubt that an appreciable rise in temperature does occur at conventional drawing speeds, and the ideas of Marshall and Thompson are very relevant to an understanding of the complex situation of fibre drawing. Calorimetric measurements by Brauer and Müller [17] have shown, however, that at slow rates of extension the increase in temperature is quite small ($\sim 10\,^{\circ}\mathrm{C}$) and not sufficient to give an explanation for necking and cold-drawing and cold-drawing in terms of adiabatic heating. In addition, Lazurkin [18] demonstrated that necking could still take place under quasi-static conditions for elastomers below their glass transition temperature cold-drawn at very low speeds. A comparable result was shown by Vincent [5] for (semicrystalline) polyethylene, which cold-draws at very slow extension rates at room temperature.

The adiabatic heating explanation arose at least in part because the initial yield process was not regarded as distinct from the drawing process. It is now recognized that up to the yield point the deformation of the sample is homo-geneous and generally quite small strains are involved, whereas once a neck forms the deformation is inhomogeneous and large strains are involved in the neck. The work of plastic deformation then can lead to a large rise in temperature in the neck. For example, Figure 11.13 shows results for the cold-drawing of poly(ethy-lene terephthalate) (PET) [19], where both the yield stress and the drawing stress were measured as a function of strain rate. It can be seen that the yield stress continues to rise with increasing strain rate, beyond the strain rate at which the drawing stress falls quite distinctly. It is argued that, provided that the drawing is carried out at a low strain rate, any heat generated will be conducted away from the neck sufficiently rapidly for no significant temperature rise to occur. As the strain rate is increased and the process becomes more nearly adiabatic, the effective temperature at which the drawing is taking place is increased. In particular, heat is conducted into the unyielded portion of the sample, and so lowers the yield stress of the undeformed material and reduces the force necessary to propagate the neck. The observed temperature rise in the neck has been found to be in approximate agreement with that calculated from the work done in drawing, assuming that no heat is generated due to crystallization. In PET, X-ray diffraction diagrams of cold-drawn fibres show that very little crystallization has occurred.

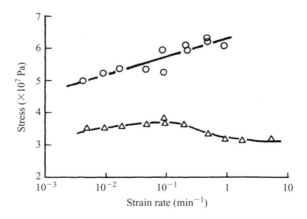

Figure 11.13   Comparison of yield stress ($\circ$) and drawing stress ($\triangle$) as a function of strain rate for PET

The work done per unit volume is given by $W = \sigma_D (\lambda_N - 1)$, where σ_D is the drawing stress and λ_N the natural draw ratio. From the results obtained $\sigma_D = 23$ MPa when $\lambda_N = 3.6$, giving $W = 4.7$ MJ m^{-3}. For PET, where the specific heat is 67 J kg^{-1}K^{-1} and the density is 1.38 Mg m^{-3}, the calculated temperature rise is 57 °C compared with the measured value of 42 °C.

11.3.2 The isothermal yield process: the nature of the load drop

There is no doubt that a temperature rise does occur in cold-drawing under many conditions of test. We have shown, however, that there is very good evidence to support the view that necking can still take place under quasi-static conditions where there is no appreciable temperature rise. Vincent [5] therefore proposed that the observed fall in load is a geometrical effect because the fall in cross-sectional area during stretching is not compensated for by an adequate degree of strain hardening. This effect, called strain softening, was attributed to the reduction in the slope of the stress–strain curve with increasing strain.

Contrary to this latter explanation of the load drop in terms of geometric softening, results reported by Whitney and Andrews [20] showed a yield drop in compression for polystyrene and poly(methyl methacrylate) (PMMA) where there are no geometrical complications. Brown and Ward [18] then made a detailed investigation of yield drops in PET, studying isotropic and oriented specimens in tension, shear and compression. They concluded that in most cases there is clear evidence for the existence of an intrinsic yield drop, i.e. that a fall in true stress can occur in polymers, as in metals. This is reflected in the work of Amoedo and Lee [7], shown above in Figure 11.5(a).

There is, however, a significant difference between polymers and many metals with regard to yield behaviour. For a polymer, as shown in Figure 11.4, stress increases continously once strain-hardening takes effect, in contrast with metals (illustrated by mild steel in Figure 11.14), where often two maxima are observed on a typical load–extension curve. The first maximum (point A in Figure 11.14), called the upper yield point, represents a fall in true stress – an intrinsic load drop – and corresponds to a sudden increase in the amount of plastic strain, which relaxes the stress. From B to C Lüders bands propagate throughout the specimen. Lüders bands have been observed also in polymers [19]. At C the specimen is

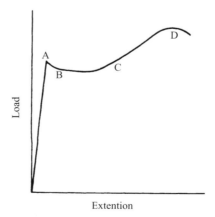

Figure 11.14   Load–extension curve in tension of mild steel

homogeneously strained and the stress begins to rise as the material work hardens uniformly. A second maximum is observed at point D and is always associated with the beginning of necking in the specimen. Necking occurs when the effects of strain hardening of the metal are overwhelmed by the geometrical softening due to the reduction in the cross-sectional area of the specimen as it is strained, i.e. the Orowan–Vincent explanation, discussed in Section 11.1.1 above.

The second maximum, as we have seen previously, is not observed if the true stress–strain curve is plotted instead of the load–extension curve. The first maximum, on the other hand, would exist on the true stress–strain curve. It is called an intrinsic yield point because it relates to the intrinsic behaviour of the material.

In polymers, as we have emphasized, necking is always associated with an initial maximum in the load–extension curve. The investigations of Whitney and Andrews [20] and Brown and Ward [21] show that this maximum combines the effect of the geometrical changes and an intrinsic load drop, and cannot be attributed to the geometrical changes alone. In particular, the cold-drawing results are not accounted for by a decrease in the slope of the true stress–strain curve, as suggested in the explanation of Vincent. It is important to note that not every element of the material follows the same true stress–strain curve, because the stress for initiation is greater than for propagation of yielding, thus confirming (as already noted in Section 11.1.3) that it will not be possible to give a complete explanation of necking and cold-drawing in terms of the Considère construction on a true stress–strain curve.

11.4 Experimental evidence for yield criteria in polymers

Many studies of the yield behaviour of polymers have bypassed the question of strain rate and temperature and sought to establish a yield criterion as discussed in Section 11.2 above. In very general terms such studies divide into two categories: (1) those that attempt to define a yield criterion on the basis of determining yield for different stress states; (2) those that confine the experimental studies to an examination of the influence of hydrostatic pressure on the yield behaviour.

11.4.1 Application of Coulomb yield criterion to yield behaviour

From the early studies of yield behaviour of polymers, one example has been selected: the plane-strain compression tests on PMMA, carried out by Bowden and Jukes [22]. The experimental set-up is shown in Figure 11.15. A particular advantage of this technique is that yield behaviour can be observed in compression for materials that normally fracture in a tensile test. In this case PMMA was

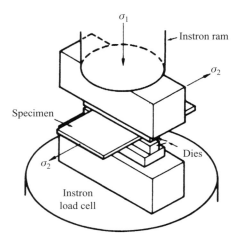

Figure 11.15 The plane-strain compression test. (Reproduced with permission from Bowden and Jukes, *J. Mater. Sci.*, **3**, 183 (1968))

studied at room temperature, i.e. below its brittle–ductile transition temperature in tension.

The yield point in compression σ_1 was measured for various values of applied tensile stress σ_2. The results, shown in Figure 11.16, give $\sigma_1 = -110.0 + 13.65\sigma_2$, where both σ_1 and σ_2 are expressed as true stresses in units of MPa. The results therefore clearly do not fit the Tresca criterion, where $\sigma_1 - \sigma_2 = $ constant at yield; neither do they fit a von Mises yield criterion. They are, however, consistent with a Coulomb yield criterion with $\tau = 47.4 - 1.58\sigma_N$.

11.4.2 Direct evidence for the influence of hydrostatic pressure on yield behaviour

There have been a number of detailed investigations of the influence of hydrostatic pressure on the yield behaviour of polymers. Because it illustrates clearly the relationship between a yield criterion, which depends on hydrostatic pressure, and the Coulomb yield criterion, an experiment will be discussed where Rabinowitz, Ward and Parry [23] determined the torsional stress–strain behaviour of isotropic PMMA under hydrostatic pressures up to 700 MPa. The results are shown in Figure 11.17.

There is a substantial increase in the shear yield stress up to a hydrostatic pressure of about 300 MPa. After this pressure brittle failure occurs, unless prevented by protecting the specimens from the hydraulic fluid [24] (e.g. by coating with a layer of solidified rubber solution). A study of polyethylene under conditions of combined pressure and tension has shown that the yield stress of

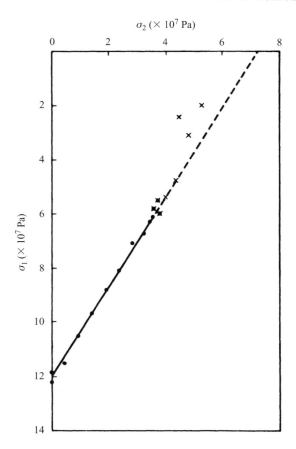

Figure 11.16 Measured values of the compressive yield stress σ_1 (true stress) plotted against applied tensile stress σ_2 (nominal stress). The full circles denote ductile yield, the crosses denote brittle fracture and the combined points denote tests where ductile yielding occurred, followed immediately by brittle fracture. (Reproduced with permission from Bowden and Jukes, *J. Mater. Sci.*, **3**, 183 (1968))

polyethylene increases approximately linearly up to pressures of 850 MPa [25]. The strain at which yield occurs also increases with increasing pressure, similar to the results of other workers for tensile tests under pressure. The shear yield stress increases linearly with pressure to an excellent approximation (Figure 11.18).

There are two other ways in which these results can be presented. First, recalling Section 11.2.6 and Figure 11.12, the Mohr circle diagram can be constructed from the data, as shown in Figure 11.19 where Bowden and Jukes's results appear as crossed points. This diagram leads naturally to a Coulomb yield criterion.

It is, however, equally reasonable to interpret Figure 11.17 directly in terms of the equation

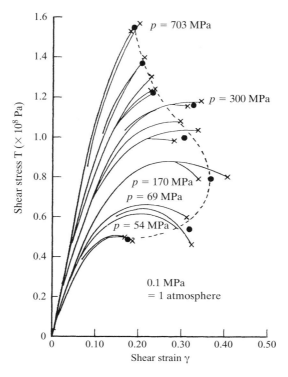

Figure 11.17   Shear stress–strain curves for PMMA showing fracture envelope. (Reproduced with permission of Rabinowitz, Ward and Parry, *J. Mater. Sci.*, **5**, 29 (1970))

$$\tau = \tau_0 + \alpha p \qquad (11.10)$$

where τ is the shear yield stress at pressure p, τ_0 is the shear yield stress at atmospheric pressure and α is the coefficient of increase of shear yield stress with hydrostatic pressure.

We will see that this simple form of pressure-dependent yield criterion is more satisfactory than the Coulomb criterion when a representation is developed that includes the effects of temperature and strain rate on the yield behaviour. In physical terms, the hydrostatic pressure can be seen as changing the state of the polymer by compressing the polymer significantly, unlike the situation in metals where the bulk moduli are much larger ($\sim$100 GPa compared with $\sim$5 GPa). Although the experimental evidence that exists is not unequivocal in this respect, it seems likely that the flow rules for the polymer subjected to hydrostatic pressure are still given by Equation (11.9), i.e. pressure has the sole effect of increasing the magnitude of the yield stresses.

Recent studies of yield behaviour, using a variety of multiaxial stressing experiments, can be described adequately by a generalization of Equation (11.10),

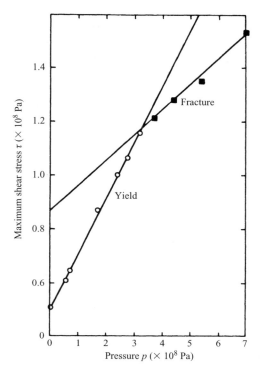

Figure 11.18 Maximum shear stress τ as a function of hydrostatic pressure p for PMMA: ($\circ$) yield; ($\blacksquare$) fracture. (Reproduced with permission of Rabinowitz, Ward and Parry, *J. Mater. Sci.*, **5**, 29 (1970))

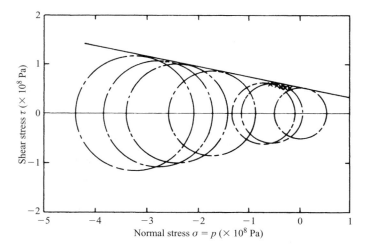

Figure 11.19 Mohr circles for yield behaviour of PMMA obtained from results of Rabinowitz, Ward and Parry. The crosses are the results of Bowden and Jukes. (Reproduced with permission from Rabinowitz, Ward and Parry, *J. Mater. Sci.*, **5**, 29 (1970))

i.e. a generalized von Mises equation where τ is replaced by the octahedral shear stress (see Section 11.6.1 for a fuller development).

Finally, it can be noted that the coefficient α in Equation (11.10) depends on the temperature of measurement and increases markedly near a viscoelastic transition. Briscoe and Tabor [26] have pointed out that α is equivalent to the coefficient of friction μ in sliding friction, and show that there is good numerical agreement between values of μ and the values of α obtained from yield stress/pressure measurements.

11.5 The molecular interpretations of yield and cold-drawing

11.5.1 Yield as an activated rate process: the Eyring equation

We have already seen in Section 10.3.4 that yield can be modelled using the Eyring process. It provides a convincing representation of both the temperature- and strain rate dependence of the yield stress. However, the discussions of Chapter 10 were confined to one-dimensional states of stress, whereas we now appreciate that yield criteria are essentially functions of the three-dimensional stress state. Also, in view of the discussion in the previous section, it is of interest to explore its applicability to pressure dependence. Both pressure dependence and the extension of the Eyring process to general stress states are considered here.

11.5.2 Alternative models: nucleation-controlled mechanisms

In the Eyring approach, stress affects the rate at which deformation occurs in a continuous manner. Although there is zero strain rate at zero stress, any non-zero value of stress, however small, gives rise to plastic deformation. The application of the stress speeds up the processes that are essentially ever-present, until the stress associated with them is equal to the applied stress; the deformation is referred to as being velocity-controlled.

There is another class of model in which the deformation is referred to as nucleation-controlled. Here, the applied stress is assumed to induce specific physical events, without which deformation cannot proceed. An example of such a model is that due to Argon [27]. He envisaged that shear strain was associated with buckling of the polymer chain via the action of a pair of opposed kinks. As pointed out by Ward [28], it is difficult to distinguish between the Argon and Eyring models by mechanical experimentation.

When considering the interactions between chains in a deforming polymer we may envisage a number of feasible deformation mechanisms; the richness of possibilities will depend on the complexity of the molecular chains. Recently, detailed computer simulations of polymer chains at the atomic level have made it possible to 'observe' the action of different mechanisms. In a simulation of the growth of plastic strain in glassy atactic polypropylene [29], no single molecular mechanism could be identified as having a major contribution to plastic flow. In a similar study of polycarbonate [30], which has a more complex chain, a greater number of deformation mechanisms were identified, including phenylene ring rotations, carbonate group reorientations and isopropylidene group rearrangements. However, as with polypropylene, no single mechanism could be identified as being dominant in the determination of plastic flow. A nucleation-controlled model involving a single deformation mechanism would therefore appear to have no natural advantage over a velocity-controlled model; both lack physical justification at the molecular level.

11.5.3   Pressure dependence and general states of stress

We have seen that the effect of pressure on the shear yield stress of a polymer can be represented very well by Equation (11.9), $\tau = \tau_0 + \alpha p$, which suggests that the Eyring equation may be modified very simply [31] to include the effect of the hydrostatic component of stress p, giving

$$\dot{\varepsilon} = \frac{\dot{\varepsilon}}{2} \exp\left(-\frac{\Delta H - \tau v + p\Omega}{RT}\right) \tag{11.11}$$

where v and Ω are the shear and pressure activation volumes, respectively. This modification of the Eyring equation can be considered to arise from a linear increase in the activation energy with increasing pressure, and so is the simplest approach.

Equation (11.11) may be expressed conveniently in terms of the octahedral yield stress

$$\tau_{\mathrm{oct}} = \tfrac{1}{3}[(\sigma_1 - \sigma_2)^2 + (\sigma_2 - \sigma_3)^2 + (\sigma_3 - \sigma_1)^2]^{1/2}$$

and the octahedral strain rate

$$\gamma_{\mathrm{oct}} = \tfrac{2}{3}[(e_1 - e_2)^2 + (e_2 - e_3)^2 + (e_3 - e_1)^2]^{1/2}$$

to give a generalized representation suitable for all stress fields

$$\dot{\gamma}_{\mathrm{oct}} = \frac{\dot{\gamma}_0}{2} \exp\left(-\frac{\Delta H - \tau_{\mathrm{oct}}v + p\Omega}{RT}\right) \tag{11.12}$$

For a constant strain rate test we therefore have

$$\Delta H - \tau_{oct}v + p\Omega = \text{constant}$$

from which an equation similar to Equation (11.10) is obtained with

$$\tau_{oct} = (\tau_{oct})_0 + \alpha p$$

where $\alpha = \Omega/v$. Figure 11.20 shows results for polycarbonate at atmospheric pressure [32] using data from torsion, tension and compression. It can be seen that, on average, the values of τ_{oct} lie in the order: compression > torsion > tension. The differences are therefore consistent with the observed linear dependence of τ_{oct} on pressure shown by direct measurement of the yield stress in torsion over a range of hydrostatic pressures (Section 11.4.1 above), and there is good numerical agreement between the two sets of measurements.

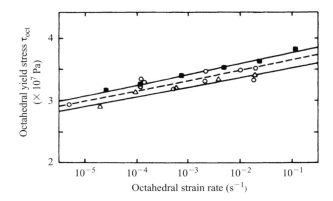

Figure 11.20   The strain rate dependence of the octahedral shear stress τ_{oct} at atmospheric pressure using data from torsion (○), tension (Δ) and compression (■). (Reproduced with permission from Duckett *et al.*, *Br. Polym. J.*, **10**, 11 (1978))

11.6   Cold-drawing

11.6.1   General considerations

The strain hardening, which is a necessary prerequisite for cold-drawing in polymers, has two possible sources:

1.   Drawing causes molecular alignment so that the drawing stress (often called the flow stress) is increased. This is a general phenomenon that is true for both crystalline and amorphous polymers.

2. Strain-induced crystallization may occur, similar to the crystallization observed in rubbers at high degrees of stretching.

In general, the morphological changes that occur in drawing are complex, and strain hardening is due to both molecular orientation and changes in morphology.

11.6.2 The natural draw ratio, maximum draw ratios and molecular networks

At room temperature, amorphous PET draws through a neck so that the undrawn material is transformed into drawn material, with a constant reduction in cross-section as it passes through the neck. If the drawing is uniaxial, as in the case of a fibre, the extension ratio λ is exactly similar to that defined for stretching a rubber (Section 2.4.3 above) and is generally called the natural draw ratio. There is some evidence to suggest that this natural draw ratio arises because cold drawing extends the network of entangled molecular chains to its limiting extensibility.

Confirmation of this idea comes from the observation that the natural draw ratio observed for melt-spun fibres is sensitive to the degree of molecular orientation introduced during the spinning process. It appears that the molecular network is formed as the polymer freezes from the melt, is subsequently stretched in the rubber-like state before the polymer cools below T_g and is eventually collected as a frozen stretched rubber. The amount of stretching in the threadline can be measured by shrinking these spun fibres back to a state of zero strain, i.e. isotropy. These results then can be combined with measurements of the natural draw ratio to give the limiting extensibility of the network [33].

Consider the cold-drawing of a sample of length l_1. If the fibre were allowed to shrink back to its isotropic state, length l_0, the shrinkage s would be defined by

$$s = \frac{l_1 - l_0}{l_1} \tag{11.13}$$

Cold-drawing to a length l_2 gives a natural draw ratio

$$\lambda_N = \frac{l_2}{l_1} \tag{11.14}$$

By combining Equations (11.13) and (11.14) we have

$$\frac{l_2}{l_0} = \frac{\lambda_N}{1 - s} \tag{11.15}$$

Table 11.1 shows collected results for a series of PET filaments. It can be seen that

λ_N varies from 4.25 to 2.58 and s varies from 0.042 to 0.378, but the ratio l_2/l_0 calculated from Equation (11.15) remains constant at a value of about 4.0.

Table 11.1   Value of $l_2/l_0 = \lambda_N/(l - s)$ for poly(ethylene terephthalate) samples of differing amounts of pre-orientation (see [33])

Initial birefringence $\times 10^3$	Natural draw ratio λ_N	Shrinkage s	$\lambda_N/(l - s)$
0.65	4.25	0.042	4.44
1.6	3.70	0.094	4.08
2.85	3.32	0.160	3.96
4.2	3.05	0.202	3.83
7.2	2.72	0.320	4.01
9.2	2.58	0.378	4.14

11.6.3   Crystalline polymers

The plastic deformation of crystalline polymers, in particular polyethylene, has been studied intensively from the viewpoint of changes in morphology. Notable contributions to this area have been made by Keller and co-workers and by Peterlin, Geil and others [34–36]. It is now evident that very drastic reorganization occurs at the morphological level, with the structure changing from a spherulitic to a fibrillar type as the degree of plastic deformation increases. The molecular reorientation processes are very far from being affine or pseudo-affine and can also involve mechanical twinning in the crystallites. It is surprising that some of the continuum ideas for mechanical anisotropy are nevertheless still relevent, although they must be appropriately modified.

In a few highly crystalline polymers, notably high-density polyethylene, extremely large draw ratios of $\sim$30 or more have been achieved by optimizing the chemical composition of the polymers and the drawing conditions [37, 38]. These high draw ratios lead to oriented polymers with very high Young's moduli, as discussed in Section 10.7.3. In spite of the much more complex deformation processes in a crystalline polymer, it has been concluded [38] that the molecular topology and the deformation of a molecular network are still the overriding considerations in determining the strain-hardening behaviour and the ultimate draw ratio achievable. For high-molecular-weight, high-density polyethylene, the key network junction points are physical entanglements, as in amorphous polymers. For low-molecular-weight, high-density polyethylene, both physical entanglements and crystallites where more than one molecular chain is incorporated can provide the network junction points. Junction points associated with the crystallites will be of a temporary nature. Very high draw ratios involve the breakdown of the crystalline structure and the unfolding of molecules, so that the simple ideas of a molecular network suggested for amorphous polymers have to be extended and modified.

References

1. Wineman, A. S. and Waldron, W. K., *Polym. Eng Sci.*, **33**, 1217 (1993).
2. Nadai, A., *Theory of Flow and Fracture of Solids*, McGraw-Hill, New York, 1963.
3. Orowan, B., *Rep. Prog. Phys.*, **12**, 185 (1949).
4. Sweeney, J., Shirataki, H., Unwin, A. P., *et al.*, *J. Appl. Polym. Sci.*, **74**, 3331 (1999).
5. Vincent, P. I., *Polymer*, **1**, 7 (1960).
6. Bridgeman, P. W., *Studies in Large Plastic Flow and Fracture*, McGraw-Hill, New York, 1952.
7. Amoedo, J. and Lee, D., *Polym. Eng. Sci.*, **32**, 1055 (1992).
8. Tresca, H., *C. R. Acad. Sci. (Paris)*, **59**, 754 (1864); **64**, 809 (1867).
9. Smith, W. F., *Principles of Materials Science and Engineering* (3rd edn), McGraw-Hill, New York, 1996.
10. Coulomb, C. A., *Mem. Math. Phys.*, **7**, 343 (1773).
11. von Mises, R., *Gottinger Nach. Math. -Phys. Kl.*, 582 (1913).
12. Levy, M., *C. R. Acad. Sci. (Paris)*, **70**, 1323 (1870).
13. Timoshenko, S. P. and Goodier, J. N., *Theory of Elasticity*, (3rd Edn), McGraw-Hill International Editions, New York, 1970.
14. Marshall, I. and Thompson, A. B., *Proc. R. Soc.*, **A221**, 541 (1954).
15. Müller, F. H., *Kolloidzeitschrift*, **114**, 59 (1949); **115**, 118 (1949); **126**, 65 (1952).
16. Hookway, D. C., *J. Text. Inst.*, **49**, 292 (1958).
17. Brauer, P. and Müller, F. H., *Kolloidzeitschrift*, **135**, 65 (1954).
18. Lazurkin, Y. S., *J. Polym. Sci.*, **30**, 595 (1958).
19. Allison, S. W., and Ward, I. M., *Br. J. Appl. Phys.*, **18**, 1151 (1967).
20. Whitney, W. and Andrews, R. D., *J. Polym. Sci. C*, **16**, 2981 (1967).
21. Brown, N. and Ward, I. M., *J. Polym. Sci. A2*, **6**, 607 (1968).
22. Bowden, P. B., and Jukes, J. A., *J. Mater. Sci.*, **3**, 183 (1968).
23. Rabinowitz, S., Ward, I. M. and Parry, J. S. C., *J. Mater. Sci.*, **5**, 29 (1970).
24. Harris, J. S., Ward, I. M. and Parry, J. S. C., *J. Mater. Sci.*, **6**, 110 (1971).
25. Sweeney, J., Duckett, R. A. and Ward, I. M., *J. Mater. Sci. Let.*, **5**, 1109 (1986).
26. Briscoe, B. J. and Tabor, D., in *Polymer Surfaces* (eds D. T. Clark and W. J. Feast), Wiley, New York, 1978, Ch. 1.
27. Argon, A. S., *Philos. Mag.*, **28**, 839 (1973).
28. Ward, I. M., *Mechanical Properties of Solid Polymers*, Wiley, Chichester, 1983.
29. Mott, P. H., Argon, A. S. and Suter, U. W., *Philos. Mag.*, **A67**, 931 (1993).
30. Hutnik, M., Argon, A. S. and Suter, U. W., *Macromolecules*, **26**, 1097 (1993).
31. Ward, I. M., *J. Mater. Sci.*, **6**, 1397 (1971).
32. Duckett, R. A., Goswami, B. C., Smith, L. S. A., *et al. Br. Polym. J.*, **10**, 11 (1978).
33. Allison, S. W., Pinnock, P. R. and Ward, I. M., *Polymer*, **7**, 66 (1966).
34. Hay, I. L. and Keller, A., *Kolloidzeitschrzft*, **204**, 43 (1965).
35. Peterlin, A., *J. Polym. Sci.*, **69**, 61 (1965).
36. Geil, P. H., *J. Polym. Sci. A*, **2**, 3835 (1964).
37. Capaccio, G. and Ward, I. M., *Nature Phys. Sci.*, **243**, 43 (1973).
38. Capaccio, G. and Ward, I. M., *Polymer*, **15**, 233 (1974).

12

Breaking Phenomena

12.1 Definition of tough and brittle behaviour in polymers

The mechanical properties of polymers are greatly affected by temperature and strain rate, and the load–elongation curve at a constant strain rate changes with increasing temperature as shown schematically (not necessarily to scale) in Figure 12.1. At low temperatures the load rises approximately linearly with increasing elongation up to the breaking point, when the polymer fractures in a brittle manner. At higher temperatures a yield point is observed and the load falls before failure, sometimes with the appearance of a neck: i.e. ductile failure, but still at quite low strains (typically 10–20 per cent). At still higher temperatures, under certain conditions, strain hardening occurs, the neck stabilizes and cold-drawing ensues. The extensions in this case are generally very large, up to 1000 per cent. Finally, at even higher temperatures, homogeneous deformation is observed, with a very large extension at break. In an amorphous polymer this rubber-like behaviour occurs above the glass transition temperature so the stress levels are very low.

For polymers the situation is clearly more complicated than that for the brittle–

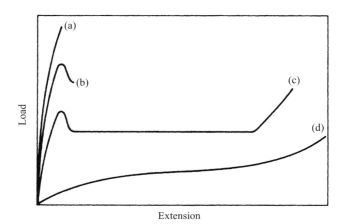

Figure 12.1 Load–extension curves for a typical polymer tested at four temperatures showing different regions of mechanical behaviour: (a) brittle fracture; (b) ductile failure; (c) necking and cold-drawing; (d) homogeneous deformation (quasi-rubber-like behaviour)

An Introduction to the Mechanical Properties of Solid Polymers I. M. Ward and J. Sweeney
© 2004 John Wiley & Sons, Ltd   ISBN: 0471 49625 1 (HB); 0471 49626 X (PB)

ductile transition in metals, as there are in general four regions of behaviour and not two. It is of considerable value to discuss the factors that influence the brittle–ductile transition, and then to consider further factors that are involved in the observation of necking and cold-drawing.

Ductile and brittle behaviour are most simply defined from the stress–strain curve. Brittle behaviour is designated when the specimen fails at its maximum load, at comparatively low strains (say < 10 per cent), whereas ductile behaviour shows a peak load followed by failure at a lower stress (Figure 12.1(a) and (b)).

The distinction between brittle and ductile failure is also manifested in two other ways: (1) the energy dissipated in fracture; and (2) the nature of the fracture surface. The energy dissipated is an important consideration for practical applications and forms the basis of the Charpy and Izod impact tests (discussed in Section 12.8 below). At the testing speeds under which the practical impact tests are conducted it is difficult to determine the stress–strain curve, so impact strengths are customarily quoted in terms of the fracture energy for a standard specimen.

The appearance of the fracture surface also can be an indication of the distinction between brittle and ductile failure, although the present state of knowledge concerning the crack propagation is not sufficiently extensive to make this distinction more than empirical.

12.2 Principles of brittle fracture of polymers

Modern understanding of the fracture behaviour of brittle materials stems from the seminal research of Griffith [1] on the brittle fracture of glass. The Griffith theory of fracture, which is the earliest statement of linear elastic fracture mechanics, has been applied extensively to the fracture of glass and metals, and more recently to polymers. Although it was conceived initially to describe the propagation of a crack in a perfectly elastic material at small elastic strains (hence linear elastic), subsequent work has shown that it is still applicable for situations including localized plastic deformation at the crack tip, which does not lead to general yielding in the specimen.

12.2.1 Griffith fracture theory

First, Griffith considered that fracture produces a new surface area and postulated that for fracture to occur the increase in energy required to produce the new surface must be balanced by a decrease in elastically stored energy.

Second, to explain the large discrepancy between the measured strength of materials and those based on theoretical considerations, he proposed that the elastically stored energy is not distributed uniformly throughout the specimen but is concentrated in the neighbourhood of small cracks. Fracture thus occurs due to the spreading of cracks that originate in pre-existing flaws.

In general the growth of a crack will be associated with an amount of work dW being done on the system by external forces and a chage dU in the elastically stored energy U. The difference between these quantities, $dW - dU$, is the energy available for the formation of the new surface. The condition for growth of a crack by a length dc is then

$$\frac{dW}{dc} - \frac{dU}{dc} \geqslant \gamma \frac{dA}{dc} \tag{12.1}$$

where γ is the surface free energy per unit area of surface and dA is the associated increment of surface. If there is no change in the overall extension Δ when the crack propagates, $dW = 0$ and

$$-\left(\frac{dU}{dc}\right)_\Delta \geqslant \gamma \frac{dA}{dc} \tag{12.1a}$$

The elastically stored energy decreases and so $-(dU/dc)_\Delta$ is essentially a positive quantity.

Griffith calculated the change in elastically stored energy using a solution obtained by Inglis [2] for the problem of a plate, pierced by a small elliptical crack, that is stressed at right angles to the major axis of the crack. Equation (12.1) then allows the fracture stress σ_B of the material to be defined in terms of the crack length $2c$ by the relationship

$$\sigma_B = (2\gamma E^*/\pi c)^{1/2} \tag{12.2}$$

where E^* is the 'reduced modulus', equal to Young's modulus E for a thin sheet in plane stress and equal to $E/(1 - \nu)$, where ν is Poisson's ratio, for a thick sheet in plane strain.

12.2.2   The Irwin model

An alternative formulation of the problem due to Irwin [3] considers the stress field near an idealized crack length $2c$ (Figure 12.2). In two-dimensional polar coordinates with the x axis as the crack axis and $r \ll c$,

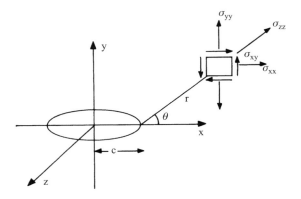

Figure 12.2   The stress field near an idealized crack of length $2c$

$$\sigma_{xx} = \frac{K_I}{(2\pi r)^{1/2}}\cos(\theta/2)[1 - \sin(\theta/2)\sin(3\theta/2)]$$

$$\sigma_{yy} = \frac{K_I}{(2\pi r)^{1/2}}\cos(\theta/2)[1 + \sin(\theta/2)\sin(3\theta/2)]$$

$$\sigma_{zz} = \nu(\sigma_{xx} + \sigma_{yy}) \text{ for plane strain}$$

$$\sigma_{zz} = 0 \text{ for plane stress} \tag{12.3}$$

$$\sigma_{xy} = \frac{K_I}{(2\pi r)^{1/2}}\cos(\theta/2)\sin(\theta/2)\sin(3\theta/2)$$

$$\sigma_{yz} = \sigma_{zx} = 0$$

In these equations θ is the angle between the axis of the crack and the radius vector.

The value of Irwin's approach is that the stress field around the crack is identical in form for all types of loading situation normal to the crack, with the magnitude of the stresses (i.e. their intensity) determined by K_I, which is constant for given loads and geometry; K_I is called the stress intensity factor, the subscript I indicating loading normal to the crack. This crack opening mode I is distinct from a sliding mode II, which is not considered here. As we approach the crack tip σ_{xx} and σ_{yy} clearly become infinite in magnitude and r tends to zero, but the products $\sigma_{xx}\sqrt{r}$ and $\sigma_{yy}\sqrt{r}$ and hence K_I remain finite.

For an infinite sheet with a central crack of length $2c$ subjected to a uniform stress σ it was shown by Irwin that

$$K_I = \sigma(\pi c)^{1/2} \tag{12.4}$$

He postulated that, when σ reaches the fracture stress σ_B, K_I has a critical value given by

$$K_{IC} = \sigma_B(\pi c)^{1/2} \tag{12.5}$$

The fracture toughness of the material then can be defined by the value of K_{IC}, termed the critical stress intensity factor, which defines the stress field at fracture.

There is clearly a link with the earlier Griffith formulation in that Equation (12.5) can be written as

$$\sigma_B = (K_{IC}^2/\pi c)^{1/2} \tag{12.6}$$

which is identical in form to Equation (12.2).

12.2.3 The strain energy release rate

In linear elastic fracture mechanics it is useful also to consider the energy G available for unit increase in crack length, which is called the 'strain energy release rate'. Following Equation (12.1) above, G is

$$G = \frac{dW}{dA} - \frac{dU}{dA} = \frac{1}{B}\left[\frac{dW}{dc} - \frac{dU}{dc}\right] \tag{12.7}$$

where B is the thickness of the specimen. It is assumed that fracture occurs when G reaches a critical value of G_c. The equivalent Equation to (12.1) is then

$$G \geqslant G_c \tag{12.8}$$

and G_c is equal to 2γ in the Griffith formulation but is generalized to include all work of fracture, not just the surface energy.

Comparison of Equations (12.2) and (12.6) shows

$$G_{IC} = K_{IC}^2/E^* \tag{12.9}$$

Although the Griffith and Irwin formulations of the fracture problems are equivalent, most recent studies of polymers have followed Irwin. Before discussing results for polymers, it is useful to show how G_c can be calculated.

Consider a sheet of polymer with a crack of length $2c$ (Figure 12.3). We now define a quantity termed the compliance of the cracked sheet, C, which is the reciprocal of the slope of the linear load–extension curve from zero load up to the point at which crack propagation begins. At the latter point the load is P and the extension is Δ, so $C = \Delta/P$.

This quantity C is not to be confused with an elastic stiffness constant as defined

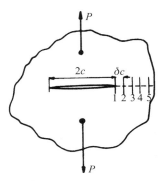

Figure 12.3   Schematic diagram of a specimen with a centre crack of length $2c$

in Section 7.1. The work done in an elemental step of crack propagation is illustrated by Figure 12.4. As the crack moves from 4 to 5, for example, the energy available for the formation of a new crack surface is the difference between the work done (45XY) and the increase in elastic stored energy (triangle 05Y – triangle 04X). This energy corresponds to the area of the shaded triangle in Figure 12.4, and for an increase of crack length dc is given by $\frac{1}{2}P^2\,dC$. Hence

$$G_c = \frac{P^2}{2B}\frac{dC}{dc} \tag{12.10}$$

which is generally known as the Irwin–Kies relationship [4].

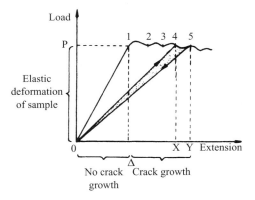

Figure 12.4   The load–extension curve for the specimen shown in Figure 12.3

Here G_c can be determined directly by combining a load–extension plot from a tensile testing machine with determination of the movement of the crack across the specimen, noting the load P for given crack lengths (points 1, 2, 3, 4, 5 in

Figure 12.4). Alternatively, test pieces of standard geometry can be used, for which the compliance is known as a function of crack length. For example, the relationship between the extension Δ (usually termed the deflection in this case) and the load P for a double cantilever beam specimen of thickness B (see Figure 12.5) is given by

$$\Delta = \frac{64c^3}{EBb^3}P$$

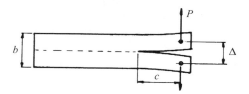

Figure 12.5   The double cantilever beam specimen

Hence

$$C = \frac{\Delta}{P} = \frac{64c^3}{EBb^3} \quad \text{and} \quad \frac{\mathrm{d}C}{\mathrm{d}c} = \frac{192c^2}{EBb^3} \tag{12.11}$$

giving

$$G_\mathrm{c} = \frac{P^2}{2B}\frac{\mathrm{d}C}{\mathrm{d}c} = \frac{P^2}{2B}\frac{192c^2}{EBb^3} \tag{12.12a}$$

or

$$G_\mathrm{c} = \frac{3\Delta^2 b^3}{128c^4}E \tag{12.12b}$$

The critical strain-energy release rate (or, in the original Griffith terminology, the fracture surface energy γ) therefore can be obtained by measurements of either the load P or the deflection Δ for given crack lengths c.

The exact equivalent formulation in terms of the critical stress intensity factor can be obtained from Equation (12.9), giving

$$K_\mathrm{IC} = 4\sqrt{6}\,\frac{P^2}{Bb^{3/2}} \tag{12.13}$$

We have discussed only the calculation for a geometrically simple specimen, so that the principles involved are not obscured by complex stress analysis. For a comprehensive discussion of the calculation of the fracture toughness parameters G_c and K_c for specimens with different geometries, see standard texts [5–7].

12.3 Controlled fracture in brittle polymers

In its simplest form the Griffith theory, and the linear elastic fracture mechanics (LEFM) that developed from it, ignore any contribution to the energy balance arising from the kinetic energy associated with movement of the crack. A basic study of the brittle fracture of polymers is therefore likely to be most rewarding if the fracture takes place slowly so that a negligible amount of energy is dissipated in this way.

With these ideas in mind, the seminal experimental studies of brittle fracture in polymers were undertaken by Benbow and Roesler at ICI in the UK [8] and by Berry at GE in the USA [9]. Benbow and Roesler devised a method of fracture in which flat strips of poly(methyl methacrylate) (PMMA) were cleaved by gradually propagating a crack down the middle, as in a cantilever double beam. Essentially, their experiments involved determining the deflection Δ for a given length c (symbols as in Figure 12.5). The results were expressed in terms of the surface energy γ. Following Equation (12.12b)

$$\frac{G_c}{E} = \frac{2\gamma}{E} = \frac{3\Delta^2 b^3}{128c^4} \tag{12.14}$$

Knowing a value for the Young's modulus, E, the surface energy γ can be found.

Berry adopted a slightly different approach to evaluate γ, and also confirmed the validity of Equation (12.2) for the fracture of PMMA and polystyrene by measuring the tensile strength of small samples containing deliberately introduced cracks of known magnitude.

Berry's summary of his own results and those of other workers is shown in Table 12.1. The very important conclusion to be drawn from these values for fracture surface energy is that they are very much greater than values estimated from the assumption that the energy required to form a new surface originates in the simultaneous breaking of chemical bonds, which might appear to provide an upper theoretical estimate. Take the bond dissociation energy as 400 kJ and the concentration of molecular chains as 1 chain per 0.2 nm^2, giving 5×10^{18} molecular chains m^{-2}. To form 1 m^2 of new surface requires about 1.5J, which

Table 12.1 Fracture surface energies (in J m$^{-2} \times 10^2$)

Method	Polymers	
	Poly(methyl methacrylate)	Polystyrene
Cleavage (Benbow [10])	4.9 ± 0.5	25.5 ± 3
Cleavage (Svensson [11])	4.5	9.0
Cleavage (Berry [9])	1.4 ± 0.07	7.13 ± 0.36
Tensile (Berry [12])	2.1 ± 0.5	17 ± 6

is two orders of magnitude less than that obtained from cleavage and tensile measurements.

12.4 Crazing in glassy polymers

The large discrepancy between experimental and theoretical values for the surface energy is comparable to that found for metals, where it was proposed by Orowan and others that the surface free energy may include a large term that arises from plastic work done in deforming the metal near the fracture surface as the crack propagates. Andrews [13] suggested that the quantity measured in the fracture of polymers should be describe by $\mathcal{J}$, the 'surface work parameter', to distinguish it from a true surface energy, and proposed a generalized theory of fracture that embraces viscoelastic as well as plastic deformation, both of which may be important in polymers.

On the basis of the results shown in Table 12.1, Berry concluded that the largest contribution to the surface energy of a glassy polymer comes from a viscous flow process that in PMMA, he suggested [14], was related to the interference bands observed on the fracture surfaces, as seen in Figure 12.6. He proposed that work was expended in the alignment of polymer chains ahead of the crack, the subsequent crack growth leaving a thin, highly oriented layer of polymeric material on the fracture surface. Following on from these ideas, Kambour [15–17] showed that a thin wedge of porous material termed a craze forms at a crack tip in a glassy polymer, as shown schematically in Figure 12.7. The craze forms under plane strain conditions, so that the polymer is not free to contract laterally and there is a consequent reduction in density. Several workers [18, 19] have attempted to determine the craze profile by examining the crack tip region in PMMA in an optical microscope. In reflected light, two sets of interference fringes were observed, which correspond to the crack and the craze, respectively. It was found that the craze profile was very similar to the plastic zone model proposed by Dugdale [20] for metals, which will now be described.

Equations (12.3) imply that there is an infinite stress at the crack tip. In practice this clearly cannot be so, and there are two possibilities. First, there can be a zone where shear yielding of the polymer occurs. In principle this can occur in both thin sheets where conditions of plane stress pertain and in thick sheets where there is a plane strain. Secondly, for thick specimens under conditions of plane strain, the stress singularity at the crack tip can be released by the formation of a craze, which is a line zone, in contrast to the approximately oval (plane stress) or kidney-shaped (plane strain) shear yield zones. As indicated, its shape approximates very well to the idealized Dugdale plastic zone where the stress singularity at the crack tip is cancelled by the superposition of a second stress field in which the stresses are compressive along the length of the crack (Fig. 12.8). A constant compressive stress is assumed and is identified with the craze stress. It is not the yield stress,

Figure 12.6 Matching fracture surfaces of a cleavage sample of poly(methyl methacrylate) showing colour alternation (green filter). (Reproduced from Berry, in *Fracture 1959* (eds B. L. Averach *et al.*), Wiley, New York, 1959, p. 263)

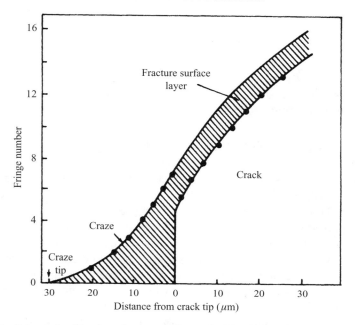

Figure 12.7 Schematic diagram of a craze. (Reproduced with permission from Brown and Ward, *Polymer*, **14**, 469 (1973). Copyright IPC Business Press Ltd)

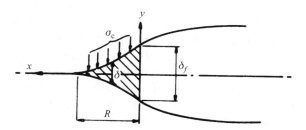

Figure 12.8 The Dugdale plastic zone model for a craze

and crazing and shear yielding are different in nature and respond differently to changes in the structure of the polymer.

Rice [21] has shown that the length of the craze for a loaded crack on the point of propagation is

$$R = \frac{\pi}{8} \frac{K_{IC}^2}{\sigma_c^2} \tag{12.15}$$

and the corresponding separation distance δ between the upper and lower surface of the craze is

$$\delta = \frac{8}{\pi E^*} \sigma_c R \left[\zeta - \frac{x}{2R} \log\left(\frac{1 + \zeta}{1 - \zeta}\right) \right] \tag{12.16}$$

where $\zeta = (1 - x/R)^{1/2}$.

The crack opening displacement (COD) δ_t is the value of the separation distance δ at the crack tip, where $x = 0$, and is therefore

$$\delta_t = 8\sigma_c R/\pi E^* = K_{IC}^2/\sigma_C E^* \tag{12.17}$$

The fracture toughness of the polymer then relates to two parameters δ_t and σ_c (the craze stress), the product of which is equal to G_{IC}, the critical strain energy release rate. Direct measurements of craze shapes for several glassy polymers, including polystyrene, poly(vinyl chloride) and polycarbonate [19, 22], have confirmed the similarity to a Dugdale plastic zone. A result of some physical significance is that the COD is often insensitive to temperature and strain rate for a given polymer, although it has been shown to depend on molecular mass. For constant COD, the true dependence of G_{IC} on strain rate and temperature is determined only by the sensitivity of the craze stress to these parameters. Because $G_{IC} = K_{IC}^2/E^*$ the fracture toughness K_{IC} will in addition be affected by E^*, which is also dependent on strain rate and temperature.

This approch offers a deeper understanding of the brittle–ductile transition in glassy polymers in terms of competition between crazing and yielding. Both are activated processes, in general with different temperature and strain rate sensitivities, and one will be favoured over the other for some conditions and vice versa for other conditions. An additional complexity can arise from the nature of the stress field, which may favour one process rather than the other, but the latter consideration does not enter into our discussion of the craze at the crack tip. The line of travel of the crack is a line of zero shear stress within the plane but maximum triaxial stress. In later discussion we will see that such a stress field favours crazing and that for long cracks where the stress field of the crack is the dominant factor, the craze length is determined solely by the requirement that the craze grows to cancel the stress singularity at the crack tip.

In several glassy polymers [22, 23], such as the polycarbonate shown in Figure 12.9, a complication occurs in that a thin line of material called a shear lip forms on the fracture surface where the polymer has yielded. Analogous to the behaviour of metals, it has been proposed that the overall strain energy release rate G_C^0 is the sum of the contribution from the craze and that from the shear lips. To a first approximation we would expect the latter to be proportional to the volume of yielded material. If the total width of the shear lip on the fracture surface is w, B is the specimen thickness and the shear lip is triangular in cross-section, then

$$G_C^0 = G_{IC}\left(\frac{B - w}{B}\right) + \frac{\phi w^2}{2B} \tag{12.18}$$

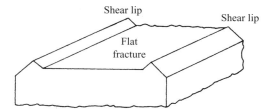

Figure 12.9 The shear lips in polycarbonate. (Reproduced with permission from Fraser and Ward, *Polymer*, **19**, 220, (1978))

where ϕ is the energy to fracture a unit volume of shear lip. It has been shown that this relationship describes results for polycarbonate and poly(ether sulphone) very well [22, 23] and that ϕ corresponds quite closely to the energy to fracture in a simple tensile extension experiment.

An alternative approach [24] assumes an additivity rule based on a plane strain K_{IC}, which pertains to fracture in the central part of the specimen and is designated K'_{IC}, and a plane stress K_{IC}, which is effective for the two surface skins of depth $w/2$ and is designated K''_{IC}. For the overall specimen it is then proposed that

$$K_{IC} = \left(\frac{B - w}{B}\right) K'_{IC} + \left(\frac{w}{B}\right) K''_{IC} \qquad (12.18a)$$

Although Equation (12.18a) is more empirically based than Equation (12.18) and is not formally equivalent, it has been shown to model fracture results very well. Moreover, in this formulation w relates to the size of the so-called Irwin plastic zone r_y, which can be defined simply on the basis of Equation (12.3) by assuming that a point r_y the stress reaches the yield stress σ_y. Hence

$$r_y = \frac{1}{2\pi} \left(\frac{K_{IC}}{\sigma_y}\right)^2$$

for plane stress and $w/2 = r_y$ in Equation (12.18a).

For PMMA Berry showed that the surface energy was strongly dependent on polymer molecular mass [25]. His results (Figure 12.10) fitted an approximately linear dependence of the fracture surface energy on reciprocal molecular mass, such that $\gamma = A' - B'/\overline{M}_v$, where $\overline{M}_v$ is the viscosity average molecular mass. Many years previously Flory [26] had proposed that the brittle strength is related to the number average molecular mass.

More recently Weidmann and Döll [27] have shown that the craze dimensions decrease markedly in PMMA at low molecular masses. In a study of the molecular mass dependence of fracture surfaces in the same polymer, Kusy and Turner [28] could observe no interference colours for a viscosity average molecular mass of less than 90 000 daltons, concluding that there was a dramatic decrease in the size

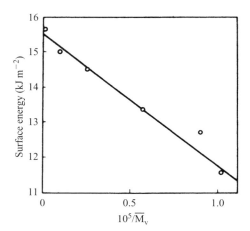

Figure 12.10 Dependence of fracture surface energy on reciprocal molecular mass) M_v is viscosity average molecular mass). (Reproduced with permission from Berry, *J. Polym. Sci.*, A2, 4069 (1964))

of the craze. Based on craze shape studies of polycarbonate, Pitman and Ward [22] reported a very high dependence of both craze stress and COD on molecular mass and observed that both would be expected to become negligibly small for $\overline{M}_w < 10^4$. Berry speculated that the smallest molecule that could contribute to the surface energy would have its end on the boundaries of the craze region, on opposite sides of the fracture plane, and be fully extended between these points. Kusy and Turner [29] presented a fracture model for PMMA in which the surface energy measured was determined by the number of chains above a critical length. Their data fitted well with their predictions, showing a limit to the surface energy at high molecular weight, but the model appeared inappropriate for the polycarbonate data of Pitman and Ward. Moreover, the extended molecular lengths, based on the extension of a random coil, would be much less than the COD (as discussed by Haward, Daniels and Treloar [30]) so that there is no direct correlation between the two quantities. The craze structure relates to the stretching of fibrils and the key molecular factors are the presence of random entanglements and the distance between these entanglements, not the extension of an isolated molecular chain.

12.5 The structure and formation of crazes

We have seen how the craze at the crack tip in a glassy polymer plays a vital role in determining its fracture toughness. Crazing in polymers also manifests itself in another way. When certain polymers, notably PMMA and polystyrene, are subjected to a tensile test in the glassy state, above a certain tensile stress opaque striations appear in planes whose normals are the direction of tensile stress, as in Figure 12.11.

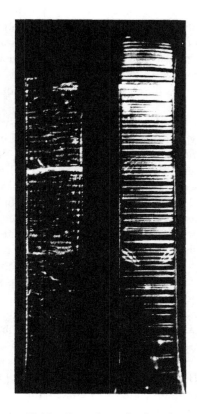

Figure 12.11   Craze formation in polystyrene

The interference bands on the fracture surfaces, which relate to the craze at the crack tip, were first observed by Berry [31] and by Higuchi [32]. Kambour confirmed that the PMMA fracture-surface layers were qualitatively similar to the internal crazes of this polymer, by showing that the refractive indices were the same [15]. Both surface layer and bulk crazes appear to be oriented polymer structures of low density, which are produced by orienting the polymer under conditions of abnormal constraint: it is not allowed to contract in the lateral direction, while being extended locally to strains of the order of unity, and so has undergone inhomogeneous cold-drawing.

Detailed studies have been made of the structure of crazes, the stress or strain criteria for their formation and environmental effects. These subjects now will be discussed in turn.

12.5.1   The structure of crazes

The structure of crazes in bulk specimens was studied by Kambour [15], who used the critical angle for total reflection at the craze/polymer interface to determine the refractive index of the craze, and showed that the craze was roughly 50 per cent polymer and 50 per cent void. Another investigation involved transmission electron microscopy of polystyrene crazes impregnated with an iodine–sulphur eutectic to maintain the craze in its extended state [33, 34]. The structure of the craze was clearly revealed as fibrils separated by the voids that are responsible for the overall low density.

Our understanding of the structure of crazes in glassy polymers developed in two stages, the first of which followed from a combination of techniques, starting with respective index measurements and transmission electron microscopy, followed by small-angle X-ray scattering (SAXS) and small-angle electron scattering (SAEX). The SAXS measurements, initiated by Parades and Fischer [35] with further contributions by Brown, Kramer and their collaborators [36], together with the SAEX measurements by Berger, Brown and Kramer [37, 38] confirmed the craze structure of a forest of cylindrical fibrils oriented normal to the craze surface. Figure 12.12 shows bright-field transmission electron microscopy and SAEX measurements by Berger [39], which were also most important in showing the presence of cross-tie fibrils between the main fibrils, suggesting the structure postulated in Figure 12.13.

Porod analysis of the SAXS and SAEX measurements provided a quantitative estimate of the mean craze fibril spacing. Brown [40] subsequently made the key observation that the presence of the cross-tie fibrils has a profound effect on the failure mechanism of a craze because they enable stress transfer between broken and unbroken fibrils. Brown [40], and then Kramer [41], followed this idea through to produce a quantitative theory of craze failure of the molecular chains at the mid-rib of the craze. Brown's theory is a very ingenious mixture of the macroscopic and the microscopic. Starting at the macroscopic level the craze can be modelled as a continuous anisotropic elastic sheet. The stress on the craze plane in front of the crack is then

$$\sigma = \frac{K_{\text{tip}}}{(2\pi r)^{1/2}}$$

where r is the distance from the crack tip and K_{tip} is the crack tip stress intensity factor. This is classical LEFM following Irwin (Equation (12.6) above). A more sophisticated analysis [41] makes use of the model of the craze as an anisotropic solid, characterized by stiffness constants c_{pq} (see Chapter 7). The dimensionless quantity α is introduced, defined as $\alpha^2 = c_{66}/c_{22}$ in a two-dimensional axis set with the 1 axis along the crack. The craze is opened by a drawing stress σ_d acting along the 2 direction normal to rigid boundaries either side of the craze. For a craze half-width h, the stress intensity at the tip is given by

(a)

(b)

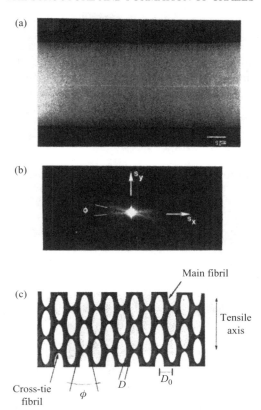

Main fibril

(c)

Tensile
axis

Cross-tie
fibril

ϕ D D_0

Figure 12.12 Bright-field TEM image of craze (A) formed in PMMA and its corresponding low-angle electron diffraction pattern (B); an idealized representation of the craze microstructure is shown in C. (Reproduced with permission from Berger, *Macromolecules*, **22**, 3162 (1989))

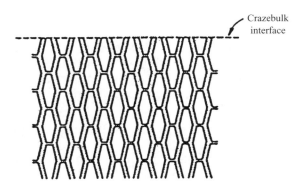

Crazebulk
interface

Figure 12.13 Schematic illustration of the fibril structure of a craze showing a regular arrangement of cross-tie fibrils. (Reproduced with permission from Brown, *Mater. Sci. Rep.*, **2**, 315 (1987))

$$K_{\text{tip}} = \sigma_{\text{d}}(2\alpha h)^{1/2} \qquad (12.19)$$

The highest stress in the craze is assumed to be in the fibril closest to the crack tip, and can be approximated by putting $r = d/2$, where d is the fibril spacing, to calculate

$$\sigma_{\text{tip}} = \frac{K_{\text{tip}}}{(\pi d)^{1/2}} \qquad (12.20)$$

Putting together Equations (12.19) and (12.20) shows the stress concentration:

$$\sigma_{\text{tip}} = \sigma_{\text{d}}\left(\frac{2\alpha h}{\pi d}\right)^{1/2} \qquad (12.21)$$

The next key idea, proposed by Brown and by Kramer, follows from the recognition that the craze is not continuous material, and that the stress is concentrated into the fibrils, which have a smaller cross-sectional area than the model continuum. This has the effect of concentrating the stress by the draw ratio λ in the fibril, to give a stress σ_{f} given by

$$\sigma_{\text{f}} = \frac{\lambda K_{\text{tip}}}{(\pi d)^{1/2}}$$

Values for the extension ratios, estimated by Kramer and colleagues [36, 42] and also by Ward and co-workers [18, 22] from analysis of optical interference patterns, compare reasonably well with estimates of the network extensibility from small-angle neutron scattering data [36] or stress-optical measurements [43]. It was therefore proposed that the criterion for craze failure, and hence crack propagation via a craze, is to assume that the entangled strands crossing the section of the craze at the crack tip break due to the development of the critical stress at the crack tip σ_{fail}.

Quantitatively, $\sigma_{\text{fail}} = \Sigma_{\text{eff}} f_{\text{b}}$, where f_{b} is the force required to break a single polymer molecule and Σ_{eff} is the effective crossing density of chains at the craze tip. If no chains are broken by forming the forest of fibrils when the craze is produced, the cross-density of strands is given by

$$\Sigma_{\text{eff}} = v d_{\text{e}}/2$$

where v is the number of chains per unit volume and d_{e} is the root-mean-square end-to-end distance of a random coil strand. This can be seen to follow from the number of entanglements in a rectangular box of cross-section v and thickness d_{e}:

$$v = k_{\text{B}}T/G_{\text{N}}^0 = \rho N/M_{\text{e}}$$

where G_N^0 is the rubbery plateau shear modulus, M_e is the entanglement molecular weight, N is Avogadro's number and ρ is the density (see Section 3.3.4 above).

If a fraction q of strands survive fibrillation it can be shown that

$$\Sigma_{eff} = (q \upsilon d_e/2)[1 - (M_e/qM_n)]$$

To calculate the fracture energy, the Dugdale model analysis is followed. Adapting Equation (12.17) using the relation $G_c = K_{IC}^2/E^*$ gives

$$G_c = \sigma_d \delta = 2h(1 - \upsilon_f)\sigma_d \qquad (12.22)$$

where $\upsilon_e = 1/\lambda$ is the volume fraction of the craze. Putting $\sigma_{tip} = \sigma_{fail} = \Sigma_{eff} f_b$ using Equation (12.21) yields an expression for h:

$$h = \frac{\pi d}{2\alpha} \left(\frac{\Sigma_{eff} f_b}{\sigma_d}\right)^2$$

which when substituted into Equation (12.22) gives

$$G_c = \frac{\pi d(1 - \upsilon_f) f_b^2 \Sigma_{eff}^2}{\alpha \sigma_d} \qquad (12.23)$$

As stated above, this calculation crosses between the macroscopic and the microscopic. It is therefore of interest to examine the numerical estimates that follow to obtain some assessment of the validity of the assumptions.

One example is the stress concentration equation (Equation (12.21)). Experimental results for PMMA suggest typical values of $h = 2$ μm and $d = 20$ nm. If α lies in the range $0.01-0.05$, σ_{tip} will be in the range $0.96\sigma_d - 2.1\sigma_d$, which gives some credence to the theory in that it is remarkably consistent with values of the craze stress obtained from Dugdale zone measurements, and will account for failure in the mid-rib of the craze.

Secondly, we can follow Brown [40] in estimating a value for f_b from the fracture toughness G_c on the basis of Equation (12.23) above. Brown takes $\Sigma_{eff} = 2.8 \times 10^{17}$ m^{-2} and a modulus ratio of 0.025 to obtain a value of 1.4×10^{-9} N for f_b. This value is within the range of 3×10^{-9} N estimated by Kausch [44] for the chain-breaking force and between 2.5 and 12×10^{-9} N estimated by Odell and Keller [45] from elongational flow experiments.

12.5.2 Craze initiation and growth

The studies of craze formation and structure described above indicate that there are clear differences between crazing and yield. Yield is essentially a shear process where the deformation occurs at constant volume (ignoring structural changes such as crystallization), but crazing occurs at a crack tip or in a solid section with a very appreciable increase in volume. It therefore appears that tensile stresses and, in particular, the hydrostatic tensile stress will be important in craze initiation and growth.

It would be desirable to obtain a stress criterion for craze initiation analogous to that for yield behaviour described in Chapter 11. Although all proposals made so far have not achieved general acceptance, it is of value to review the most important findings. Sternstein, Ongchin and Silverman [46] examined the formation of crazes in the vicinity of a small circular hole (1.59 mm diameter) punched in the centre of PMMA strips (12.7 mm $\times$ 50.8 mm $\times$ 0.79 mm) when the latter are pulled in tension. A typical pattern is shown in Figure 12.14(a). When the solutions for the elastic stress field in the vicinity of the hole were compared with the craze pattern it was found that the crazes grew parallel to the minor principal stress vector. Because the contours of the minor principal stress vector are orthogonal to those of the major principal stress vector, this result shows that the major principal stress acts along the craze plane normal and therefore parallel to the molecular orientation axis of the crazed material.

The boundary of the crazed region coincided to a good approximation with contour plots showing lines of constant major principal stress σ_1, as shown in Figure 12.14(b) where the contour numbers are per unit of applied stress. At low applied stresses it is not possible to discriminate between the contours of constant σ_1 and contours showing constant values of the first stress invariant $I_1 = \sigma_1 + \sigma_2$. However, the consensus of the results is in accord with a craze-stress criterion based on the former rather than on the latter and, as we have seen, the direction of the crazes is consistent with the former.

Sternstein and Ongchin [47] extended this investigation by examining the formation of crazes under biaxial stress conditions, and found that the stress conditions for crazing involved both the principal stresses σ_1 and σ_2. The most physically acceptable explanation of these results was proposed by Bowden and Oxborough [48], who suggested that crazing occurs when the extensional strain in any direction reaches a critical value e_1, which depends on the hydrostatic component of stress.

For small strains, for the two-dimensional stress field, e_1 is given by

$$e_1 = \frac{1}{E}(\sigma_1 - \nu\sigma_2)$$

where E is Young's modulus and ν is Poisson's ratio.

It was proposed that the crazing criterion was

$$Ee_1 = \sigma_1 - \nu\sigma_2 = A + B/I_1 \tag{12.24}$$

where $I_1 = \sigma_1 + \sigma_2$. Equation (12.24) predicts that the stress required to initiate a

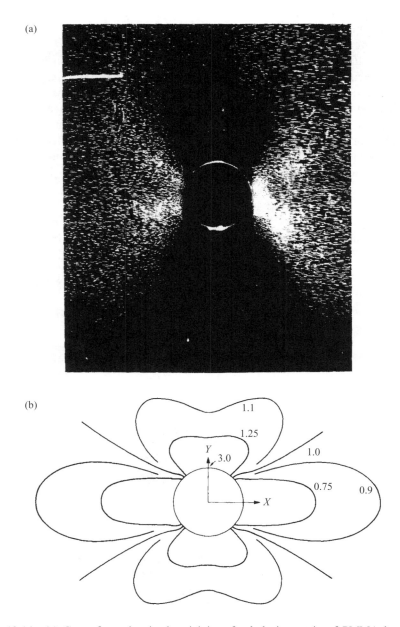

Figure 12.14 (a) Craze formation in the vicinity of a hole in a strip of PMMA loaded in tension. (Result obtained by L. S. A. Smith.) (b) Major principal stress contours (σ_1) for an elastic solid containing a hole. The specimen is loaded in tension in the x direction. Contour numbers are per unit of applied tensile stress. (Reproduced with permission from Sternstein, Ongchin and Silverman, *Appl. Polym. Symp.*, **7**, 175 (1968))

craze becomes infinite when $I_1 = 0$, i.e. crazing requires a dilational stress field. Unfortunately there are several pieces of experimental evidence [49–51] that contradict this assumption, so there is still no completely satisfactory stress criterion for craze initiation.

There is, however, a theory for the growth of crazes that is consistent with all the experimental evidence. Argon, Hannoosh and Salama [52] have proposed that the craze front advances by a meniscus instability mechanism in which craze tufts are produced by the repeated break-up of the concave air/polymer interface at the crack tip, as illustrated in Figure 12.15. A theoretical treatment of this model predicted that the steady-state craze velocity would relate to the five-sixths power of the maximum principal tensile stress, and support for this result was obtained from experimental results on polystyrene and PMMA [52].

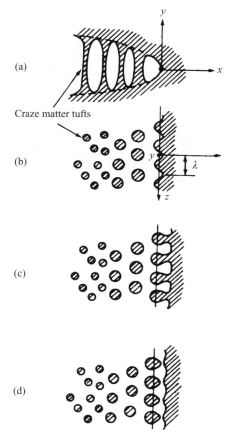

Figure 12.15 Schematic diagram showing craze matter production by the mechanism of meniscus instability: (a) outline of a craze tip; (b) cross-section in the craze plane across craze matter tufts; (c, d) advance of the craze front by a completed period of interface convolution. (Reproduced with permission from Argon, Hanncosh and Salama, in *Fracture 1977*, Vol. 1, Waterloo, 1977, p. 445)

12.5.3   Crazing in the presence of fluids and gases: environmental crazing

The crazing of polymers by environmental agents is of considerable practical importance and has been studied extensively, with notable contributions from Kambour [16, 53–55], Andrews and Bevan [56], Williams and co-workers [57, 58] and Brown [59–61]. The subject has been reviewed by Kambour [62] and by Brown [63]. In general environmental agents, which can be fluids or solids, reduce the stress or strain required to initiate crazing.

Kambour and co-workers [16, 53–55] showed that the critical strain for crazing decreased as the solubility of the environmental agent was increased. It was also found that the critical strain decreased as the glass transition temperature of the solvated polymer decreased. Andrews and Bevan [56], adopting a more formal approach and applying the ideas of fracture mechanics, performed fracture tests on single-edge notched tensile specimens, where a central edge crack of length c is introduced into a large sheet of polymer that is then loaded in tension. The fracture stress is related to the surface work parameter $\mathcal{J}$ of Andrews (or the strain energy release rate $G_c = 2\gamma$) by an equation identical in form to Equation (12.2) above. The critical stress for crack and craze propagation σ_c was indeed proportional to $c^{-1/2}$, so the $\mathcal{J}$ values could be determined. For constant experimental conditions, a range of values of $\mathcal{J}$ was obtained from which a minimum value $\mathcal{J}_0$ was estimated. From tests in a given solvent over a range of temperatures it was found that values of $\mathcal{J}_0$ decreased with increasing temperatures up to a characteristic temperature $\mathcal{J}_c$, above which $\mathcal{J}_0$ remained constant at value $\mathcal{J}_0^*$. The values of $\mathcal{J}_0^*$ for the different solvents were shown to be a smooth function of the difference between the solubility parameters of the solvent and the polymer, reaching a minimum when this difference was zero (Figure 12.16).

These findings were explained on the basis that the work done in producing the craze can be modelled by the expansion of a spherical cavity of radius r under a negative hydrostatic pressure p, which has two terms so that

$$p = \frac{2\gamma_\tau}{r} + \frac{2\sigma_Y}{3}\psi \qquad (12.25)$$

where γ_τ is the surface tension between the solvent in the void and the surrounding polymer, σ_Y is the yield stress and ψ is a factor close to unity. The effect of temperature is to change the yield stress, so that with increasing temperature σ_Y falls eventually to zero at T_c, which is the glass transition temperature of the plasticized polymer. Above T_c the fracture surface energy $\mathcal{J}_0^*$ relates solely to the intermolecular forces represented by the surface tension γ_τ.

Brown has pointed out that gases at sufficiently low temperatures make almost all linear polymers craze [59–61, 63]. Parameters such as the density of the crazes and the craze velocity increase with the pressure of the gas and decrease with

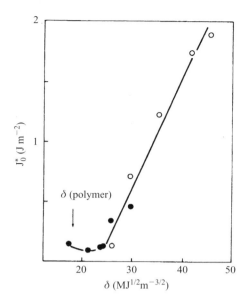

Figure 12.16   Variation of J_0^* for PMMA with the solubility parameter of the solvent: (●) pure solvents; (○) water isopropanol mixtures. (Reproduced with permission from Andrews and Bevan, *Polymer*, **13**, 337 (1972), © IPC Business Press Ltd)

increasing temperature. It was concluded that the surface concentration of the absorbed gas was a key factor in determining its effectiveness as a crazing agent.

In a related but somewhat different development, Williams and co-workers [57, 58] studied the rate of craze growth in PMMA in methanol. In all cases the craze growth depended on the initial stress-intensity factor K_0, calculated from the load and the initial notch length. Below a specific value of K_0 termed K_0^*, the craze would decelerate and finally arrest. For $K_0 > K_0^*$ the craze would decelerate initially and finally propagate at constant speed.

It was argued that the controlling factor determining craze growth was the diffusion of methanol into the craze. Where $K_0 < K_0^*$, the methanol is considered to diffuse along the length of the craze, and it may be shown that the length of the craze x is proportional to the square root of the time of growth (Figure 12.17). In the second type of growth, where $K_0 > K_0^*$, it is considered that the methanol diffuses through the surface of the specimens, maintaining the pressure gradient in the craze and producing craze growth at constant velocity.

12.6   Controlled fracture in tough polymers

The development of brittle fracture, as outlined in Sections 12.2.1–12.2.3 above, is directly applicable to the failure of many glassy polymers, including PMMA

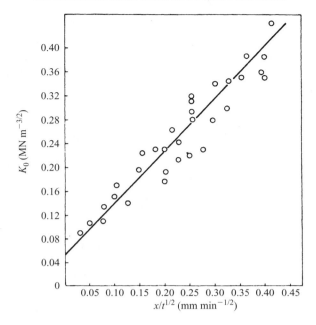

Figure 12.17   Craze growth behaviour for poly(methyl methacrylate) in methanol at 20 °C (Reproduced with permission from Williams and Marshall, *Proc. R. Soc. A*, **342**, 55 (1975))

and polystyrene, which have been studied intensively by LEFM. A key criterion for quantitative analysis is that extensive yielding should not occur, either at the crack tip or in the body of the specimen, for which explicit rules have been proposed. In essence, the load–extension curves should be of the form shown in Figure 12.4, i.e elastic deformation up to the point of initiation of crack growth. In practice, this means that fracture occurs under conditions of plane strain for specimens of minimum thickness, whereas for thin specimens plane stress conditions apply and the stress–strain curve will show a yield point as in Figure 12.1, curves (a) and (c).

The fracture of most semicrystalline polymers, notably polyethylene, polypropylene and nylon, cannot be described by LEFM based on the theory of Griffith and Irwin because large-scale yielding occurs at the crack tip prior to failure. For these materials and for toughened polymers and polymer blends other approaches have been developed, three of which will be discussed in detail:

1. The *J*-integral.

2. Essential work of fracture.

3. Crack opening displacement (COD).

12.6.1   The *J*-integral

The *J*-integral approach was initiated by Rice [64] and developed by Begley and
Landes [65]. It is most instructive to follow the exposition adopted by Landes and
Begley [66] and later by Chan and Williams [67].

Rice defined a quantity termed the *J*-integral, which describes the flow of
energy into the crack tip region. It is defined for the two-dimensional problem of a
straight crack in the x direction (Figure 12.18), and Γ is any contour surrounding
the crack tip. Formally

$$J = \int W \, \mathrm{d}y - T_i \frac{\mathrm{d}u_i}{\mathrm{d}x} \, \mathrm{d}s$$

where $W = \int \sigma_{ij} \, \mathrm{d}\varepsilon_{ij}$ is the strain energy density relating to the stress and strain
components σ_{ij} and ε_{ij} in the crack tip region and $T_i \, \mathrm{d}u_i$ are the work terms when
components of the surface tractions T_i on the contour path move through
displacements $\mathrm{d}u_i$. Rice showed that J is independent of the path chosen for
integration of the total energy. In the case of linear elasticity, J equates the strain
energy release rate G.

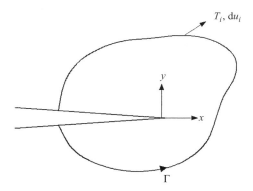

Figure 12.18   Line integral contour

When the displacements are prescribed, the J integral is more simply defined as
the rate of decrease of potential energy U with crack length. As shown in Figure
12.19, the *J*-integral is then given by the shaded area and

$$J = -\frac{1}{B} \frac{\mathrm{d}U}{\mathrm{d}a} \qquad (12.26)$$

where B is the specimen thickness and a is the crack length. It can be seen that the

(a)

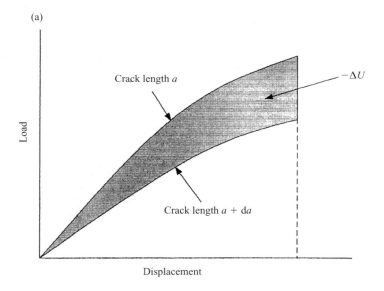

(b)

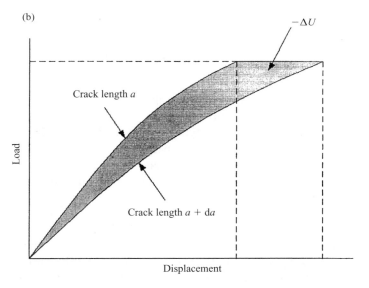

Figure 12.19 Load displacement curves for crack growth: (a) constant displacement; (b) constant load.

difference between constant load and constant displacement is second order and can be neglected.

Sumpter and Turner [68] have expressed J as the sum of elastic and plastic components J_e and J_p, respectively:

$$J = J_e + J_p$$

where

$$J_e = \frac{\eta_e U_e}{B(W - a)}, \quad \text{and} \quad J_p = \frac{\eta_p U_p}{B(W - a)}$$

and η_e and η_p are elastic and plastic work factors dependent on the specimen geometry.

For single-edge-notch blend specimens when

$$0.4 < \frac{a}{w} < 0.6 \quad \text{and} \quad \eta_e = \eta_p = 2$$

then

$$J = \frac{2U}{B(W - a)} \tag{12.27}$$

The accepted procedure is to introduce a sharp crack into a single-edge-notch sample. This is usually done by machining a notch, which is then sharpened by insertion of a razor blade, either by tapping (for brittle polymers) or by slicing (for ductile polymers). After each loading, the specimens (either in tension or bending) are broken up so that the amount of crack extension Δa can be measured. This can be done by cooling the specimen to low temperatures and then breaking it to observe the difference in the fracture surface that occurs at the point of initial crack extension.

The value of J is calculated from Equation (12.27) and plotted against Δa as shown in Figure 12.20(b). Initially the crack extends by blunting the initial sharp crack. If it is assumed that the blunted crack has a semi-circular profile (this is not precisely the case for polymers) so that $\Delta a = \delta/2$, where δ is the COD, the crack blunting line is given by

$$J = \sigma_y \delta = 2\sigma_y \Delta a \tag{12.28}$$

where σ_y is the yield stress or craze stress.

The J resistance curve shows a point of inflection where true crack growth starts, and this defines an equivalent quantity to the G_{IC} of LEFM for plane strain

(a)

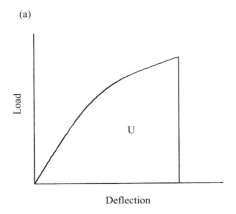

(b)

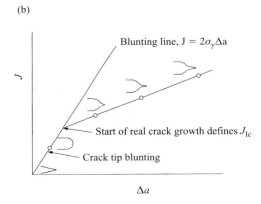

Figure 12.20 Procedure for J_{IC} measurement: (a) load identical specimens to a range of deflections; (b) plot of J against Δa, showing crack tip blunting and crack growth. (Reproduced with permission from Chan and Williams, *Int. J. Fracture*, **19**, 145 (1983))

brittle fracture, which is termed J_{IC}. The sample size requirements for a valid J_{IC} are given by

$$a, W - a \quad \text{and} \quad B \geqslant 25\left(\frac{J_{IC}}{\sigma_y}\right)$$

In very tough polymers, an idealized plot like that of Figure 12.20 is not obtained and a more arbitrary procedure has been proposed [69]. The J resistance curve is represented by a power law

$$J = C_1 \Delta a^{C_2} \tag{12.29}$$

where C_1 and C_2 are fitting parameters, and a crack initiation value $J_{0.2}$ is defined that is determined from the J value for a crack extension of 0.2 mm.

12.6.2   Essential work of fracture

Another approach to the fracture of ductile polymers stems from the recognition that for such materials the crack tip deformation zone has two components, as shown in Figure 12.21. There is an inner zone where the fracture process occurs – which could involve a combination of shear yielding and crazing – and an outer zone where extensive yielding and plastic deformation occur. This approach was originally proposed by Broberg [70], and has been developed by Mai and Cottrell [71], Hashemi and Williams [72], Mai [73] and others.

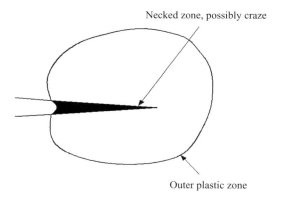

Figure 12.21   Schematic diagram of process zone in ductile fracture specimen

The measurements are carried out on deeply notched specimens, either single-edge-notched tension (SENT) or double-edge-notched tension (DENT) (see Figure 12.22). The total work of fracture W_f consists of two components. First, there is the work expended in the inner zone W_e, which is called the essential work of fracture. This relates directly to the energy required to fracture the sample and is therefore proportional to the ligament length l. Secondly, there is what is termed the non-essential work of fracture W_p, the energy dissipated in the outer plastic zone, where shear yielding and other forms of plastic deformation can occur. This component is proportional to the second power of the ligament length.

We have

$$W_e + W_p \tag{12.30}$$

which can be written as

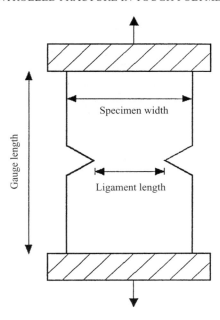

Figure 12.22   Schematic diagram of test specimen for essential work of fracture test

$$W_f = w_e B l + \beta w_p B l^2 \qquad (12.31)$$

where w_e is the specific essential work of fracture, w_p is the specific non-essential work of fracture, B is the specimen thickness and β is a shape factor for the plastic zone. More conveniently the specific total work of fracture w_f is given by

$$w_f = \frac{W_f}{lB} = w_e + \beta w_p l \qquad (12.32)$$

It is necessary to consider whether the fracture occurs under conditions of plane stress or of plane strain. Figure 12.23 shows schematically how the total specific work of fracture varies with ligament length. At high ligament length, conditions of plane stress pertain and Equation (12.32) applies to give an extrapolated value of W_e, the essential work of fracture for plane stress fracture; βw_p is the work dissipation in the outer plastic zone. At low ligament length there is a transition from plane stress to plane strain and extrapolation to zero ligament length gives W_{1e}, the plane strain essential work of fracture.

Wu and Mai [73] have examined the relationship between the essential work of fracture method and the *J*-integral method. Figure 12.24 shows w_f as function of ligament length for a DENT specimen of thickness 0.285 mm. At large ligament lengths failure occurs under conditions of plane stress; at low ligament lengths failure occurs under plane strain. The corresponding values for w_e are 46.93 and

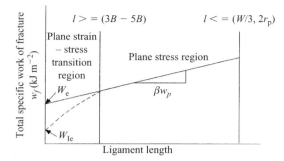

Figure 12.23   Schematic diagram showing specific total work of fracture against ligament length

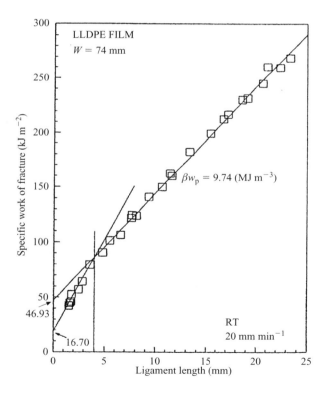

Figure 12.24   Specific work of fracture versus ligament length for linear low-density polyethylene films. (Reproduced with permission from Win and Mai, *Polym. Eng. Sci.*, **36**, 2275 (1996))

16.70 kJ m^{-2}. Wu and Mai concluded that the plane stress w_e obtained by linear extrapolation is equivalent to the plane strain J_{IC}, as proposed earlier by Mai and Cotterell [71].

12.6.3   Crack opening displacement

For LEFM there is an explicit relationship between K_{IC} and the COD, which is exemplified by the Dugdale plastic zone model (Equation (12.17)). For tough polymers it cannot be assumed that this is still the case. Nevertheless, the measurement of COD can still be a valuable tool and it has been used extensively for assessing the toughness of polyethylene gas pipe, especially by Brown and co-workers [74, 75]. The COD is measured under plane strain conditions or nearly plane strain conditions (which cannot always be assumed for very tough samples), where a damage zone forms at the root of the notch, a craze similar to that observed in a glassy polymer (Section 12.4 above). These tests are conducted under conditions of slow crack growth, and usually under constant stress, at elevated temperatures to accelerate the crack growth. Figure 12.25 shows a typical experimental set-up.

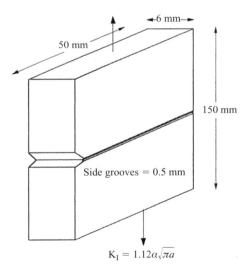

Figure 12.25   Single-edge-notched fracture specimen. (Reproduced with permission from O'Connell, Bonner, Duckett, *et al.*, *Polymer*, **36**, 2355 (1995))

For tough polyethylenes, the craze angle remains approximately constant as the damage zone grows so the growth of the craze in the crack direction is linearly related to the COD. Figure 12.26 shows the COD versus time for a polyethylene

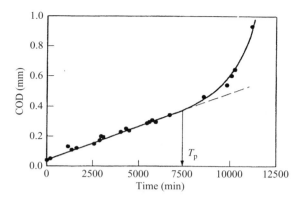

Figure 12.26 Crack opening displacement (COD) for polyethylene copolymer at a bulk stress of 3 MPa. (Reproduced with permission from O'Connell, Bonner, Duckett, *et al.*, *Polymer*, **36**, 2355 (1995))

copolymer. There is an initial linear portion followed by an accelerating rate. Experimental observations show that the point at which the COD rate starts to accelerate T_p, is associated with the first signs of fracture at the base of the craze, with failure occurring in the mid-rib of the fibrils of the craze, and this gives the failure time for slow crack growth. The situation as envisaged is shown schematically in Figure 12.27.

Ward and co-workers [76] showed that slow crack growth data obtained from measurements of COD can be related to creep of the fibrils in the craze. The COD rate and hence the fracture time are dominated by creep to fracture of the fibrils. This result gives support to the previous observations by Capaccio and co-workers [77], who showed that there was a direct correlation between slow crack growth and the gradient of a Sherby–Dorn creep plot for log strain rate versus strain (see Section 10.3.2 above), which they called the creep rate deceleration factor (CRDF). The larger the value of CRDF, i.e. the greater the reduction in creep rate with strain, the more resistance a polymer has to slow crack propagation. The creep rates were reduced in polyethylene by copolymerization (introduction of side groups into the polyethylene chains) and increasing molecular weight, as shown previously by Wilding and Ward [78]. These results are consistent with the conclusions Brown and co-workers [79] on the basis of their very extensive COD measurements of slow crack growth in a range of polyethylenes.

Ward and co-workers used extensive creep data to construct master plots of log strain rate versus true stress at constant plastic strain (draw ratio) [76]. One such plot for a polyethylene copolymer is shown in Figure 12.28. It was shown that slow crack propagation data were consistent with the proposition that this related to creep to failure of oriented fibrils in the craze. These fibrils then followed the computed route to failure shown in Figure 12.28.

In several publications, Brown and co-workers [79] developed the idea that the

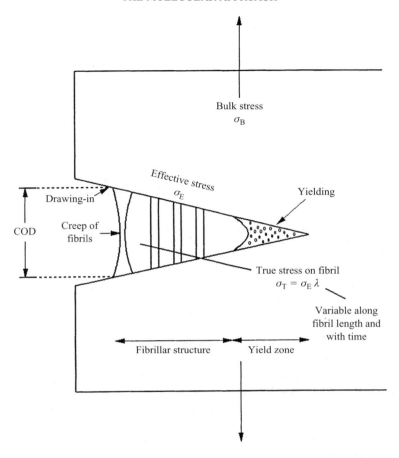

Figure 12.27 Schematic diagram showing model for craze growth and processes involved. (Reproduced with permission from O'Connell, Bonner, Duckett, *et al.*, *Polymer*, **36**, 2355 (1995))

reduction in crack growth rate in polyethylene due to incorporation of branches relates to a difference in the fraction of tie molecules in the initial structure. The extensive studies of Capaccio and co-workers [80], confirmed by the computer model of Ward *et al.*, suggest that the critical factor is the creep to failure of the oriented fibrils in the craze and is not related directly to the initial morphology [81].

12.7 The molecular approach

It has long been recognized that oriented polymers (i.e. fibres) are much less strong than would be predicted on the basis of elementary assumptions that fracture involves simultaneously breaking the bonds in the molecular chains across

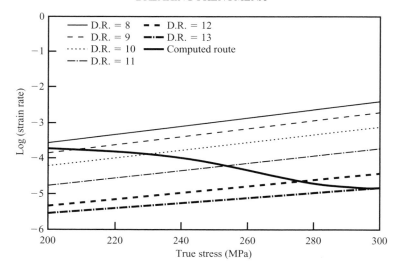

Figure 12.28 Log_{10} (strain rate) versus true stress at constant draw ratio for polyethylene copolymer at indicated draw ratios (D. R.). (Reproduced with permission from O'Connell, Bonner, Duckett, *et al.*, *Polymer*, **36**, 2355 (1995))

the section perpendicular to the applied stress. Calculations of this nature were originally undertaken by Mark [82] and rather more recently by Vincent [83] on polyethylene. It was found that in both cases the measured tensile strength was at least an order of magnitude less than that calculated.

We have seen one possible explanation of this descrepancy—the Griffith flaw theory of fracture. It has been considered also that there may be a general analogy between this difference betweeen measured and calculated strengths and the difference between measured and calculated stiffness for oriented polymers. A general argument for both discrepancies could be that only a small fraction of the molecular chains are supporting the applied load. In Chapter 8 we discussed how the tie molecules or crystalline bridges that connect adjacent crystalline blocks play a key role in determining the axial stiffness of an oriented semicrystalline polymer. There has therefore been considerable interest in examining chain fracture in oriented polymers, using electron paramagnetic resonance to observe the free radicals produced or infrared spectroscopy to identify such entities as aldehyde end groups, which suggest chain scission. A very comprehensive survey of the results of such studies has been given by Kausch [84]. Kausch and Becht [85] have emphasized that the total number of broken chains is much too small for their load-carrying capacity to account for the measured reductions in macroscopic stress. We must therefore conclude that the tie molecules that eventually break are not the main source of strength of highly oriented polymers, a conclusion confirmed by the lack of any positive correlations between the strength of fibres and the radical concentration at break.

Although these strong reservations have to be borne in mind, studies using molecular methods are relevant to the deformation of polymers. Examination of the infrared and Raman spectra of oriented polymers under stress show that there are distinct shifts in frequency from the unstressed state [86, 87] indicative of a distortion of bonds in the chain due to stress. Furthermore, changes in the shape of the spectrum lines are observed, which is interpreted as implying that certain bonds are much more highly stressed than the average.

Recent Raman spectroscopy studies, notably by Young and co-workers [88, 89] have shown that the shifts in the Raman frequency per unit strain for a range of oriented fibres are proportional to the fibre tensile moduli. This is consistent with a series aggregate model for the fibre structure (Section 7.5 above). For this model, strain and stress σ are related by

$$\varepsilon = \sigma/E_3$$

The Raman shift Δv with stress is a constant so that

$$\frac{d\Delta v}{d\sigma} = \alpha$$

and the Raman shift with strain is given by

$$\frac{d\Delta v}{d\varepsilon} = \frac{d\Delta v}{d\sigma} \cdot \frac{d\sigma}{d\varepsilon} = \alpha E_3$$

A positive attempt to obtain a molecular understanding of fracture took as its starting point the time and temperature dependence of the fracture process. This approach dates back to the early work of Bueche [90] and Zhurkov and co-workers [91]. It is assumed that the fracture process relates to the rate of bond breakage v_B at high stress, via an Eyring-type thermally activated process, so that

$$v_B = v_{B0} \exp[-(U_0 - v\sigma_B)/kT]$$

where U_0 is the activation energy and v is an activation volume. The time to failure τ under an applied stress σ_B is then given by

$$\tau = \tau_0 \exp[(U_0 - v\sigma_B)/kT]$$

This equation was shown to hold for a wide range of polymers and, moreover, the values obtained for U_0 correlated very well with values obtained for the activation energy for thermal degradation.

The existence of submicrocracks in polymers has already been mentioned in connection with the argon theory of craze initiation. Zhurkov, Kuksenko and Slutsker [92] have use small-angle X-ray scattering to establish the presence of such submicroscopic cracks. Although it has been proposed by Zakrevskii [93] that the formation of these submicrocracks is associated with a cluster of free radicals and the associated ends of molecular chains, Peterlin [94] has argued that the cracks occur at the ends of microfibrils, and Kausch [84] has concluded that the submicrocrack formation is essentially independent of chain scission.

12.8 Factors influencing brittle–ductile behaviour: brittle–ductile transitions

12.8.1 The Ludwig–Davidenkov–Orowan hypothesis

Many aspects of the brittle–ductile transition in metals, including the effect of notching, which we will discuss separately, have been discussed in terms of the Ludwig–Davidenkov–Orowan hypothesis that brittle fracture occurs when the yield stress exceeds a critical value [95], as illustrated in Figure 12.29(a). It is assumed that brittle fracture and plastic flow are independent processes, giving separate characteristic curves for the brittle fracture stress σ_B and the yield stress σ_Y as a function of temperature at constant strain rate (as shown in Figure 12.29(b)). Changing strain rate will produce a shift in these curves. It is then argued that whichever process, either fracture or yield, can occur at the lower stress will be the operative one. Thus the intersection of the σ_B/σ_Y curves defines the brittle–ductile transition and the material is ductile at all temperatures above this point.

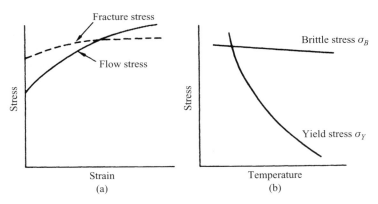

Figure 12.29 Diagrams illustrating the Ludwig–Davidenkov–Orowan theories of brittle–ductile transitions

The influence of chemical and physical structure on the brittle–ductile transition can be analysed by considering how these factors affect the brittle stress curve and the yield stress curve, respectively. As will be appreciated, this approach bypasses the relevance of fracture mechanics to brittle failure. If, however, we consider fracture initiation (as distinct from propagation of a crack) as governed by a fracture stress σ_B, the concept of regarding yield and fracture as competitive processes provides a useful starting point.

Vincent and others [96–98] have shown that the brittle stress is not much affected by strain and temperature (e.g. by a factor of 2 in the temperature range -180 to $+20\,°C$). The yield stress, on the other hand, is greatly affected by strain rate and temperature, increasing with increasing strain rate and decreasing with increasing temperature. (A typical figure would be a factor of 10 over the temperature range -180 to $+20\,°C$.) These ideas are clearly illustrated by results for PMMA shown in Figure 12.30(a). The brittle–ductile transition will therefore be expected to move to higher temperatures with increasing strain rate (Figure 12.30(b)). The effect can be illustrated by varying the strain rate in a tensile test on a sample of nylon at room temperature: at low strain rates the sample is ductile and cold-draws, whereas at high strain rates it fractures in a brittle manner.

A further complication in varying strain rate occurs at low speeds, where within a certain temperature range cold-drawing occurs. It is possible that at high speeds the heat is not conducted away rapidly enough, so that strain hardening is prevented and the specimen fails in a ductile manner. Such an isothermal–adiabatic transition does not affect the yield stress and therefore does not affect the brittle–ductile transition; but it does cause a considerable reduction in the energy to break and may be operative in impact tests, even if brittle fracture does not intervene. It has been proposed therefore that there are two critical velocities at which the fracture energy drops sharply as the strain rate is increased: the isothermal–adiabatic transition and, at higher strain rates, the brittle–ductile transition. Changes in ambient temperature have very little effect on the position of the isothermal–adiabatic transition but have a large effect on the brittle–ductile transition.

It was thought at first that the brittle–ductile transition was related to mechanical relaxation and in particular to the glass transition, which is true for natural rubber, polyisobutylene and polystyrene but is not the case for most thermoplastics. It was then proposed [99] that where there is more than one mechanical relaxation the brittle–ductile transition may be associated with a lower temperature relaxation. Although again it appeared that there might be cases where this is correct, it was soon shown that this hypothesis has no general validity. Because the brittle–ductile transition occurs at fairly high strains, whereas the dynamic mechanical behaviour is measured in the linear, low strain region, it is unreasonable to expect that the two can be linked directly. It is certain that fracture, for example, depends on several other factors such as the presence of flaws, which will not affect the low-strain dynamic mechanical behaviour. The subject has been discussed extensively by Boyer [100] and by Heijboer [101].

12.8.2 Notch sensitivity and Vincent's σ_B–σ_Y diagram

As for metals the presence of a sharp notch can change the fracture of a polymer from ductile to brittle. For this reason a standard impact test for a polymer is the

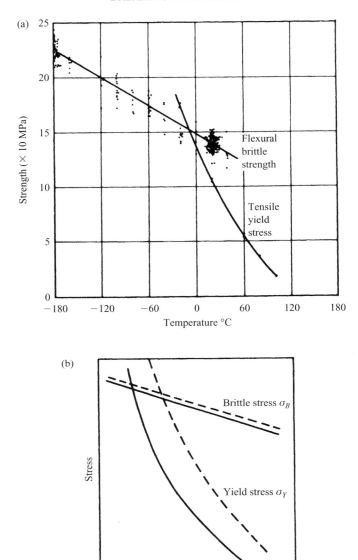

Figure 12.30 (a) Effect of temperature on brittle strength and tensile yield stress of PMMA. (Reproduced with permission from Vincent. *Plastics*, **26**, 141 (1961).) (b) Diagram illustrating the effect of strain rate on the brittle–ductile transition: (—) low strain rate; (- - -) high strain rate

Charpy or Izod test, where a notched bar of polymer is struck by a pendulum and the energy dissipated in fracture is calculated.

A very simple explanation of the effect of notching has been given by Orowan [95]. For a deep, symmetrical tensile notch, the distribution of stress is identical to that for a flat frictionless punch indenting a plate under conditions of plane strain [102] (Figure 12.31). The compressive stress on the punch required to produce plastic deformation can be shown to be $(2 + \pi)K$, where K is the shear yield stress. For the Tresca yield criterion the value is $2.57\sigma_Y$ and for the von Mises yield criterion the value is $2.82\sigma_Y$, where σ_Y is the tensile yield stress. Hence for an ideally deep and sharp notch in an infinite solid the plastic constraint raises the yield stress to a value of approximately $3\sigma_Y$ which leads to the following classification for brittle–ductile behaviour first proposed by Orowan [95]:

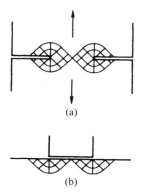

(a)

(b)

Figure 12.31   The slip-line field for a deep symmetrical notch (a) is identical to that for the frictionless punch indenting a plate under conditions of plane strain (b). (Reproduced with permission from Cottrell, *The Mechanical Properties of Matter*, Wiley, New York, 1964)

1.   If $\sigma_B < \sigma_Y$, the material is brittle.

2.   If $\sigma_Y < \sigma_B < 3\sigma_Y$, the material is ductile in an unnotched tensile test but brittle when a sharp notch is introduced.

3.   If $\sigma_B < 3\sigma_Y$, the material is fully ductile, i.e. ductile in all tests, including those in notched specimens.

Vincent σ_B–σ_Y diagram

We may ask how relevant the above ideas are to the known behaviour of polymers. Vincent [103] has constructed a σ_B–σ_Y diagram that is very instructive in this regard (Figure 12.32).

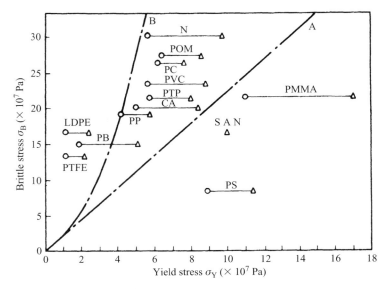

Figure 12.32   Plot of brittle stress at about $-180\,^{\circ}\text{C}$ against a line joining yield-stress values at $-20\,^{\circ}\text{C}$ ($\circ$), respectively, for various polymers. Line A divides polymers that are brittle unnotched from those that are ductile unnotched but brittle notched, and line B divides polymers that are brittle notched but ductile unnotched from those that are ductile even when notched. PMMA, poly(methyl methacrylate); PVC, poly(vinyl chloride); PS, polystyrene; PET, poly(ethylene terephthalate); SAN, copolymer of styrene and acrylonitrile; CA, cellulose acetate; PP, polypropylene; N, nylon 6:6; LDPE, low-density polyethylene; POM, polyoxymethylene; PB, polybutene-1; PC, polycarbonate; PTFE, polytetrafluoroethylene. (Reproduced with permission from Vincent, *Plastics*, **29**, 79 (1964))

Where possible, the value of σ_Y was taken as the yield stress in a tensile test at a strain rate of about 50 per cent per minute; for polymers that were brittle in tension, σ_Y was the yield stress in uniaxial compression and σ_B was the fracture strength measured in flexure at a strain rate of $18\ \text{min}^{-1}$ at $-180\,^{\circ}\text{C}$.

The yield stresses were measured at $+20$ and $-20\,^{\circ}\text{C}$, the idea being that the $-20\,^{\circ}\text{C}$ values would give a rough indication of the behaviour in impact at $+20\,^{\circ}\text{C}$, i.e. lowering the temperature by $40\,^{\circ}\text{C}$ is assumed to be equivalent to increasing the strain rate by a factor of about 10^5.

In the diagram the circles represent σ_B and σ_Y at $+20\,^{\circ}\text{C}$; the triangles represent σ_B and σ_Y at $-20\,^{\circ}\text{C}$. Both σ_Y and σ_B are affected by subsidiary factors such as molecular mass and the degree of crystallinity so that each point can be regarded only as of first-order significance.

From the known behaviour of the 13 polymers shown in this diagram, two characteristic lines can be drawn. Line A divides the brittle materials on the right, which are brittle when notched, from those on the left, which are ductile even when notched. Both of these lines are approximations, but they do summarize the existing knowledge.

For line A the ratio $\sigma_B/\sigma_Y \sim 2$ rather than unity, but the difference may be accounted for by the measurement of σ_B at very low temperatures and possibly by the measurement of σ_B in flexure rather than in tension. (The latter may reduce the possibility of fracture at serious flaws in the surface.) It is encouraging that even an approximate relationship holds along the lines of the Ludwig–Davidenkov–Orowan hypothesis. Even more encouraging is the fact that line B has a slope $\sigma_B/\sigma_Y \sim 6$, which is three times that of A, as expected on the basis of the plastic constraint theory.

The principal value of the σ_B–σ_Y diagram is that it may guide the development of modified polymers or new polymers. Together with the ideas of the previous section on the influence of material variables on the brittle stength and yield stress, it can lead to a systematic search for improvements in toughness.

12.9 The impact strength of polymers

The ability of a structural part to maintain its integrity and to absorb a sudden impact is often a relevant issue when selecting a suitable material. Impact testing of polymers is thus a subject of some importance and is extensively employed, although many of the results obtained are of an empirical and hence comparative nature.

The two major types of impact test are categorized as flexed beam and falling weight.

12.9.1 Flexed-beam impact

Examples of flexed-beam impact are the Izod and Charpy impact test, in which a small bar of polymer is struck with a heavy pendulum. In the Izod test the bar is held vertically by gripping one end in a vice and the other free end is struck by the pendulum. In the Charpy test the bar is supported near its ends in a horizontal plane and struck either by a single-pronged or two-pronged hammer so as to simulate a rapid three-point or four-point bend test, respectively (Figure 12.33(a)), It is customary to introduce a centre notch into the specimen so as to add to the severity of the test, as discussed in Section 12.5.1 above. The standard Charpy impact specimen has a 90° V-notch with a tip radius of 0.25 mm. For polymers a very much sharper notch is often adopted by tapping a razor blade into a machined crack tip, which has important consequences for interpretation of the subsequent impact test.

The interpretation of impact tests is not straightforward and it is necessary to consider several alternatives, as follows:

1. It was proposed independently by Brown [104], and by Marshall, Williams and Turner [105], that Charpy impact tests on sharply notched specimens can

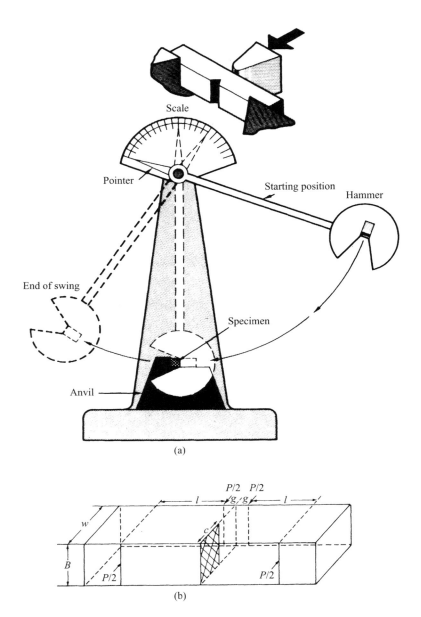

(a)

(b)

Figure 12.33 (a) Schematic drawing of a Charpy impact tester. (b) The notched Charpy impact specimen

be analysed quantitatively in terms of linear elastic fracture mechanics. It is assumed that the polymer deforms in a linear elastic fashion up to the point of failure, which occurs when the change in stored elastic energy due to crack growth satisfies the Irwin-Kies relationship (Equation (12.10) above). So that

$$G_c = \frac{K_c^2}{E^*} = \frac{P_0^2}{2B} \frac{dC}{dc}$$

where P_0 is the load immediately prior to fracture, C and B are the specimen compliance and thickness, respectively, and c is the crack length as in Equation (12.12a) in Section 12.2.3. Because the elastically stored energy in the specimens immediately prior to failure is $U_0 = P_0^2 C/2$,

$$G_c = \frac{U_0}{B} \frac{1}{C} \frac{dC}{dc} \qquad (12.33)$$

where U_0 is determined in a commercial impact tester from the potential energy lost due to impact. The total measured impact energy U_I must be reduced by the kinetic energy of the sample U_k to give $U_0 = U_I - U_k$.

It is conventional to follow Williams and co-workers [106] and express Equation (12.33) as

$$G_c = \frac{U_0}{BW} \frac{1}{C} \frac{dC}{d(c/W)} \qquad (12.34)$$

We then have

$$U_I = BW\phi(c/W)G_c + U_k \qquad (12.35)$$

where $\phi(c/W) = \frac{c}{dC/d(c/W)}$ can be calculated (see for example Williams [7], Chapter 4), and a plot of U_I versus $BW\phi$ produces a straight line with G_0 as slope and U_k as the intercept. This approach has been shown to give values for G_c that are independent of specimen geometry for impact tests on razor-notched samples of several glassy polymers, including PMMA, polycarbonate [107] and poly(ether sulphone) [108]. Similar results have been obtained also for razor-notched samples of polyethylene [109].

2. Vincent [110] and others have recognized that the impact strength depends on the geometry of the notch, which led Fraser and Ward [111] to propose that for comparatively blunt notches (i.e. those not introduced by a razor blade or a sharp cutting tool) failure occurs when the stress at the root of the notch reaches a critical value. This stress, which in a glassy polymer marks the stress required to initiate a craze, can be calculated by assuming that the deformation is elastic. On this hypothesis, the Charpy test, as undertaken in

the Hounsfield impact tester, can be regarded as a four-point bend test with the bending moment $M = Pl/2$, where P is the applied load and l is a sample dimension (Figure 12.33(b)). Immediately prior to fracture, $M = M_0$, $P = P_0$ and the elastically stored energy is $U_0 = \frac{1}{2}(2M_0/l)^2 C$, where C is the sample compliance. Hence

$$M_0 = \frac{l}{2}\sqrt{\frac{2U_0}{C}}$$

where C is calculable from specimen geometry.

For pure bending, the nominal stress at the root of the notch σ_n is given by $\sigma_n = (M/I)y$, where I is the second moment of area ($= Bt^3/12$ for a rectangular beam) and y is the distance to the neutral axis.

Using the linear stress assumption the maximum stress at the root of the notch is the product of the nominal stress and the stress concentration factor α_k. Calculations of α_k for general shapes of notch are available in the literature, but when the crack length c is much greater than the notch tip radius ρ, α_k reduces to the simple expression $\alpha_k = 2\sqrt{c/\rho}$.

It has been shown that the impact behaviour of blunt notched specimens of PMMA is consistent with a critical stress at the root of the notch [111], and similar considerations apply to polycarbonate [107] and poly(ether sulphone) [108] in the absence of shear lips. In these instances it appears therefore that the maximum local stress is the fracture criterion, independent of specimen geometry.

3. The most unsophisticated interpretation of the flexed bend impact test is that it is a measure of the energy required to propagate the crack across the specimen, irrespective of whether the specimen is notched or unnotched. Notch sensitivity is ignored and only the energy of propagation is involved. In this case

$$G_c = \frac{U_0}{A} = \frac{U_0}{BW(1 - c/W)} \qquad (12.36)$$

where the area of the uncracked cross-section is $A = B(W - c)$.

Justification for this approach has been given by Plati and Williams [106], where

$$J_c = \sigma_y u \qquad (12.37)$$

If full yielding is assumed in bending impact

$$U_0 = \frac{u}{2}\sigma_y B(W - c) = \frac{J_c B(W - c)}{2}$$

Because the ligament area $A = B(W - c)$, then

$$J_c = \frac{2U_0}{A}$$

This differs only by a factor of 2 from Equation (12.36), which arises because the average displacement in bending is $u/2$, compared with u in tension. Plati and Williams showed results from high-impact polystyrene (HIPS) and acrylonitrile–butadiene–styrene (ABS) polymers that agreed with values of J_c from impact tests.

More recent work has extended the J-integral and essential work of fracture methods described in Section 12.6 above to impact tests.

Bramuzzo [112] used high-speed photography to monitor the crack propagation in a three-point bend test in parallel with determining the force–time curve. In this way resistance curves were obtained for polypropylene copolymers by plotting the J-integral as fraction of crack length. Martinatti and Riccio [113] used the multispecimen technique to determine the J_R curves for rubber-toughened polypropylenes using an instrumented Charpy test where the hammer of the pendulum could be stopped at different displacements of the specimen. The crack advancement of each loading was measured after successive fractures at low temperatures using an optical microscope. Crouch and Huang [114] produced multispecimen resistance curves for toughened nylon by impacting SENT three-point bend specimens to different levels of crack growth using a falling-weight impact tower. Force–time curves were determined to obtain the total energy up to maximum deflection.

In further recent work, Ramsteiner [115] determined $J_{0.2}$ values by constructing the J values as a function of crack length, impacting specimens of HDPE with different masses from the same height to give a constant impact velocity of 2 m s^{-1}. Finally, Fasce and co-workers [116] have attempted to apply the essential work of fracture methodology to impact testing of two PP copolymers and ABS using pre-cracked specimens of different notch deeply double-edge-notched tension (DENT) and single-edged-notched bend (SENB) depth. In both cases the specific total work of fracture was plotted against ligament length (Figure 12.34).

12.9.2 Falling-weight impact

In the falling-weight impact test a circular disc of material (typically 6 cm diameter and 2 mm thickness, freely supported on an annulus of 4 cm diameter) is impacted by a metal dart with a hemispherical tip (typically of radius 1 cm). The tests are carried out either under conditions where the impact energy is far in excess of that required to break the specimen or at low levels of impact energy so that damage tolerance and the possible initiation of a crack can be observed.

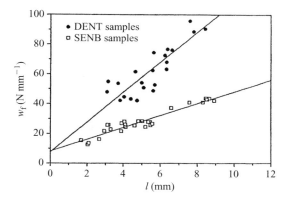

Figure 12.34 Specific total work of fracture w_f versus ligament length l for deeply double-edge-notched tension (DENT) (●) and single-edge-notched bend (SENB) (□) 3 mm thick polypropylene random copolymer samples. (Reproduced with permission from Fasce, Bernal, Frontini, *et al.*, *Polym. Eng. Sci.*, **41**, 1 (2001))

Moore and his colleagues have described the application of such tests to polymers and to polymer composites [117–119]. It is emphasized that for any reasonable attempt at interpretation the following must be carried out:

1. Measurement of the force–time curve so that the input energy to maximum force can be determined, as well as the total impact energy.

2. Photography of the tension surface during the impact event.

For fibre composites Moore and colleagues showed that the peak on the force–time curve corresponds well with the energy required to initiate a crack. It was shown also that for both composites and polymers the total fracture energy corresponded quite well with that detemined from notched Charpy tests.

Only for the Charpy test, and to a rather lesser extent the Izod test, has a satisfactory theoretical analysis been achieved. Even for these tests, however, there is still a gap between the engineering analysis and any accepted interpretation in physical terms. For example, although it seems likely that the brittle failure of razor-notched impact specimens is associated with the craze at the crack tip, there is no convincing numerical link between craze parameters and the fracture toughness K_{IC}, as exists for the cleavage fracture of compact tension specimens (see Section 12.2 above). Again, although the mechanics point to a critical stress criterion for some blunt notched specimens and there is an empirical correlation with the craze stress determined in other ways, the magnitude of the critical stress is very great and suggests that a more sophisticated explanation may be required. For the brittle epoxy resins, which do not show a craze at the crack tip, Kinloch and Williams [120] have suggested that the fracture of both razor-notched and blunt-notched specimens can be described by a critical stress at a critical distance ($\sim$10 μm) below the root of the notch.

As the temperature and strain rate in a polymer change, the nature of the stress–strain curve can alter remarkably. It is therefore natural to seek correlations between the area beneath the stress–strain curve and the impact strength, and between dynamic mechanical behaviour and the impact strength. Attempts to make such correlations directly have met with mixed success [121], which is not surprising in view of the complex quantitative interpretations of impact strength suggested above.

Vincent [122] has examined the statistical significance of a possible inverse correlation between impact strength and dynamic modulus and concluded that, at best, this correlation only accounts for about two-thirds of the variance in impact strength. Factors such as the influence of molecular mass, and details of molecular structure such as the presence of bulky side groups, are not accounted for. He also reported impact tests over a wide temperature range on some polymers, notably polytetrafluorethylene and polysulphone, where peaks in brittle impact strength were observed at temperatures close to dynamic loss peaks, suggesting that in some instances it may be necessary to consider the relevance of a more general-ized form of fracture mechanics [123], where the viscoelastic losses occurring during loading and unloading must be taken into account.

12.9.3 Toughened polymers: high-impact polyblends

The comparatively low impact strength of many well-known polymers, such as PMMA, polystyrene and PVC, led to the production of rubber-modified thermo-plastics with high impact strength. The best known examples are high-impact polystyrene (HIPS) and ABS copolymer, where the rubbery phase is dispersed throughout the polymer in the form of small aggregates or balls. Othe polymers that have been toughened in this way include PMMA, PVC, polypropylene, polycarbonate, nylons and thermosets such as epoxies, polyesters and polyimides.

In an excellent review Bucknall [124] explains that rubber toughening involves three principal deformation mechanisms: shear yielding, crazing and rubber particle cavitation. The rubber particles, with a much lower stiffness than the matrix polymer, give rise to stress concentrations for the initiation of shear yielding and crazing.

Nielson [125] lists three conditions that are required for an effective polyblend:

1. The glass temperature of the rubber must be well below the test temperature.

2. The rubber must form a second phase and not be soluble in the rigid polymer.

3. The two polymers should be similar enough in solubility behaviour for good adhesion between the phases.

In rubber-toughened ABS, shear yielding is dominant. Optical microscopy examination by Newman and Strella [126] showed that plastic deformation had occurred in the matrix around the rubber particles. Later studies, notably by Kramer and co-workers, suggested that the rubber particles initiate microshear bands. Donald and Kramer [127] showed that cavitation in the rubber particles initiates shear yielding of the matrix and that shear deformation occurs when the particles are small, and crazing when the particles are large.

In rubber-toughened HIPS, Bucknall and Smith [128] showed that the improved toughness was related to crazing and stress whitening. The crazes are initiated at points of maximum triaxial stress concentration produced by incorporation of the rubber particles. The rubber particles also act as craze terminators so that a large number of small crazes is produced to give high energy absorption and extensive stress whitening. Work by Yang and Bucknall [129] suggests that cavitation in the rubber particles precedes crazing.

Bucknall [130] and Bucknall and Smith compared the force–time curves for impact specimens over a range of temperatures, with both the notched Izod impact strength and the falling-weight impact strength and the nature of the fracture surface. The force–time curves, such as in Figure 13.35(a), show regions similar to those observed for a homopolymer as discussed in the introduction above. Both impact strength tests also showed three regions (Figure 12.35(b) and (c)). The fracture surfaces at the lowest temperature were quite clear, whereas at high temperatures stress whitening or craze formation occurred. Three temperature regions were considered:

1. *Low temperature*. The rubber is unable to relax at any stage of fracture. There is no craze formation and brittle fracture occurs.

2. *Intermediate temperature*. The rubber is able to relax during the relatively slow build-up of stress at the base of the notch, but not during the fast crack propagation stage. Stress whitening occurs only in the first (precrack) stage of fracture and is therefore confined to the region near the notch.

3. *High temperature*. The rubber is able to relax even in the rapidly forming stress field ahead of the travelling crack. Stress whitening occurs over the whole of the fracture surface. Bucknall and Smith [128] report similar results for other rubber-modified impact polymers.

12.9.4   Crazing and stress whitening

Bucknall and Smith [128] remarked on the connection between crazing and stress whitening. It was observed that the fracture of high-impact polystyrene, which incorporates rubber particles into the polystyrene, is usually preceded by opaque whitening of the stress area. Figure 12.11 shows a stress-whitened

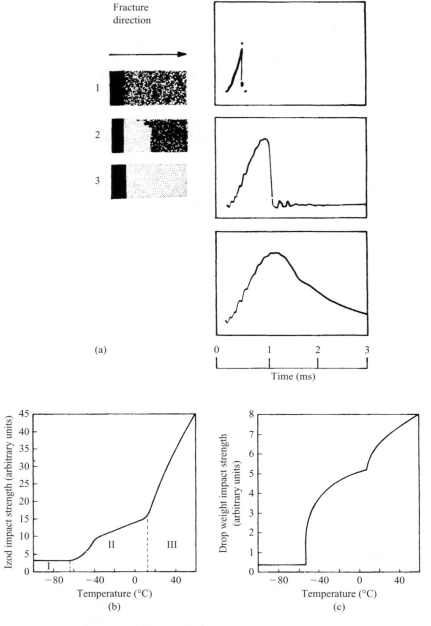

Figure 12.35 (a) Fracture surfaces of modified polystyrene notched Izod impact specimens: top, broken at −70 °C, type I fracture; centre, broken at 40 °C, type II fracture; bottom, broken at 150 °C, type III fracture. (b) Notched Izod impact strength of modified polystyrene as a funtion of temperature, showing the limits of the three types of fracture behaviour. (c) Falling-weight impact strength of 2.03 mm high-impact polystyrene sheet as a function of temperature. (Reproduced with permission from Bucknall, *Br. Plast.*, **40**, 84 (1967))

bar of high-impact polystyrene that failed at an elongation of 35 per cent. A combination of different types of optical measurements (polarized light to measure molecular orientation and phase contrast microscopy to determine refractive index) showed that these stress-whitened regions are similar to the crazes formed in unmodified polystyrene. They are birefringent, of low refractive index, capable of bearing load and are healed by annealing treatements. The difference between stress whitening and crazing exists merely in the size and concentration of the craze bands, which are of much smaller size and greater quantity in stress whitening. Thus the higher conversion of the polymer into crazes accounts for the high breaking elongation of toughened polystyrene. It is suggested that the effect of the rubber particles is to lower the craze initiation stress relative to the fracture stress, thereby prolonging the crazing stage of deformation. The crazing stage appears to require the relaxation of the rubber phase, so that it behaves like a rubber and not a glass. The function of the rubber particles is not, however, merely to provide points of stress concentration, and there must be a good bond between the rubber and polystyrene, which is achieved by chemical grafting. The rubber must bear part of the load at the stage when the polymer has crazed but not fractured. Bucknall and Smith suggested that the rubber particles may be constrained by the surrounding polystyrene matrix so that their stiffness remains high. These ideas lead directly to an explanation of the three regimes for impact testing, as discussed above. At low temperatures there is no stress whitening because the rubber does not relax during, the fracture process, giving low impact strengths. At intermediate temperatures, stress whitening occurs near the notch, where the crack initiates and is travelling sufficiently slowly compared with the relaxation of the rubber. Here the impact strength increases. Finally, at high temperatures, stress whitening is observed along the whole of the crack and the impact strength is high.

12.9.5 Dilatation bands

Lazzeri and Bucknall [131] have proposed that the pressure dependence of yield behaviour caused by the presence of microvoids can explain the observation of dilatation bands in rubber-toughened epoxy resins [132], rubber-toughened polycarbonate [133] and styrene–butadiene diblock copolymers [134]. These dilatation bands combine in-plane shear with dilatation normal to the shear plane. Whereas true crazes contain interconnecting strands, as described in Section 12.5.1 above, dilatation bands contain discrete voids that, for rubber-toughened polymers, are confined to the rubber phase.

12.10   The tensile strength and tearing of polymers in the rubbery state

12.10.1   The tearing of rubbers: extension of Griffith theory

The Griffith theory of fracture implies that the quasi-static propagation of a crack is a reversible process. Rivlin and Thomas [135, 136] recognized, however, that this may be unnecessarily restrictive, and also that the reduction in elastically stored energy due to the crack propagation may be balanced by changes in energy other than that due to an increase in surface energy. Their approach was to define a quantity termed the 'tearing energy', which is the energy expended per unit thickness per unit increase in crack length. The tearing energy includes surface energy, energy dissipated in plastic flow processes and energy dissipated irreversibly in viscoelastic processes. Provided that all these changes in energy are proportional to the increase in crack length and are primarily determined by the state of deformation in the neighbourhood of the tip of the crack, then the total energy will still be independent of the shape of the test piece and the manner in which the deforming forces is applied.

In formal mathematical terms, if the crack increases in length by an amount $\mathrm{d}c$, an amount of work $TB\,\mathrm{d}c$ must be done, where T is the tearing energy per unit area and B is the thickness of the sheet. Assuming that no external work is due, this can be equated to the change in elastically stored energy, giving

$$-\left[\frac{\partial U}{\partial c}\right]_l = TB \tag{12.38}$$

The suffix l indicates that differentiation is carried out under conditions of constant displacement of the parts of the boundary that are not force-free. Equation (12.38) is similar in form to Equation (12.1) above but T is defined for unit thickness of specimen and is therefore equivalent to 2γ in Equation (12.1). As in the case of glassy polymers, T is not to be interpreted as a surface free energy, but involves the total deformation in the crack tip region as the crack propagates.

The so-called 'trouser tear' experiment shown in Figure 12.36 is a particularly

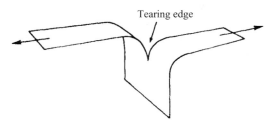

Tearing edge

Figure 12.36   The standard 'trouser tear' experiment

simple case where the equation can be evaluated immediately. After making a uniform cut in a rubber sheet the sample is subjected to tear under the applied forces P. The stress distribution at the tip of the tear is complex, but provided that the legs are long it is independent of the depth of the tear.

If the sample tears a distance Δc under the force F, and changes in extension of the material between the tip of the tear and the legs are ignored, the work done is given by $\Delta W = 2F\Delta c$.

Because the tearing energy $T = \Delta W / B\Delta c$, $T = 2F/B$ and can be measured easily.

Rivlin and Thomas [135] found that two characteristic tearing energies could be defined, one for very slow rates of tearing ($T = 37 \text{ kJ m}^{-2}$) and one for catastrophic growth ($T = 130 \text{ kJ m}^{-2}$), and that both of these quantities were independent of the shape of the test piece.

The tearing energy is the energy required to extend the rubber to its maximum elongation and does not relate directly to tensile strength but depends on the shape of the stress–strain curve together with the viscoelastic nature of the rubber. For example, we may contrast two different rubbers, the first possessing a high tensile strength but a very low elongation to fracture and very low viscoelastic losses, and the second possessing a low tensile strength but a high elongation to fracture and high viscoelastic losses. In spite of its comparatively low tensile strength the second rubber may still possess a high tearing energy.

12.10.2 Molecular theories of the tensile strength of rubbers

Most molecular theories of the strength of rubber treat rupture as a critical stress phenomenon. It is accepted that the strength of the rubber is reduced from its theoretical strength in a perfect sample by the presence of flaws. Moreover, it is assumed that the strength is reduced from that of a flawless sample by approximately the same factor for different rubbers of the same basic chemical composition. It is then possible to consider the influence on the strength of such factors as the degree of cross-linking and the primary molecular mass.

Bueche [137] has considered the tensile strength of a model network consisting of a three-dimensional net of cross-linked chains. Figure 12.37 illustrates a unit cube whose edges are parallel to the three chain directions in the idealized network. Assume that there are N chains in this unit cube and that the number of chains in each strand of the network is n. There are then n^2 strands passing through each face of the cube. To relate the number n to the number of chains per unit volume of the network (and so form a link with rubber elasticity theory) we note that the product of the number of strands passing through each cube face and the number of chains in each strand will be $\frac{1}{3}N$ because there are three strand directions. Thus

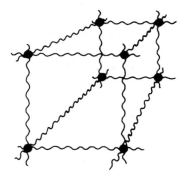

Figure 12.37   Model network of cross-linked chains

$$n^3 = \tfrac{1}{3}N, \qquad n = (N/3)^{1/3} \tag{12.39}$$

Apply a stress σ parallel to one of the three strand directions and assume that the strands break simultaneously at an individual fracture stress σ_c. Then

$$\sigma_B = n^2 \sigma_c$$

which from Equation (12.39) can be written as

$$\sigma_B = (N/3)^{2/3} \sigma_c$$

For a real network, N is the number of effective chains per unit volume and is given in terms of the actual number of chains per unit volume N_a by the Flory relationship

$$N = N_a[1 - 2\overline{M}_c/M_n]$$

where $\overline{M}_c$ and $\overline{M}_n$ are the average molecular mass between cross-links and the number average molecular mass of the polymer, respectively. (Note that for a network there must be at least two cross-links per chain, i.e. $\overline{M}_n > 3\overline{M}_c$.)

    This substitution gives

$$\sigma_B \alpha [1 - 2\overline{M}_c/\overline{M}_n]^{2/3} \sigma_c$$

Flory [138] found that the variation of tensile strength with the polymer molecular mass $\overline{M}_n$, for butyl rubber, follows the predicted $[1 - 2\overline{M}_c/\overline{M}_n]^{2/3}$ relationship, but for natural rubber [139] an initial increase in tensile strength with increasing degree of cross-linking was followed by a decrease at very high degrees of cross-linking. Flory attributed this decrease to the influence of cross-links in the crystallization of the rubber. However, a similar effect was observed for the non-crystallizing styrene–butadiene rubber by Taylor and Darin [140], which led Bueche [141] to propose that the simple model described above fails because of the assumption that each chain holds the load at fracture, which may be a good approximation at low degrees of cross-linking but is less probable at high degrees of cross-linking.

It is of considerable technological importance that the tensile strength of rubbers can be much increased by the inclusion of reinforcing fillers such as carbon black and silicone, which increase the tensile strength by allowing the applied load to be shared among a group of chains, thus decreasing the chance that a break will propagate [142].

12.11 Effect of strain rate and temperature

The influence of strain rate and temperature on the tensile properties of elastomers and amorphous polymers has been studied extensively, particularly by Smith and co-workers [143–145], who measured the variation of tensile strength and ultimate stain as a function of strain rate for a number of elastomers. The results for different temperatures could be superimposed, by shifts along the strain rate axis, to give master curves for tensile strength and ultimate strain as a function of strain rate. Results of this nature are shown in Figure 12.38, which summarizes Smith's data for an unfilled styrene–butadiene rubber. Remarkably, the shift factors obtained from superposition of both tensile strength and ultimate strain took the form predicted by the WLF equation (see Section 6.3.2) for the superposition of low-strain linear viscoelastic behaviour of amorphous polymers (Figure 12.39). The actual value for T_g agreed well with that obained from dilatometric measurements.

This result suggests that, except at very low strain rates and high temperatures where the molecular chains have complete mobility, the fracture process is dominated by viscoelastic effects. Bueche [146] has treated this problem theoretically and obtained the observed form of the dependence of tensile strength on strain rate and temperature. Later theories have attempted to obtain the time dependence for both tensile strength and ultimate strain, or the time to break at a constant strain rate [147, 148].

Smith plotted $\log \sigma_B / T$ against $\log e$ for the above and similar data to obtain a unique curve for all strain rates and test temperatures, which he termed the 'failure envelope' for elastomers. It was also found [145] that the failure envelope can represent failure under more complex conditions such as creep and stress relaxation. In Figure 12.40 such failure can take place by starting from the initial stage G and progressing parallel to the abscissa (constant stress, i.e. creep) or parallel to the ordinate (constant strain, i.e. stress relaxation) until a point is reached on the failure envelope ABC, as indicated by the progress along the dotted lines.

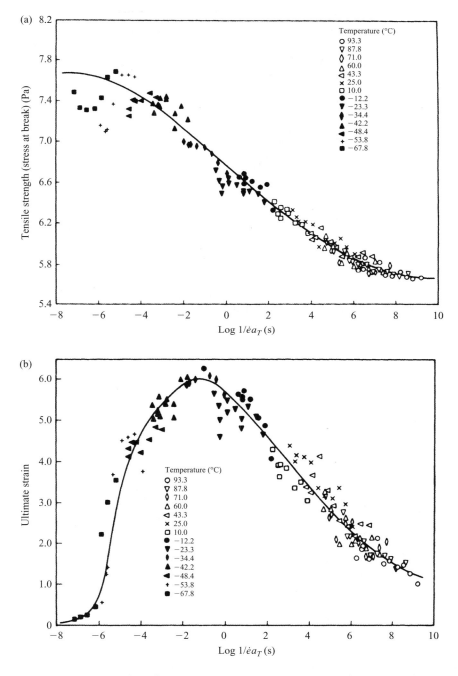

Figure 12.38 Variation of tensile strength (a) and ultimate strain (b) of a rubber with reduced strain rate $\dot{e}a_T$. Values were measured at various temperatures and rates and reduced to a temperature of 263 K. (Reproduced with permission from Smith, *J. Polym. Sci.*, **32** 99 (1958))

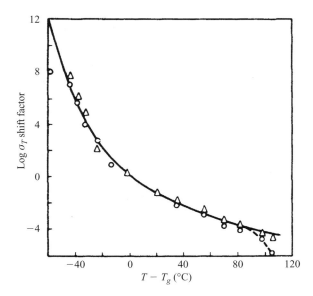

Figure 12.39 Experimental values of $\log a_T$ shift factor obtained from measurement of ultimate properties compared with those predicted using the WLF equation: ($\triangle$) from tensile strength; ($\circ$) from ultimate strain; (——) WLF equations with $T_g = 263$ K (Reproduced with permission from Smith, *J. Polym. Sci.*, **32**, 99 (1958))

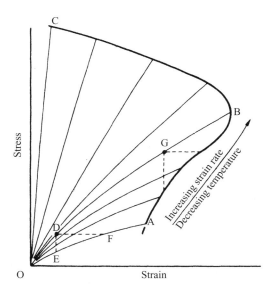

Figure 12.40 Schematic representation of the variation of stress–strain curves with the strain rate and temperature. Envelope connects rupture point and the dotted lines illustrate stress relaxation and creep under different conditions (Reproduced with permission from Smith and Stedry, *J. Appl. Phys.*, **31**, 1892 (1960))

12.12 Fatigue in polymers

Materials frequently fail by fatigue due to the cyclic application of stress below that required to cause yield or fracture when a continuously rising stress is applied. The effect of such cyclic stresses is to initiate microscopic cracks at centres of stress concentration within the material or on the surface, and subsequently to enable these cracks to propagate, leading to eventual failure.

Early studies of fatigue in polymers concentrate on stress cycling of unnotched samples, to produce S versus N plots similar to those that have proved so useful for characterizing fatigue in metals (S being the maximum loading stress and N the number of cycles to failure). An example of this type of plot for PVC [149] is shown in Figure 12.41. A major aspect of such a test is the question of adiabatic heating, which can lead to failure by thermal melting. Clearly there will be a critical frequency above which thermal effects become important.

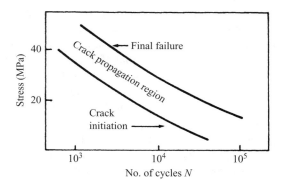

Figure 12.41   Fatigue response of PVC: relationship between applied stress σ and number of cycles to failure N, for both initiation of fatigue cracks and final failure. (Reproduced with permission from Manson and Hertzberg, *CRC Crit. Rev. Macromol Sci.*, **1**, 433 (1973))

Stress cycling tests in unnotched samples do not readily distinguish between crack initiation and crack propagation. Further progress requires a similar approach to that adopted in fracture studies, namely the introduction of very sharp initial cracks in order to examine crack propagation utilizing fracture mechanics concepts.

The first quantitative studies of fatigue in polmers, which concentrated on rubbers [150–152], applied the tearing energy concept of fracture proposed by Rivlin and Thomas to fatigue crack propagation. Thomas [150] showed that the fatigue crack growth rate could be expressed in the form of an empirical relationship

$$\frac{\mathrm{d}c}{\mathrm{d}N} = A\mathcal{J}^n \tag{12.40}$$

where c is the crack length, N is the number of cycles and $\mathcal{J}$ is the surface work parameter, which is analogous to the strain energy release rate G in linear elastic fracture mechanics. For a SENT specimen

$$\mathcal{J} = 2k_1 c U \qquad (12.41)$$

where $U = \sigma^2/2E$ is the stored energy density for a linear elastic material and k_1 is a constant that varies from π at small extensions (the linear elastic value) to approximately unity at large extensions [153]. Here A and n are constants that are dependent on the material and generally vary with test conditions such as temperature. The exponent n usually lies between 1 and 6 and for rubber is approximately 2 for anything other than very small dc/dN.

As expressed in Equation (12.40), $\mathcal{J}$ is essentially a positive quantity and can be considered to vary during the test cycle from zero ($\mathcal{J} = \mathcal{J}_{\min} = 0$) to a finite value ($\mathcal{J} = \mathcal{J}_{\max}$). It has been found that where $\mathcal{J}_{\min}$ is increased there is a corresponding decrease in A, which has been attributed to reduced crack propagation where strain-induced crystallization occurs. Furthermore, it has been shown that there is a fatigue limit $\mathcal{J} = \mathcal{J}_0$ below which a fatigue crack will not be propagated. Lake and Thomas showed that $\mathcal{J}_0$ corresponds to the minimum energy required per unit area to extend the rubber at the crack tip to its breaking point. Andrews [154] pointed out that initiation requires either intrinsic flaws of magnitude c_0 or that flaws of this size are produced during the test itself, with c_0 defined by Equation (12.41), where $\mathcal{J}_0 = k_1 c_0 U$. Andrews and Walker [155] carried this approach one stage further, incorporating a generalized form of fracture mechanics to analyse the fatigue behaviour of low-density polyethylene, which was viscoelastic in the range of interest so the more generalized fracture mechanics was required to deal with unloading as well as loading during crack propagation. The fatigue characteistics were predicted from the crack growth data using a single fitting constant, the intrinsic flaw size c_0, which it was suggested corresponded to the spherulite dimensions so that interspherulite boundary cracks constituted the the the intrinsic flaws.

For glassy polymers, fracture mechanics has been the usual starting point [156–159], with the fatigue crack growth rate usually expressed as an empirical relationship

$$\frac{dc}{dN} = A'(\Delta K)^m \qquad (12.42)$$

where c is the crack length, N is the number of cycles, ΔK is the range of the stress intensity factor (i.e. $K_{\max} - K_{\min}$, where $K_{\min}$ is generally zero) and A' and m are constants depending on the material and test conditions.

For $K_{\min} = 0$, Equation (12.42), is clearly identical in form to Equation (12.40), which is generally adopted for rubbers. Recall from Section 12.2.3 that the strain energy release rate $G = K^2/E$ for plane stress. Then

$$G = 2\mathcal{J} = K_{\text{max}}^2/2E = (\Delta K)^2/2E$$

and Equations (12.40) and (12.42) are formally equivalent if $m = 2n$.

Equation (12.42) is also the most general form of the law proposed by Paris [160, 161] for predicting fatigue crack growth rates in metals. The general situation for glassy polymers is illustrated in Figure 12.42(a), with some typical results shown in Figure 12.42(b). The data differ in two respects from the Paris equation: first, analogous to the case of rubbers, there is a distinct threshold value of ΔK, denoted by ΔK_{th}, below which no crack growth is observed; second, as ΔK approaches the critical stress intensity factor K_{c}, the crack accelerates. A further criticism of Equation (12.42) is that it allows for the influence of the stress intensity factor but not for the mean stress, which usually has an important influence on the crack growth rate. The latter consideration led Arad, Radon and Culver [162] to suggest an equation of the form

$$\frac{\mathrm{d}c}{\mathrm{d}N} = \beta\lambda^n \tag{12.43}$$

where $\lambda = (K_{\text{max}}^2 - K_{\text{min}}^2)$. This relation is equivalent to Equation (12.42) because the cycle strain energy release rate ΔG is given by

$$\Delta G = \frac{1}{E}(K_{\text{max}}^2 - K_{\text{min}}^2)$$

A comprehensive review of the application of the Paris equation and its modified form (Equation (12.43)) to the fatigue behaviour of polymers has been given by Manson and Hertzberg [149], who considered the effect of physical variables such as crystallinity and molecular mass. They noted a strong sensitivity of fatigue crack growth to molecular mass: in polystyrene a fivefold increase in molecular mass resulted in a more than tenfold increase in fatigue life. A general correlation was observed between the fracture toughness K_{c} and the fatigue behaviour, expressed as the stress intensity range ΔK corresponding to an arbitrary value of $\mathrm{d}c/\mathrm{d}N$ (chosen as 7.6×10^{-7} m cycle^{-1}), as is shown in Figure 12.43. A study of fatigue behaviour in polycarbonate by Pitman and Ward [163] also brought out the similarity between fatigue and fracture, so that the fatigue behaviour can be analysed in terms of mixed mode failure. Similar to the fracture behaviour described in Section 12.2, changing molecular mass again changed the balance between energy dissipated in propagating the craze and shear lips, respectively. A development by Williams [164, 165] attempts to model fatigue crack propagation behaviour in terms of the Dugdale plastic zone analysis of the crack tip. Each fatigue cycle is considered to reduce the craze stress in one part of craze, so that a two-stage plastic zone is established, leading to an equation for crack growth of the form

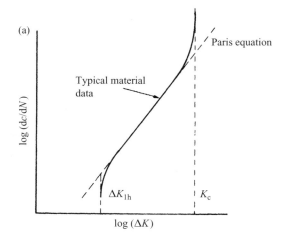

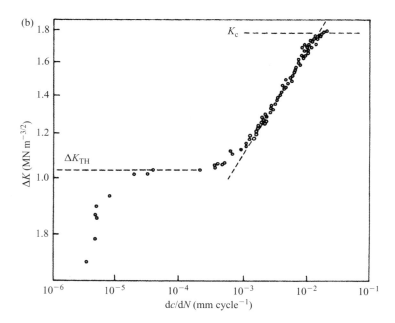

Figure 12.42 (a) Schematic diagram of fatigue crack growth rate dc/dN as a function of the range of stress intensity factor ΔK. (b) Fatigue crack growth characteristics for a vinyl urethane polymer. (Reproduced with permission from Harris and Ward, *J. Mater. Sci.*, **8**, 1655 (1973))

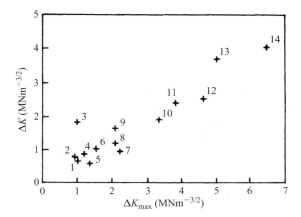

Figure 12.43 Relationship between the stress intensity range ΔK, corresponding to an arbitary value of dc/dN 7.6×10^{-7} m cycle^{-1}, and the maximum stress intensity factor range ΔK_{max} observed at failure for a group of polymers. The polymers are: (1) cross-linked polystyrene, (2) PMMA, (3) PVC, (4) LDPE, (5) polystyrene, (6) polysulphone, (7) high-impact polystyrene, (8) ABS resin, (9) chlorinated polyether, (10) poly(phenylene oxide), (11) nylon 6, (12) polycarbonate, (13) nylon 6:6, (14) poly(vinylidene fluoride) (Reproduced with permission from Manson and Hertzberg, *CRC Crit. Rev. Macromol. Sci.*, **1**, 433 (1973))

$$\frac{dc}{dN} = \beta'(K^2 - \alpha K_c^2) \tag{12.44}$$

which gives a good fit to experimental data for polystyrene over a substantial range of temperatures.

Both Williams and Pitman and Ward conclude that it is difficult to assign physical significance to the parameters in the Paris equation. Further developments in this area will require a more distinctly physical approach.

References

1. Griffith, A. A., *Philos. Trans. R. Soc.*, **221**, 163 (1921).
2. Inglis, G. E., *Trans. Inst. Nav. Architec.*, **55**, 219 (1913).
3. Irwin, G. R., *J. Appl. Mech.*, **24**, 361 (1957).
4. Irwin, G. R. and Kies, J. A., *Welding J. Res. Suppl.*, **33**, 1935 (1954).
5. Brown, W. F. and Srawley, J. F., *ASTM STP 410*, 1966.
6. Srawley, J. F. and Gross, B., *NASA Report E-3701*, 1967.
7. Williams, J. G., *Stress Analysis of Polymers* (2nd edn), Ellis Horwood, Chichester, 1980.
8. Benbow, J. J. and Roesler, F. C., *Proc. Phys. Soc.*, **B70**, 201 (1957).
9. Berry, J. P., *J. Appl. Phys.*, **34**, 62 (1963).
10. Benbow, J. J., *Proc. Phys. Soc.*, **78**, 970 (1961).
11. Svensson, N. L., *Proc. Phys. Soc.*, **77**, 876 (1961).

12. Berry, J. P., *J. Polym. Sci.*, **50**, 313 (1961).
13. Andrews, E. H., in *Proceedings of the Conference on the Physical Basis of Yield and Fracture*, Oxford, 1966, p. 127.
14. Berry, J. P., in *Fracture* (eds B. L. Auerbach *et al.*), Wiley, New York, 1959, p. 263.
15. Kambour, R. P., *Polymer*, **5**, 143 (1964).
16. Kambour, R. P., *J. Polym. Sci., A2*, **4**, 349, (1966).
17. Kambour, R. P., *Macromol. Rev.*, **7**, 1 (1973).
18. Brown, H. R. and Ward, I. M., *Polymer*, **14**, 469 (1973).
19. Doll, W. and Weidmann, G. W., *Colloid Polym. Sci.*, **254**, 205 (1976).
20. Dugdale, D. S., *J. Mech. Phys. Solids*, **8**, 100 (1960).
21. Rice, J. R., in *Fracture—An Advanced Treatise* (ed. H. Liebowitz.), Academic Press, New York, 1968, Ch. 3.
22. Pitman, G. L. and Ward, I. M., *Polymer*, **20**, 895 (1979).
23. Hine, P. J., Duckett, R. A. and Ward, I. M., *Polymer*, **22**, 1745 (1981).
24. Williams, J. G. and Parvin, M., *J. Mater. Sci.*, **10**, 1883 (1975).
25. Berry, J. P., *J. Polym. Sci., A2*, **2**, 4069 (1964).
26. Flory, P. J., *J. Am. Chem. Soc.*, **67**, 2048 (1945).
27. Doll, W. and Weidmann, G. W., *Prog. Colloid Polym. Sci.*, **66**, 291 (1979).
28. Kusy, R. P. and Turner, D. T., *Polymer*, **18**, 391 (1977).
29. Kusy, R. P. and Turner, D. T., *Polymer*, **17**, 161 (1976).
30. Haward, R. N., Daniels, H. E. and Treloar, L. R. G., *J. Polym. Sci. Polym. Phys. Ed.*, **16**, 1169 (1978).
31. Berry, J. P., in *Fracture*, (eds B. L. Averback *et al.*), Wiley, New York, 1959, p. 263.
32. Higuchi, M., *Rep. Res. Inst. Appl. Mech. Jpn.*, **6**, 173 (1959).
33. Kambour, R. P. and Holik, A. S., *J. Polym. Sci., A2*, **7**, 1393 (1969).
34. Kambour, R. P. and Russell, R. R., *Polymer*, **12**, 237 (1971).
35. Parades, E. and Fischer, E. W., *Makromol. Chem.*, **180**, 2707 (1979).
36. Brown, H. R. and Kramer, E. J., *J. Macromol. Sci. Phys.*, **B19**, 487 (1981).
37. Yang, A. C.-M. and Kramer, E. J., *J. Mater. Sci.*, **21**, 3601 (1986).
38. Berger, L. L., Buckley, D. J., Kramer, E. J., *et al.*, *J. Polym. Sci. Polym. Phys. Ed.*, **25**, 1679 (1987).
39. Berger, L. L., *Macromolecules*, **22** 3162 (1989).
40. Brown, H. R., *Macromolecules*, **24**, 2752 (1991).
41. Sha, Y., Hui, C. Y., Ruina, A., *et al.*, *Macromolecules*, **28**, 2450 (1995).
42. Donald, A. M. and Kramer, E. J., *Polymer*, **23**, 457 (1982).
43. Rietsch, F., Duckett, R. A. and Ward, I. M., *Polymer*, **20**, 2235 (1979).
44. Kausch, H. H., *Polymer Fracture* (2nd edn), Springer Verlag, Berlin, 1987.
45. Odell, J. A. and Keller, A., *J. Polym. Sci. Polym. Phys. Ed.*, **24**, 1889 (1986).
46. Sternstein, S. S., Ongchin, L. and Silverman, A., *Appl. Polym. Symp.*, **7**, 175 (1968).
47. Sternstein, S. S. and Ongchin, L., *Am. Chem. Soc. Polym. Prepr.*, **10**, 1117 (1969).
48. Bowden, P. B. and Oxborough, R. J., *Philos. Mag.*, **28**, 547 (1973).
49. Matsushige, K., Radcliffe, S. V. and Baer, E., *J. Mater. Sci.*, **10**, 833 (1974).
50. Duckett, R. A., Goswami, B. C., Smith, L. S. A., *et al.*, *Br. Polym. J.*, **10**, 11 (1975).
51. Kitagawa, M., *J. Polym. Sci. Polym. Phys. Ed.*, **14**, 2095 (1976).
52. Argon, A. S., Hannoosh, J. C. and Salama, M. M., in *Fracture 1977*, Vol. 1, Waterloo, Canada, 1977, p. 445.
53. Bernier, G. A. and Kambour, R. P., *Macromolecules*, **1**, 393 (1968).

54. Kambour, R. P., Gruner, C. L. and Romagosa, E. E., *J. Polym. Sci.*, **11** 1879 (1973).
55. Kambour, R. P., Gruner, C. L. and Romagosa, E. E., *Macromolecules*, **7**, 248 (1974).
56. Andrews, E. H. and Bevan, L., *Polymer*, **13**, 337 (1972).
57. Marshall, G. P., Culver, L. E. and Williams, J. G., *Proc. R. Soc.*, **A319**, 165 (1970).
58. Williams, J. G. and Marshall, G. P., *Proc. R. Soc., A*, **342**, 55 (1975).
59. Brown, N. and Imai, Y., *J. Appl. Phys.*, **46**, 4130 (1975).
60. Imai, Y. and Brown, N., *J. Mater. Sci.*, **11** 417 (1976).
61. Brown, N., Metzger, B. D. and Imai, Y., *J. Polym. Sci. Polym. Phys. Ed.*, **16**, 1085 (1978).
62. Kambour, R. P., in *Proceedings of the International Conference on the Mechanics of Environment Sensitive Cracking Materials*, University of Surrey, Guildford, UK 1977, p. 213.
63. Brown, N., in *Methods of Experimental Physics*, Vol. 16, Part C (ed. R. A. Fava), Academic Press, New York, 1980, p. 233.
64. Rice, J. R., *J. Appl. Mechan.*, **35**, 379 (1968).
65. Begley, J. A. and Landes, J. D., *ASTM STP 514*, 1972.
66. Landes, J. D. and Begley, J. A., *Post Yield Fracture Mechanics* (ed. D. G. A. Latzko), Applied Science, London, 1979.
67. Chan, M. K. V. and Williams, J. G., *Int. J. Fracture*, **19**, 145 (1983).
68. Sumpter, J. D. and Turner, C. E., *Int. J. Fracture*, **9**, 320 (1973).
69. ESIS Technical Committee on Polymers and Composites, *A Testing Protocol for Conducting J Crack Growth Resistance Curve Tests on Plastics*, May 1995.
70. Broberg, K. B., *Int. J. Fracture*, **4**, 11 (1986).
71. Mai, Y. W. and Cotterell, B., *Int. J. Fracture*, **32**, 105 (1986).
72. Hashemi, S. and Williams, J. G., *Plast., Rubber Comp.*, **29**, 294 (2000).
73. Wu, J. and Mai, Y.-W., *Polym. Eng. Sci.*, **36**, 2275 (1996).
74. Lu, X. and Brown, N., *J. Mater. Sci.*, **21**, 2423 (1986).
75. Huang, Y.-L. and Brown, N., *J. Polym. Sci. Polym. Phys. Ed.*, **28**, 2007 (1990).
76. O'Connell, P. A., Bonner, M. J., Duckett, R. A., *et al.*, *Polymer*, **36**, 2255 (1995).
77. Cawood, M. J,. Channell, A. D. and Capaccio, G., *Polymer*, **34** 323 (1993).
78. Wilding, M. A. and Ward, I. M., *Polymer*, **19**, 969 (1978).
79. Huang, Y.-L. and Brown, N., *J. Polym. Sci., Polym. Phys.*, **29**, 129 (1991).
80. Clutton, E. Q., Rose, L. J. and Capaccio, G., *Plast. Rubber Comp. Proc. Appl.*, **27**, 478 (1998).
81. O'Connell, P. A., Bonner, M. J., Duckett, R. A., *et al.*, *J. Appl. Polym. Sci.*, **89**, 1663 (2003).
82. Mark, H., *Cellulose and its Derivatives*, Interscience, New York, 1943.
83. Vincent, P. I., *Proc. R. Soc., A*, **282**, 113 (1964).
84. Kausch, H. H., *Polymer Fracture*, Springer-Verlag, Berlin, 1978.
85. Kausch, H. H. and Becht, J., *Rheol. Acta*, **9**, 137 (1970).
86. Zhurkov, S. N., Novak, I. I., Slutsker, A. I., *et al.*, in *Proceedings of the Conference on the Yield, Deformation and Fracture of Polymers*, Cambridge, 1970, Session 3, Talk 3, pp. 1–6.
87. Wool, R. P., *J. Polym. Sci.*, **13**, 1795 (1975).
88. Yeh, W.-Y. and Young, R. J., *Polymer*, **40**, 857 (1999).
89. Ward, Y. and Young, R. J., *Polymer*, **42**, 7857 (2001).
90. Bueche, F., *J. Appl. Phys.*, **26**, 11343 (1955).

91. Zhurkov, S. N. and Tomashevsky, E. E., in *Proceedings of the Conference on the Physical Basis of Yield and Fracture*, Oxford, 1966, p. 200.
92. Zhurkov, S. N., Kuksenko, V. S. and Slutsker, A. I., in *Proceedings of the Second International Conference on Fracture*, Brighton. 1969, p. 531.
93. Zakrevskii, V. A. and Korsukov, V. Ye., *Polym. Sci. USSR*, **14**, 1064 (1972).
94. Peterlin, A., *Int. J. Fracture*, **11**, 761 (1975).
95. Orowan, E., *Rep. Prog. Phys.*, **12**, 185 (1949).
96. Vincent, P. I., *Polymer*, **1**, 425 (1960).
97. Stearne, J. M. and Ward, I. M., *J. Mater. Sci.*, **4**, 1088 (1969).
98. Clarke, P. L., *PhD Thesis*, Leeds University, 1981.
99. Hoff, E. A. W. and Turner, S., *Bull. Am. Soc. Test. Mater.*, **225**, TP208 (1957).
100. Boyer, R. F., *Polym. Eng. Sci.*, **8**, 161 (1968).
101. Heijboer, J., *J. Polym. Sci., C*, **16**, 3755 (1968).
102. Cottrell, A. H., *The Mechanical Properties of Matter*, Wiley, New York, 1964, p. 327.
103. Vincent, P. I., *Plastics*, **29**, 79 (1964).
104. Brown, H. R., *J. Mater. Sci.*, **8**, 941 (1973).
105. Marshall, G. P., Williams, J. G. and Turner, C. E., *J. Mater. Sci.*, **8**, 949 (1973).
106. Plati, E. and Williams, J. G., *Polym. Eng. Sci.*, **15**, 470 (1975).
107. Fraser, R. A. W. and Ward, I. M., *J. Mater. Sci.*, **12**, 459 (1977).
108. Hine, P. J., *PhD. Thesis*, Leeds University, 1981.
109. Truss, R. W., Duckett, R. A. and Ward, I. M., *Polym. Eng. Sci.*, **23**, 708 (1983).
110. Vincent, P. I., *Impact Tests and Service Performance of Thermoplastics*, Plastics and Rubber Institute, London, 1971.
111. Fraser, R. A. W. and Ward, I. M., *J. Mater. Sci.*, **9**, 1624 (1974).
112. Bramuzzo, M., *Polym. Eng. Sci.*, **29**, 1077 (1989).
113. Martinatti, F. and Ricco, T., *Polym. Testing*, **13**, 405 (1994).
114. Crouch, B. A and Huang, D. P., *J. Mater. Sci.*, **29**, 861 (1994).
115. Ramsteiner, F., *Polym. Testing*, **18**, 641 (1999).
116. Fasce, L., Bernal, C., Frontini, P., *et al.*, *Polym. Eng. Sci.*, **41**, 1 (2001).
117. Johnson, A. E., Moore, D. R., Prediger, R. S., *et al.*, *J. Mater. Sci.*, **21**, 3153 (1986); **22**, 1724 (1987).
118. Moore, D. R. and Prediger, R. S., *Polym. Compos.*, **9**, 330 (1988).
119. Jones, D. P., Leach, D. C. and Moore, D. R., *Plast. Rubber, Proc. Appl.*, **6**, 67 (1986).
120. Kinlock, A. J. and Williams, J. G., *J. Mater. Sci.*, **15**, 987 (1980).
121. Evans, R. M., Nara, H. R. and Bobalek, R. G., *Soc. Plast. Eng. J.*, **16**, 76 (1960).
122. Vincent, P. I., *Polymer*, **15**, 111 (1974).
123. Andrews, E. H., *J. Mater. Sci.*, **9**, 887 (1974).
124. Bucknall, C. B., in *Physics of Glassy Polymers* (2nd edn) (eds R. N. Haward and R. J. Young), Chapman & Hall, London, 1977, Ch. 8.
125. Nielsen, L. E., *Mechanical Properties of Polymers*, Reinhold, New York, 1962.
126. Newman, S. and Strella, S., *J. Appl. Polym. Sci.*, **9**, 2297 (1965).
127. Donald, A. M. and Kramer, E. J., *J. Mater. Sci.*, **17**, 1765 (1982).
128. Bucknall, C. B. and Smith, R. R., *Polymer*, **6**, 437 (1965).
129. Yang, H. H. and Bucknall, C. B., *10th International Conference on Deformation, Yield & Fracture of Polymers, Churchill College, Cambridge*, Institute of Materials, London, 1997, p. 458.
130. Bucknall, C. B., *Br. Plast.*, **40**, 84 (1967).

131. Lazzeri, A. and Bucknall, C. B., *Polymer*, **36**, 2895 (1995).

132. Yee, A. F. and Pearson, R. A., in *Fractography and Failure Mechanisms of Polymers and Composites* (ed. A. C. Roulin-Moloney), Elsevier, London, 1989, Ch. 8, pp. 291–350.

133. Cheng, J., Hiltner, A., Baer, E., *et al.*, *J. Appl. Sci.*, **55**, 1691 (1995).

134. Argon, A. S. and Cohen, R. E., *Adv. Polym. Sci.*, **91/92**, 301 (1990).

135. Rivlin, R. S. and Thomas A. G., *J. Polym. Sci.*, **10**, 291 (1953).

136. Thomas, A. G., *J. Polym. Sci.*, **18**, 177 (1955).

137. Bueche, F., *Physical Properties of Polymers*, Interscience, New York, 1962, 1962, p. 237.

138. Flory, P. J., *Ind. Eng. Chem.*, **38**, 417 (1946).

139. Flory, P. J., Rabjohn, N. and Shaffer, M. C., *J. Polym. Sci.*, **4**, 435 (1949).

140. Taylor, G. R. and Darin, S., *J. Polym. Sci.*, **17**, 5 (1955).

141. Bueche, F., *J. Polym. Sci.*, **24**, 189 (1957).

142. Bueche, F., *J. Polym. Sci.*, **33**, 259 (1958).

143. Smith, T. L., *J. Polym. Sci.*, **32**, 99 (1958).

144. Smith, T. L., *Soc. Plast. Eng. J.*, **16**, 1211 (1960).

145. Smith, T. L. and Stedry, P. J., *J. Appl. Phys.*, **31**, 1892 (1960).

146. Bueche, F., *J. Appl. Sci.*, **26**, 1133 (1955).

147. Bueche, F. and Halpin, J. C., *J. Appl. Phys.*, **35**, 36 (1964).

148. Halpin, J. C., *J. Appl. Phys.*, **35**, 3133 (1964).

149. Manson, J. A. and Hertzberg, R. W., *CRC Crit. Rev. Macromol. Sci.*, **1**, 433 (1973).

150. Thomas, A. G., *J. Polym. Sci.*, **31**, 467 (1958).

151. Lake, G. J. and Thomas, A. G., *Proc. R. Soc., A*, **300**, 108 (1967).

152. Lake, G. J. and Lindley, P. B., in *Proceedings of the Conference on the Physical Basis of Yield and Fracture*, Oxford, 1966, p. 176.

153. Greensmith, H. W., *J. Appl. Polym. Sci.*, **7**, 993 (1963).

154. Andrews, E. H., in *Testing of Polymers*, Vol. 4 (ed. W. E. Brown), Wiley, New York, 1968, p. 237.

155. Andrews, E. H. and Walker, B. J., *Proc. R. Soc., A*, **325**, 57 (1971).

156. Borduas, H. F., Culver, L. E. and Burns, D. J., *J. Strain Anal.*, **3**, 193 (1968).

157. Hertzberg, R. W., Nordberg, H. and Manson, J. A., *J. Mater. Sci.*, **5**, 521 (1970).

158. Arad, S., Radon, J. C. and Culver, L. E., *J. Mech. Eng. Sci.*, **13**, 75 (1971).

159. Harris, J. S. and Ward, I. M., *J. Mater. Sci.*, **8**, 1655 (1973).

160. Paris, P. C., in *Fatigue, an Interdisciplinary Approach*, Syracuse University Press, Syracuse, NY, 1964, p. 107.

161. Paris, P. C. and Erdogan, F., *J. Basic Eng., Trans. ASME*, **85**, 528 (1963).

162. Arad, S., Radon, J. C. and Culver, L. E., *Polym. Eng. Sci.*, **12**, 193 (1972).

163. Pitman, G. L. and Ward, I. M., *J. Mater. Sci.*, **15**, 635 (1980).

164. Williams, J. G., *J. Mater. Sci.*, **12**, 2525 (1980).

165. Mai, Y. W. and Williams, J. G., *J. Mater. Sci.*, **14**, 1933 (1979).

Problems for Chapters 11 and 12

1. The critical shear stress τ for yielding of a certain polymer is given by $\tau = \tau_0 + \mu\sigma_N$, where σ_N is the compressive stress on the yield plane and τ_0 and μ are constants.

If $\tau_0 = 10^7$ Pa and $\mu = 0.1$, calculate the magnitude of the yield stress compression, showing that yield occurs on the plane whose normal make angle 47°51' with the compressive stress direction.

2. A thin walled cylinder with closed ends of radius r and wall thickness is fabricated from a polymer with a yield stress in pure shear of k. Calculate the internal pressure required to produce yielding of the cylinder walls if the yield criterion under appropriate conditions of temperature and strain rate may be written as

$$(\sigma_1 - \sigma_2)^2 + (\sigma_2 - \sigma_3)^2 + (\sigma_3 - \sigma_1)^2 = 6k^2 + \mu(\sigma_1 + \sigma_2 + \sigma_3)$$

(Hint: the hoop stress, radial stress and axial stress due to an internal pressure p are pt/t, 0 and $pr/2t$, respectively.)

3. A batch of isotropic polymer is observed to yield at 40 MPa in compression and 35 MPa in tension, both tests at a uniaxial strain rate of 10^{-3} s^{-1}. Suggest a reason for the difference in these two yield stresses and estimate the yield stress in equal biaxial tension ($\sigma_1 = \sigma_2 > 0$, $\sigma_3 = 0$), explaining carefully the assumptions you have made. What strain rate ($\dot{\varepsilon}_1 = \dot{\varepsilon}_2 = \dot{\varepsilon}_3/2$) would you use in biaxial tension to obtain the same effective strain rate as in the uniaxial tests?

4. The compressive yield stress of a fictional polymer decreases linearly from 200 MPa at 50 K to 100 MPa at room temperature when tested at a strain rate of 10^{-3} s^{-1}. The room temperature compressive yield stress increases to 120 MPa at a strain rate of 10^{-1} s^{-1}.

    According to the Eyring theory of flow these observations can be analysed in terms of the activation volume $V = kT\Delta(\ln \dot{\varepsilon})/\Delta\sigma$, the activation energy $\Delta\mu$ and a pre-exponential factor $\dot{\varepsilon}_0$. Estimate the values of these parameters using the given data. (Boltzmann constant = 1.38×10^{-23} JK^{-1}.)

5. A large sheet of PMMA has a central elliptical crack of length 1 cm. Given that K_{IC} is 1 MNm$^{-3/2}$, calculate the fracture stress.

6. Show how the Irwin–Kies relationship can be derived by considering the load–extension curve for a specimen with a central crack. Use this relationship to calculate the load that must be applied to a double cantilever beam specimen of half-width 1 cm and thickness 3 mm to cause a crack of initial length 8 cm to begin to propagate, given that the fracture toughness K_{IC} is 2 MN m$^{-3/2}$.

7. A craze at the crack tip in glassy polymer has a crack opening displacement of 4 μm. Given that the fracture toughness is 1 MN m$^{-3/2}$ and the plane strain modulus is 1.5 GPa, calculate the length of the craze and the craze stress.

Appendix 1

A1.1 Scalars, vectors and tensors

Quantities such as mass and temperature are scalars, whose magnitude does not depend on direction. Force, however, is a vector, which relative to a given set of axes may be resolved into three components parallel to the coordinate axes. Referred to a new set of axes obtained by rotating the initial set, the three axial components of the force will be changed. It may be advantageous to perform a rotation of axes so that the total force is directed along one axis. Area may also be represented as a vector, whose direction is that of the outward-facing normal to the surface and whose length is proportional to the area.

When each component of one vector is linearly related to each component of another vector the coefficients of proportionality are the components of a second-rank tensor. Stress, defined as force per unit area, is the quotient of two vectors, and is an example of a second-rank tensor. Note that the condition defining positive in the direction of the outward normal means that a hydrostatic pressure must be negative. Strain is also a tensor, and in the most general case both stress and strain can be expressed in terms of nine tensor components. Stress and strain are both examples of second-rank tensors, which have nine components; vectors, which have three components, are sometimes referred to as first-rank tensors; single-valued scalars are zero rank tensors.

A1.2 Tensor components of stress

The components of stress have already been defined in Section 2.1 in terms of the equilibrium of a cube and these form the elements of a symmetric second-rank tensor

$$\sigma_{ij} = \begin{bmatrix} \sigma_{xx} & \sigma_{xy} & \sigma_{xz} \\ \sigma_{xy} & \sigma_{yy} & \sigma_{yz} \\ \sigma_{xz} & \sigma_{yz} & \sigma_{zz} \end{bmatrix}$$

An Introduction to the Mechanical Properties of Solid Polymers I. M. Ward and J. Sweeney
© 2004 John Wiley & Sons, Ltd   ISBN: 0471 49625 1 (HB); 0471 49626 X (PB)

A1.3 Tensor components of strain

It was emphasized in Section 2.2 above that the engineering components of strain were defined. The nine tensor components of strain are defined by the symmetric strain tensor

$$
\varepsilon_{ij} =
\begin{bmatrix}
\dfrac{\partial u}{\partial x} & \dfrac{1}{2}\left(\dfrac{\partial v}{\partial x}+\dfrac{\partial u}{\partial y}\right) & \dfrac{1}{2}\left(\dfrac{\partial w}{\partial x}+\dfrac{\partial u}{\partial z}\right) \\[3mm]
\dfrac{1}{2}\left(\dfrac{\partial v}{\partial x}+\dfrac{\partial u}{\partial y}\right) & \dfrac{\partial v}{\partial y} & \dfrac{1}{2}\left(\dfrac{\partial v}{\partial z}+\dfrac{\partial w}{\partial y}\right) \\[3mm]
\dfrac{1}{2}\left(\dfrac{\partial w}{\partial x}+\dfrac{\partial u}{\partial z}\right) & \dfrac{1}{2}\left(\dfrac{\partial v}{\partial z}+\dfrac{\partial w}{\partial y}\right) & \dfrac{\partial w}{\partial z}
\end{bmatrix}
$$

It can be seen that this follows from a general definition of the strain ε_{ij} as

$$
\varepsilon_{ij} = \frac{1}{2}\left(\frac{\partial u_i}{\partial x_j}+\frac{\partial u_j}{\partial x_i}\right)
$$

where i, j, take the values 1, 2 and 3 in turn and we write $x_1 = x, x_2 = y, x_3 = z$ and $u_1 = u, u_2 = v, u_3 = w$.

In terms of the engineering components of strain e_{xx}, etc. defined in Section 2.2 we have

$$
\varepsilon_{ij} =
\begin{bmatrix}
e_{xx} & \tfrac{1}{2}e_{xy} & \tfrac{1}{2}e_{xz} \\[2mm]
\tfrac{1}{2}e_{xy} & e_{yy} & \tfrac{1}{2}e_{yz} \\[2mm]
\tfrac{1}{2}e_{xz} & \tfrac{1}{2}e_{yz} & e_{zz}
\end{bmatrix}
$$

i.e. the engineering and tensor extensional strains are identical but the engineering shear strains are twice the tensor shear strains.

A1.4 Generalized Hooke's law

The generalized Hooke's law assumes that each of the nine components of the stress tensor is linearly related to each of the nine components of the strain tensor and vice versa. For example

$$
\sigma_{xx} = a\varepsilon_{xx} + b\varepsilon_{xy} + c\varepsilon_{xz} + d\varepsilon_{yx} + e\varepsilon_{yy} + f\varepsilon_{yz} + g\varepsilon_{zx} + h\varepsilon_{zy} + i\varepsilon_{zz}
$$

and

$$
\varepsilon_{xx} = a'\sigma_{xx} + b'\sigma_{xy} + c'\sigma_{xz} + d'\sigma_{yx} + e'\sigma_{yy} + f'\sigma_{yz} + g'\sigma_{zx} + h'\sigma_{zy} + i'\sigma_{zz},
$$

where a, b, ... and a', b' ... are constants.

In one dimension Hooke's law defines stiffness (or modulus) as stress divided by strain; alternatively, compliance is strain divided by stress. In three dimensions the stress and strain tensors are related through stiffness $[c]$ and compliance $[s]$ tensors (**note the confusing nomenclature**). As stress and strain are each second-rank symmetric tensors, $[c]$ and $[s]$ are each fourth-rank symmetric tensors: each component of strain is linearly related to all nine components of stress, and vice versa, so there are 81 components in the stiffness and compliance tensors, which when written out in full form a 9×9 array.

$$\sigma_{ij} = c_{ijkl}\varepsilon_{kl}$$

or equivalently

$$\varepsilon_{ij} = s_{ijkl}\sigma_{kl}.$$

In terms of earlier nomenclature $\sigma_{ij} = \sigma_{xx}$, σ_{yy}, etc. and $\varepsilon_{ij} = \varepsilon_{xx}$, ε_{yy}, etc.

The fourth-rank tensors s_{ijkl} and c_{ijkl} define the compliance and stiffness constants, with i, j, k, l taking values 1, 2, 3 in turn. In these equations the use of 1, 2, 3 is synonymous with the x, y, z used to define stress and strain components.

In a fourth-rank symmetric tensor not all 81 terms are independent: in general

$$s_{ijkl} = s_{ijlk} = s_{jilk} = s_{jikl},$$

and similarly for stiffness components. When crystalline symmetry is taken into account the number of independent terms is reduced still further.

A1.5 Engineering strains and matrix notation

It is often more convenient to work in terms of engineering strains rather than use tensor strain components. Such an approach leads to a more compact notation, in which a generalized Hooke's law relates the six independent components of stress to the six independent components of engineering strain:

$$\sigma_p = c_{pq}e_q \quad \text{and} \quad e_p = s_{pq}\sigma_q,$$

where σ_p represents σ_{xx}, σ_{yy}, σ_{zz}, σ_{xz}, σ_{yz} or σ_{xy} and e_q represents e_{xx}, e_{yy}, e_{zz}, e_{xz}, e_{yz} or e_{xy}; c_{pq} and s_{pq} now form 6×6 matrices, in which p and q take the value 1, 2, ... 6.

For stiffness constants the following substitution is used for obtaining p and q in terms of i, j, k and l:

Tensor subscript	11	22	33	23 or 32	31 or 13	12 or 21
Matrix subscript	1	2	3	4	5	6

For compliance constants the substitution is based on the above conversion for stiffness constants, but additional rules apply because of the factor 2 difference between the definition of engineering and tensor shear strains:

$$s_{ijkl} = s_{pq} \text{ when } p \text{ and } q \text{ are 1, 2 or 3,}$$

$$2s_{ijkl} = s_{pq} \text{ when either } p \text{ or } q \text{ are 4, 5 or 6,}$$

$$4s_{ijkl} = s_{pq} \text{ when both } p \text{ and } q \text{ are 4, 5 or 6.}$$

A typical relation between strain and stress is thus changed from

$$\varepsilon_{xx} = s_{1111}\sigma_{xx} + s_{1112}\sigma_{xy} + s_{1113}\sigma_{xz} + s_{1121}\sigma_{yx} + s_{1122}\sigma_{yy}$$
$$+ s_{1123}\sigma_{yz} + s_{1131}\sigma_{zx} + s_{1132}\sigma_{zy} + s_{1133}\sigma_{zz},$$

which equals

$$\varepsilon_{xx} = s_{1111}\sigma_{xx} + s_{1122}\sigma_{yy} + s_{1133}\sigma_{zz} + 2s_{1112}\sigma_{xy}$$
$$+ 2s_{1113}\sigma_{xz} + 2s_{1123}\sigma_{yz},$$

because of the symmetries already discussed, to

$$e_{xx} = s_{11}\sigma_{xx} + s_{12}\sigma_{yy} + s_{13}\sigma_{zz} + s_{14}\sigma_{xz} + s_{15}\sigma_{yz} + s_{16}\sigma_{xy},$$

which is sometimes abbreviated further as

$$e_1 = s_{11}\sigma_1 + s_{12}\sigma_2 + s_{13}\sigma_3 + s_{14}\sigma_4 + s_{15}\sigma_5 + s_{16}\sigma_6.$$

The existence of a strain-energy function ([2], p. 49) means that

$$c_{pq} = c_{qp} \quad \text{and} \quad s_{pq} = s_{qp},$$

so that the number of independent constants is reduced from 36 to 21, before taking crystal symmetry into account, i.e.

$$c_{pq} = \begin{bmatrix} c_{11} & c_{12} & c_{13} & c_{14} & c_{15} & c_{16} \\ c_{12} & c_{22} & c_{23} & c_{24} & c_{25} & c_{26} \\ c_{13} & c_{23} & c_{33} & c_{34} & c_{35} & c_{36} \\ c_{14} & c_{24} & c_{34} & c_{44} & c_{45} & c_{46} \\ c_{15} & c_{25} & c_{35} & c_{45} & c_{55} & c_{56} \\ c_{16} & c_{26} & c_{36} & c_{46} & c_{56} & c_{66} \end{bmatrix}$$

Similarly

$$s_{pq} = \begin{bmatrix} s_{11} & s_{12} & s_{13} & s_{14} & s_{15} & s_{16} \\ s_{12} & s_{22} & s_{23} & s_{24} & s_{25} & s_{26} \\ s_{13} & s_{23} & s_{33} & s_{34} & s_{35} & s_{36} \\ s_{14} & s_{24} & s_{34} & s_{44} & s_{45} & s_{46} \\ s_{15} & s_{25} & s_{35} & s_{45} & s_{55} & s_{56} \\ s_{16} & s_{26} & s_{36} & s_{46} & s_{56} & s_{66} \end{bmatrix}$$

Because the simplified notation involves a matrix rather than a tensor it is necessary to convert back into tensor notation in order to calculate, for instance, a compliance in terms of a rotated set of coordinate axes, using the tensor relationship

$$s'_{mnop} = a_{mi} a_{nj} a_{ok} a_{pl} s_{ijkl},$$

where a_{mi}, etc. represent direction cosines of the angles between the two sets of axes. An example is given in A1.7.

A1.6 The elastic moduli of isotropic materials

Most of the present book is concerned with isotropic polymers, for which measured properties, such as the Young's modulus E, Poisson's ratio ν and the shear modulus G, relate directly to the constants of the compliance matrix.

For an isotropic solid, the matrix s_{pq} reduces to

$$s_{pq} = \begin{pmatrix} s_{11} & s_{12} & s_{12} & 0 & 0 & 0 \\ s_{12} & s_{11} & s_{12} & 0 & 0 & 0 \\ s_{12} & s_{12} & s_{11} & 0 & 0 & 0 \\ 0 & 0 & 0 & 2(s_{11} - s_{12}) & 0 & 0 \\ 0 & 0 & 0 & 0 & 2(s_{11} - s_{12}) & 0 \\ 0 & 0 & 0 & 0 & 0 & 2(s_{11} - s_{12}) \end{pmatrix}$$

It can be shown that the Young's modulus is given by

$$E = 1/s_{11},$$

the Poisson's ratio by

$$\nu = -s_{12}/s_{11}$$

and the torsional modulus by

$$G = \frac{1}{2(s_{11} - s_{12})}.$$

Thus we obtain the stress–strain relationships derived more simply in Section 2.3:

$$e_{xx} = \frac{1}{E}\sigma_{xx} - \frac{\nu}{E}(\sigma_{yy} + \sigma_{zz}),$$

$$e_{yy} = \frac{1}{E}\sigma_{yy} - \frac{\nu}{E}(\sigma_{xx} + \sigma_{zz}),$$

$$e_{zz} = \frac{1}{E}\sigma_{zz} - \frac{\nu}{E}(\sigma_{xx} + \sigma_{yy}),$$

$$e_{xz} = \frac{1}{G}\sigma_{xz},$$

$$e_{yz} = \frac{1}{G}\sigma_{yz},$$

$$e_{xy} = \frac{1}{G}\sigma_{xy},$$

where

$$G = \frac{E}{2(1 + \nu)}$$

Another basic quantity is the bulk modulus K, which determines the dilation $\Delta = e_{xx} + e_{yy} + e_{zz}$ produced by a uniform hydrostatic pressure. Using the stress–strain relationships above, it can be shown that the strains produced by a uniform hydrostatic pressure p are given by

$$e_{xx} = (s_{11} + 2s_{12})p,$$

$$e_{yy} = (s_{11} + 2s_{12})p,$$

$$e_{zz} = (s_{11} + 2s_{12})p.$$

Then

$$K = \frac{p}{\Delta} = \frac{1}{3(s_{11} + 2s_{12})} = \frac{E}{3(1 - 2\nu)}$$

A1.7 Transformation of tensors from one set of coordinate axes to another

If the stress or strain components are known in terms of one set of initially perpendicular axes, they can readily be determined for another set of perpendicular axes by the tensor transformation rule

$$\sigma'_{ij} = a_{ik} a_{jl} \sigma_{kl},$$

$$\varepsilon'_{ij} = a_{ik} a_{jl} \varepsilon_{kl},$$

where $i, j, k, l = x, y, z$ in turn and a_{ik}, a_{jl}, etc. represent the direction cosines relating the two sets of axes.

Similarly if the compliances and stiffnesses are defined for one set of axes (usually the symmetry axes of the polymer) their values can be determined in another set of axes by the tensor transformation rule

$$s'_{mnop} = a_{mi} a_{nj} a_{ok} a_{pl} s_{ijkl},$$

$$c'_{mnop} = a_{mi} a_{nj} a_{ok} a_{pl} c_{ijkl}.$$

For example, consider a film of polymer with orthorhombic symmetry, with the x and z axes lying in the plane of the film and the y axis normal to that plane. The compliance matrix is

$$\begin{bmatrix} s_{11} & s_{12} & s_{13} & 0 & 0 & 0 \\ s_{12} & s_{22} & s_{23} & 0 & 0 & 0 \\ s_{13} & s_{23} & s_{33} & 0 & 0 & 0 \\ 0 & 0 & 0 & s_{44} & 0 & 0 \\ 0 & 0 & 0 & 0 & s_{55} & 0 \\ 0 & 0 & 0 & 0 & 0 & s_{66} \end{bmatrix}$$

Determine Young's modulus (E_θ) for a strip of material cut in a direction making an angle θ with the z axis (Figure A1.1). Using $E_\theta = 1/s_\theta$ take the direction of the strip as the z' axis of a rotated set of Cartesian axes. Then $s_\theta = s_{3'3'3'3'}$.

In the most general case of rotation of axes the expansion will involve 81 terms. Here, however, the z' axis is perpendicular to the y axis, so that $a_{3'2}$ will be zero,

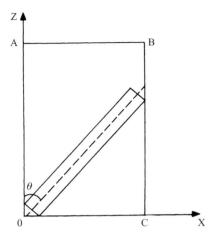

Figure A1.1   Strip cut at θ to the z axis from a polymer sheet with orthorhombic symmetry

immediately reducing the number of terms to 16. Crystal symmetry is now used to determine which of these 16 terms can be neglected because the relevant compliance constant is zero. The compliance constants are of the following forms:

$$s_{1111}, \; s_{1113}, \; s_{1133}, \; s_{1333}, \; s_{3333}.$$

Terms which include an odd number of threes relate to terms in the compliance matrix of the form s_{p5}, where $p = 5$, and so are zero. The complete expansion therefore contains only eight terms, which correspond with the compliance matrix constants s_{11}, s_{33}, s_{13} and s_{55} (Table A1.1).

$$s_{3'3'3'3'} = a_{3'1}a_{3'1}a_{3'1}a_{3'1}s_{1111} + a_{3'3}a_{3'3}a_{3'3}a_{3'3}s_{3333}$$

$$+ \; a_{3'1}a_{3'1}a_{3'3}a_{3'3}s_{1133} + a_{3'3}a_{3'3}a_{3'1}a_{3'1}s_{3311}$$

$$+ \; a_{3'1}a_{3'3}a_{3'3}a_{3'1}s_{1331} + a_{3'3}a_{3'1}a_{3'1}a_{3'3}s_{3113}$$

$$+ \; a_{3'3}a_{3'1}a_{3'3}a_{3'1}s_{3131} + a_{3'1}a_{3'3}a_{3'1}a_{3'3}s_{1313},$$

but

$$a_{3'3} \; = \; \cos\theta \quad \text{and} \quad a_{3'1} = \cos\left(\pi/2 - \theta\right) = \sin\theta.$$

$$\therefore \quad s_{3'3'3'3'} = \sin^4\theta \, s_{1111} + \cos^4\theta \, s_{3333} + 2\sin^2\theta\cos^2\theta \, s_{1133}$$

$$+ \; 4\sin^2\theta\cos^2\theta s_{1313},$$

Table A1.1 Conversion from tensor to matrix format

Tensor term	Matrix equivalent	Does it occur?
1111	11	Yes
1113	15	No
1131	15	No
1311	51	No
3111	51	No
1133	13	Yes
1331	55	Yes
3311	31	Yes
1313	55	Yes
3131	55	Yes
3113	55	Yes
1333	53	No
3331	35	No
3313	35	No
3133	53	No
3333	33	Yes

as

$$s_{1133} = s_{3311} \quad \text{and} \quad s_{1331} = s_{3113} = s_{3131} = s_{1313}.$$

Converting back to the abbreviated matrix notation

$$s_\theta = s_{3'3'} = \sin^4\theta\, s_{11} + \cos^4\theta\, s_{33} + \sin^2\theta \cos^2\theta\,(2s_{13} + s_{55}).$$

Note the factor of 4 that is required when converting s_{1313} to the matrix notation s_{55}, as in this case both p and q are greater than 3.

Important practical cases are those for strips cut along the z axis and at 45° and 90° to the z axis. Substituting in the general expression above we obtain the appropriate Young's moduli:

$$E_0 = \frac{1}{s_0} = \frac{1}{s_{33}}, \qquad E_{90} = \frac{1}{s_{90}} = \frac{1}{s_{11}}$$

and

$$\frac{1}{E_{45}} = s_{45} = \frac{1}{4}[s_{11} + s_{33} + (2s_{13} + s_{55})].$$

A1.8 The Mohr circle construction

This construction enables the components of a two-dimensional stress (or strain), expressed in terms of a given set of perpendicular axes, to be converted into the components relative to any other set of perpendicular axes. In particular it provides a simple method for determining the principal axes of stress and strain. For fibre symmetry all planes that contain the fibre axis as one of the perpendicular axes are identical: thus the construction is again applicable.

Let the components of stress be σ_{xx}, σ_{yy}, $\sigma_{xy} = \sigma_{yx}$ in terms of perpendicular axes OX, OY. Now consider an identical total stress in terms of axes OX', OY', obtained by rotating OX and OY through an angle θ in the anticlockwise direction. It can be shown that the shear stress components σ'_{xy} and σ'_{yx} in terms of the new axes are

$$\sigma'_{xy} = \sigma'_{xy} = \tfrac{1}{2}(\sigma_{yy} - \sigma_{xx})\sin 2\theta + \sigma_{xy}\cos 2\theta. \qquad (A1.1)$$

For a proof of (A1.1) using elementary mathematics see Hall [1]. The expression may be obtained directly from the general tensor relation $\sigma'_{ij} = a_{ik}a_{jl}\sigma_{kl}$, where a_{ik}, etc. represent direction cosines: $a_{11} = \cos\theta$, $a_{12} = \sin\theta$, $a_{21} = -\sin\theta$, $a_{22} = \cos\theta$.

A particular angle θ, where $0 < \theta < \pi/2$, can always be found for which σ'_{xy} in (A1.1) is zero. The condition is

$$\tan 2\theta = \frac{2\sigma_{xy}}{\sigma_{xx} - \sigma_{yy}} \qquad (A1.2)$$

When stress components are referred to axes rotated through that value of θ compared with the original axes, the shear stress components vanish, and we have defined the principal axes of stress. Principal axes of strain are defined similarly. In Figure A1.2 extensional stresses σ_{xx}, σ_{yy} are plotted along the OX axis, and shear stresses $\sigma_{xy} = \sigma_{yx}$ are plotted along the OY axis.

Here P is the point $(\sigma_{xx}, -\sigma_{xy})$; Q is the point $(\sigma_{yy}, \sigma_{xy})$. A circle with PQ as diameter will cut the OX axis at A. Call the angle between PQ and the OX axis 2θ. Then $\tan 2\theta = MP/AM$.

$$AM = \sigma_{xx} - OA = \sigma_{xx} - \tfrac{1}{2}(\sigma_{xx} + \sigma_{yy}) = \tfrac{1}{2}(\sigma_{xx} - \sigma_{yy}). \qquad MP = \sigma_{xy}.$$

Hence

$$\tan 2\theta = \frac{2\sigma_{xy}}{(\sigma_{xx} - \sigma_{yy})},$$

which is identical with equation (A1.2), used above to define the principal axes of stress. Hence θ (*half* the above angle) gives the angle between the original axes

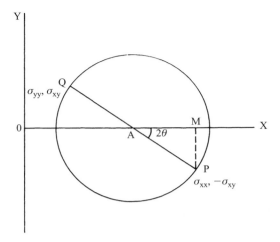

Figure A1.2   The Mohr circle construction

and the principal axes. Similarly the stress components referred to any set of axes rotated through an angle ϕ compared with the original axes, can be determined from the coordinates of the ends of the diameter of the Mohr circle obtained by rotating PQ through an angle 2ϕ. The construction holds also for strain components.

References

1. I. H. Hall, *The Deformation of Solids*, Nelson, London 1968.
2. A. E. H. Love, *A Treatise on the Mathematical Theory of Elasticity*, 4th Edn, Macmillan, New York, 1944.

Appendix 2

A2.1 Rivlin, Mooney, Ogden

A very useful and simple form for the strain-energy function has already been given as follows:

$$U = \tfrac{1}{2}NkT(\lambda_1^2 + \lambda_2^2 + \lambda_3^2 - 3), \tag{3.23a}$$

but more complex forms have been proposed. Because a rubber is an isotropic material whose properties are the same in all directions, it is plausible that any function of λ_1, λ_2, λ_3 which is invariant to a permutation of λ_1, λ_2, λ_3 and also becomes zero for $\lambda_1 = 1$, $\lambda_2 = 1$, $\lambda_3 = 1$, might provide a satisfactory form for the strain energy function. Functions of λ_1, λ_2, λ_3 which are independent of the choice of axes are termed strain invariants. Rivlin [1], and Rivlin and Saunders [2] examined the behaviour of vulcanized rubber and explored the use of a strain energy function of the form

$$U = C_1(I_1 - 3) + f(I_2 - 3), \tag{A2.1}$$

where $I_1 = \lambda_2^2 + \lambda_2^2 + \lambda_3^2$ is called the first strain invariant so that the first term in equation (A2.1) is identical to equation (3.30), $I_2 = \lambda_1^2\lambda_2^2 + \lambda_2^2\lambda_3^2 + \lambda_3^2\lambda_1^2$ is called the second strain invariant and f indicates that we can envisage a power law series

$$f(I_2 - 3) = C_2(I_2 - 3) + C_3(I_2 - 3)^2 + C_4(I_2 - 3)^3,$$

where C_2, C_3 and C_4 are constants.

A further simple strain invariant is $I_3 = \lambda_1^2\lambda_2^2\lambda_3^2$, which for an incompressible rubber is always unity, because $\lambda_1\lambda_2\lambda_3 = 1$. Therefore I_2 can also be written as

$$I_2 = \frac{1}{\lambda_1^2} + \frac{1}{\lambda_2^2} + \frac{1}{\lambda_3^2}$$

This leads us to a practically useful form of the strain-energy function,

An Introduction to the Mechanical Properties of Solid Polymers I. M. Ward and J. Sweeney
© 2004 John Wiley & Sons, Ltd   ISBN: 0471 49625 1 (HB); 0471 49626 X (PB)

originally proposed by Mooney [3], and shown by him to give a good fit to experimental data for the uniaxial extension of rubbers. The Mooney equation is

$$U = C_1(\lambda_1^2 + \lambda_2^2 + \lambda_3^2 - 3) + C_2\left(\frac{1}{\lambda_1^2} + \frac{1}{\lambda_2^2} + \frac{1}{\lambda_3^2}\right) \qquad (A2.2)$$

Using the incompressibility relationship $\lambda_1\lambda_2\lambda_3 = 1$, for uniaxial extension with $\lambda_1 = \lambda$, we have $\lambda_2^2 = \lambda_3^2 = 1/\lambda$ and

$$U = C_1(\lambda^2 - 2/\lambda - 3) + C_2\left(\frac{1}{\lambda^2} + 2\lambda - 3\right). \qquad (A2.3)$$

Differentiation with respect to λ gives the force per unit unstrained area as

$$f = 2\left(\lambda - \frac{1}{\lambda^2}\right)\left(C_1 + \frac{C_2}{\lambda}\right) \qquad (A2.4)$$

This equation can be rearranged to give

$$\frac{f}{2(\lambda - 1/\lambda^2)} = C_1 + \frac{C_2}{\lambda}$$

A plot of

$$\frac{f}{2(\lambda - 1/\lambda^2)}$$

(often called the 'reduced stress') versus $1/\lambda$ is called a Mooney plot, and should be linear with a gradient of C_2 and an ordinate of $(C_1 + C_2)$ at $\lambda = 1$.

In practice, it is found that the results for rubbers are typically similar to those shown in Figure A2.1 where the Mooney equation is adequate for $1/\lambda > 0.45$. In spite of this limitation, Mooney plots are a very useful way of dealing with data for the deformation of rubber networks. Such plots are often used as a first step to fitting results to more complicated relationships such as those proposed by Edwards and Vilgis [4], which provide a molecular understanding of rubber elasticity.

Rivlin's choice of a strain-energy function U that involved the squares of the extension ratios arose because he assumed that negative values of the extension ratios λ were a mathematical possibility, whereas it was necessary for U always to be greater than zero. We have seen, however, that by choosing suitable rotations of coordinate axes the most general deformation can be described in terms of pure strain, i.e. three principal extension ratios λ_1, λ_2, λ_3 which are all positive (although some are necessarily less than unity, because $\lambda_1\lambda_2\lambda_3 = 1$).

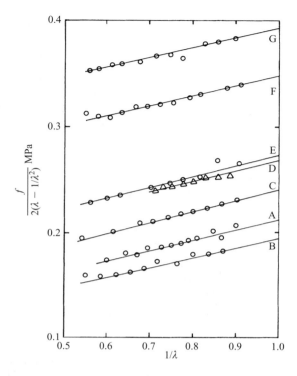

Figure A2.1 Mooney plots for rubbers covering a wide degree of vulcanization (Gumbrell, Mullins, and Rivlin 1953). Reproduced with permission from Treloar, *the Physics of Rubber Elasticity*, 3rd edn., Clarendon Press, Oxford, 1975

In addition to these theoretical considerations, which suggest that we need not be restricted to squares of extension ratios in formulating the strain energy function, it has been found by experimentalists that there is high sensitivity to experimental error when small values of the strain invariants I_1 and I_2 are involved. It is therefore natural to postulate that the only constraint on the form U is that imposed by the invariance of U with respect to the axis lables, which implies that $U(\lambda_1, \lambda_2, \lambda_3)$ should be a symmetric function of the extension ratios, i.e. invariant to any permutation of the indices 1, 2, 3.

Ogden, Chadwick and Haddon [5] have proposed that a useful form is

$$U = \sum_n \frac{\mu_n}{\alpha_n} (\lambda_1^{\alpha_n} + \lambda_2^{\alpha_n} + \lambda_3^{\alpha_n} - 3).$$

They showed that experimental data of Treloar [6] for tension, pure shear and equibiaxial tension could be fitted very well indeed (Figure A2.2) by assuming a

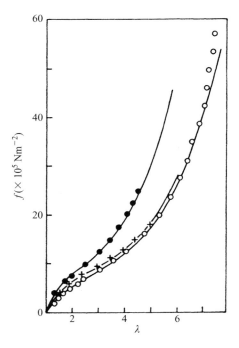

Figure A2.2   The three-term Ogden representation compared with the Treloar data for simple tension (○); pure shear (+) and equibiaxial tension (○). (Reproduced with permission from Treloar, *Proc. Roy. Soc.* **A351**, 301 (1976) and Ogden, *Proc. Roy. Soc.* **A326**, 565 (1972))

three-form expression for U with $\alpha_1 = 1.3$, $\alpha_2 = 5.0$, $\alpha_3 = -2.0$, $\mu_1 = 6.2 \times 10^5$ Pa, $\mu_2 = 1.2 \times 10^3$ Pa, $\mu_3 = -1 \times 10^3$ Pa.

References

1. R. S. Rivlin, *Phil. Trans. Roy. Soc.*, **A240**, 459, 491 (1948), **241**, 379 (1949).
2. R. S. Rivlin and D. W. Saunders., *Phil. Trans. Roy. Soc.*, **A243**, 251 (1951); *Trans. Faraday. Soc.*, **48**, 200 (1952).
3. M. Mooney, *J. Appl. Phys.*, **11**, 582 (1940)
4. S. F. Edwards and Th. Vilgis, *Polymer*, **27**, 483 (1986).
5. R. W. Ogden, P. Chadwick and E. W. Haddon, *Quart. J. Mech. Appl. Math.*, **26**, 23 (1973).
6. L. R. G. Treloar, *Trans Faraday. Soc.*, **40**, 59 (1944).

Answers to Problems

Chapters 2 and 3

1. An ideal rubber is assumed to be composed of very long molecules which can adopt a variety of conformations. The individual chains are interlinked by cross-links or junction points. The number of cross-links per unit volume is low and there is no interference with the motion of the chains.

    In the absence of external stress the chain between junction points adopts a maximum entropy state. On straining, the junction points deform affinely resulting in the orientation of the molecular segments joining them and hence a reduction in entropy.

    The entropy of a single chain can be calculated from the number of conformations which are possible for the given position of the junction points. The entropy of the whole system is obtained by summing over the network.

    The dependence of the entropy on strain in the network leads directly to the dependence of the free energy on strain and the differential of the free energy with respect to strain is the force developed in the network.

    For an ideal rubber, the shear modulus G at a temperature T is given by $G = NkT$, where N is the number of chains per unit volume so

$$N = \frac{G}{kT}$$

Also we have $\rho = N \times$ mean mass of chain, $\rho = N \times n_m \times$ mass of monomer (M), where n_m is the mean number of monomers per chain, i.e.

$$N = \frac{\rho N_A}{n_m M}$$

Combining these equations gives

An Introduction to the Mechanical Properties of Solid Polymers I. M. Ward and J. Sweeney
© 2004 John Wiley & Sons, Ltd   ISBN: 0471 49625 1 (HB); 0471 49626 X (PB)

$$n_{\text{m}} = \frac{\rho N_A}{M} \frac{kT}{G}$$

$$= \frac{1300 \times 6.022 \times 10^{26} \times 1.38 \times 10^{-23} \times 293}{68 \times 4 \times 10^5}$$

$$= 116.$$

2. Volume before deformation $V_0 = \pi r_0^2 l$. Volume after deformation

$$V_0 = \pi r_0^2 l (1 + \varepsilon_x)^2 (1 + \varepsilon_z)$$

$$= \pi r_0^2 l (1 + 2\varepsilon_x + \varepsilon_z + \varepsilon_x^2 + 2\varepsilon_x \varepsilon_z + \varepsilon_x^2 \varepsilon_z).$$

At low strains second-order terms can be neglected, i.e.

$$\pi r_0^2 l = \pi r_0^2 l (1 + 2\varepsilon_x + \varepsilon_z),$$

$$\therefore \quad \varepsilon_x = -\tfrac{1}{2}\varepsilon_z \quad \text{or} \quad \nu = 0.5000.$$

At higher strains second-order terms cannot be neglected.

$$\therefore \quad \pi r_0^2 l = \pi r_0^2 l (1 + \varepsilon_x)^2 (1 + \varepsilon_z)$$

or

$$(1 + \varepsilon_x) = (1 + \varepsilon_z)^{-1/2} = 1 - \varepsilon_{z/2} + \tfrac{3}{8}\varepsilon_z^2$$

$$\nu = -\varepsilon_x/\varepsilon_z = \tfrac{1}{2} - \tfrac{3}{8}\varepsilon_z.$$

Required value is given by

$$0.49 = \tfrac{1}{2} - \tfrac{3}{8}\varepsilon_z \quad \text{or} \quad \tfrac{3}{8}\varepsilon_z = 0.01$$

Therefore $\varepsilon_z = 0.027$ or 2.7 per cent.

3. The root mean square distance

$$(\bar{r}^2)^{1/2} = \sqrt{n}l,$$

where n is the number of links, length l, in the chain

$$(\bar{r}^2)^{1/2} = \sqrt{1000} \times 1.53 \text{ Å},$$

$$= 48.4 \text{ Å}.$$

4.
$$U = C_1(\lambda_1^{1.3} + \lambda_2^{1.3} + \lambda_3^{1.3} - 3).$$

For this deformation let λ be the extension ratio in the direction of the applied force, i.e. $\lambda_1 = \lambda$ and $\lambda_2 = \lambda_3 = \lambda^{-1/2}$ from conservation of volume. Because of incompressibility, stress can only be defined relative to an unknown hydrostatic pressure p. Along the 1 and 3 directions,

$$\sigma_1 = \lambda_1 \frac{\partial U}{\partial \lambda_1} - p = 1.3 C_1 \lambda_1^{1.3} - p$$

$$\sigma_3 = \lambda_3 \frac{\partial U}{\partial \lambda_3} - p = 1.3 C_1 \lambda_3^{1.3} - p$$

Since $\sigma_3 = 0$, the second equation above gives $p = 1.3 C_1 \lambda_3^{1.3}$ and so the first equation becomes

$$\sigma_1 = 1.3 C_1 \left(\lambda_1^{1.3} - \lambda_3^{1.3} \right)$$

to give

$$\sigma = 1.3 C_1 (\lambda^{1.3} - \lambda^{-1.3/2})$$

The force per unit undeformed area f is given by σ/λ:

$$f = 1.3 C_1 (\lambda^{0.3} - \lambda^{-1.65})$$

For $\lambda = 3$

$$f = 1.3 C_1 (3^{0.3} - 3^{-1.65}).$$

The load required to produce this extension is just

$$Af = 1.3 C_1 A (1.390 - 0.163)$$

$$= 382 \text{ N}$$

$$= 39 \text{ kg}.$$

5.
$$U = C_1(\lambda_1^2 + \lambda_2^2 + \lambda_3^2 - 3).$$

Owing to material incompressibility, stresses can only be defined to within an unknown hydrostatic pressure p. Along 1 and 3 directions,

$$\sigma_1 = \lambda_1 \frac{\partial U}{\partial \lambda_1} - p = 2C_1\lambda_1^2 - p$$

$$\sigma_3 = \lambda_3 \frac{\partial U}{\partial \lambda_3} - p = 2C_1\lambda_3^2 - p \tag{1}$$

In both cases, $\sigma_3 = 0$ and the second of the above equations (1) gives

$$p = 2C_1\lambda_3^2$$

the first equation (1) now giving

$$\sigma_1 = 2C_1\left(\lambda_1^2 - \lambda_3^2\right). \tag{2}$$

In uniaxial extension, $\lambda_1 = \lambda$ and $\lambda_3 = \lambda^{-1/2}$ and we obtain

$$\sigma = 2C_1\left(\lambda^2 - 1/\lambda\right).$$

or in terms of strain e, where $1 + 2e = \lambda^2$

$$\sigma = 2C_1\{1 + 2e - (1 + 2e)^{-1/2}\}$$

$$= 2C_1\{1 + 2e - \{1 - \tfrac{1}{2}2e + \tfrac{1}{2}\cdot\tfrac{1}{2}\cdot\tfrac{3}{2}(2e)^2 \ldots\}\}$$

$$= 2C_1\{1 + 2e - 1 + e\} \text{ for small strains,}$$

i.e. $\sigma = 6\,C_1 e$

For biaxial stretching we have:

$$\lambda_1 = \lambda_2 = \lambda \text{ and } \lambda_3 = 1/\lambda^2$$
$$\therefore U = C_1(2\lambda^2 + 1/\lambda^2 - 3).$$

Then equation (2) gives

$$\sigma = 2C_1\left(\lambda^2 - 1/\lambda^4\right)$$

$$= 2C_1\left((1 + 2e) - (1 + 2e)^{-2}\right)$$

$$= 2C_1\left(1 + 2e - (1 - 4e + \frac{2.3}{2}(2e)^2 - \ldots)\right)$$

$$= 2C_1(1 + 2e - (1 - 4e)) \text{ at small strains}$$

$$= 12C_1 e$$

6.
$$U = C_1(\lambda_1^2 + \lambda_2^2 + \lambda_3^2 - 3) + C_2(\lambda_1^2\lambda_2^2 + \lambda_2^2\lambda_3^2 + \lambda_3^2\lambda_1^2 - 3)$$

Following the method of Question 5,

$$\sigma_1 = \lambda_1 \frac{\partial U}{\partial \lambda_1} - p = 2C_1\lambda_1^2 + 2C_2\left(\lambda_1^2\lambda_2^2 + \lambda_3^2\lambda_1^2\right) - p$$

$$\sigma_3 = \lambda_3 \frac{\partial U}{\partial \lambda_3} - p = 2C_1\lambda_3^2 + 2C_2\left(\lambda_2^2\lambda_3^2 + \lambda_3^2\lambda_1^2\right) - p$$

Putting $\sigma_3 = 0$ then gives

$$p = 2C_1\lambda_3^2 + 2C_2\left(\lambda_2^2\lambda_3^2 + \lambda_3^2\lambda_1^2\right)$$

and σ_1 becomes

$$\sigma_1 = 2C_1\left(\lambda_1^2 - \lambda_3^2\right) + 2C_2\lambda_2^2\left(\lambda_1^2 - \lambda_3^2\right).$$

In uniaxial conditions, $\lambda_1 = \lambda$, $\lambda_2 = \lambda_3 = \lambda^{-1/2}$ and

$$\sigma = 2C_1\left(\lambda^2 - 1/\lambda\right) + 2C_2\left(\lambda - 1/\lambda^2\right)$$

or in terms of the strain e where $1 + 2e = \lambda^2$

$$\sigma = 2C_1[1 + 2e - \{1 - \tfrac{1}{2}2e + \tfrac{1}{2}\cdot\tfrac{1}{2}\cdot\tfrac{3}{2}(2e)^2 \ldots\}]$$

$$+ 2C_2\left[1 + \frac{1}{2}2e + \frac{1}{2}\cdot\frac{1}{2}\left(-\frac{3}{2}\right)(2e)^2 \ldots\right) - \left(1 - 2e + \frac{1.2}{2}(2e)^2 \ldots\right)\right]$$

$$= 2C_1[1 + 2e - 1 + e] + 2C_2[1 + e - 1 + 2e] \quad \text{for small strains}$$

$$= 2C_1(3e) + 2C_2(3e)$$

$$= 6e(C_1 + C_2).$$

Hence the modulus $= \sigma/e = 6(C_1 + C_2)$.

7. For a rubber we can use the equation

$$f = NkT\left(\lambda - \frac{1}{\lambda^2}\right)$$

where f is the force per unit undeformed area, N is the number of chains per unit volume (terminated by cross-links) and λ is the extension ratio.

Here a load of 1 kg on an initial cross-sectional area of 10 mm^2 is equivalent to a value f given by

$$f = \frac{9.81}{10^{-5}} \text{ Pa},$$

$$\lambda = 2 \quad \text{and} \quad T = 300 \text{ K},$$

$$\therefore \quad N = \frac{f}{kT}\left(\frac{1}{2 - \frac{1}{4}}\right) = \frac{4f}{7kT}$$

$$= \frac{4 \times 9.81 \times 10^5}{7 \times (1.38 \times 10^{-23}) \times 300}$$

$$= 1.35 \times 10^{26} \text{ chains per cubic metre.}$$

8.   The force f per unit area of undeformed material is given by

$$f = NkT\left(\lambda - \frac{1}{\lambda^2}\right).$$

The density ρ is given by

$$\rho = \text{no. of chains per unit vol} \times \text{mean mass of the chains} = N \times M/N_A,$$

where N_A is Avogadro's number and M the mean molecular mass between cross-links, i.e.

$$M = \frac{\rho N_A}{N} = \frac{\rho N_A kT(\lambda - 1/\lambda^2)}{f}$$

$$= \frac{900 \times 6.022 \times 10^{26} \times 1.38 \times 10^{-23} \times 300 \times (3 \times \frac{1}{9})}{9.81/(15 \times 10^{-3} \times 1.5 \times 10^{-3})}$$

$$= 14\,900.$$

Chapter 4

1.   The form of the stress relaxation is that for a Maxwell element

$$G(t) = E \exp\left(\frac{-t}{\tau}\right)$$

where $\tau = \eta/E$.

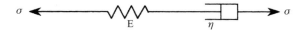

At $t = 0$

$$G(t) = E = 2 \text{ GPa}.$$

At $t = 10^4$

$$G(t) = 2 \exp\left(\frac{-10^4}{\tau}\right) = 1.0 \text{ GPa},$$

i.e.

$$\exp\left(\frac{10^4}{\tau}\right) = 2 \quad \text{or} \quad \frac{10^4}{\tau} = \ln_e 2 = 0.693,$$

$$\therefore \quad \tau = \frac{10^4}{0.693} = \frac{\eta}{E} \text{ s}, \ \eta = \frac{2 \times 10^4}{0.693} \text{ GPa s}.$$

Total strain in the Maxwell element

$$\varepsilon = \varepsilon_E + \varepsilon_v.$$

For the viscous element we can write

$$\frac{d\varepsilon_v}{dt} = \frac{\sigma}{\eta} \quad \text{or} \quad \int_0^{\varepsilon_v} d\varepsilon_v = \frac{\sigma}{\eta} \int_0^t dt$$

giving $\varepsilon_v = \dfrac{\sigma t}{\eta}$. Total strain therefore is

$$\varepsilon(t) = \frac{\sigma}{E} + \frac{\sigma t}{\eta} = \sigma\left(\frac{1}{E} + \frac{t}{\eta}\right)$$

putting $J_0 = 1/E$.

$$\therefore \quad J(t) = J_0 + \frac{t}{\eta}$$

$$\varepsilon(1000) = 0.1 \left(\frac{1}{2} + \frac{10^3 \times 0.693}{2 \times 10^4} \right) = 0.1(0.5 + 0.0347)$$

$$= 5.35 \text{ per cent.}$$

2.

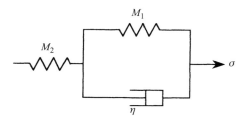

When the load is applied spring 2 extends by σ/M_2 and remains stretched. Time dependency is due entirely to the Kelvin unit, in which the total strain $\varepsilon = $ viscous strain, $\varepsilon_v = $ elastic strain ε_E.

In the Kelvin unit the stress is shared between the two components: $\sigma = \sigma_E + \sigma_v$

$$\sigma_E = M_1\varepsilon; \qquad \sigma_v = \eta\frac{d\varepsilon}{dt}$$

$$\therefore \quad \sigma = M_1\varepsilon + \eta\frac{d\varepsilon}{dt} \quad \text{or} \quad \frac{\sigma}{M_1} = \varepsilon + \frac{\eta}{M_1}\frac{d\varepsilon}{dt}$$

Rearranging

$$\int_0^\varepsilon \frac{d\varepsilon}{(\sigma/M_1 - \varepsilon)} = \frac{M_1}{\eta}\int_0^t dt$$

$$-\ln\left[\frac{\sigma}{M_1} - \varepsilon\right]_0^\varepsilon = \frac{M_1}{\eta}t$$

or

$$\ln\left(1 - \frac{\varepsilon}{\sigma/M_1}\right) = -\frac{M_1}{\eta}t,$$

where η/M_1 has the dimensions of time. It is defined as the retardation time τ_2.

$$\therefore \quad 1 - \frac{\varepsilon}{\sigma/M_1} = \exp(-t/\tau_2).$$

Rearranging

$$\varepsilon = \frac{\sigma}{M_1}[1 - \exp(-t/\tau_2)],$$

for total strain add on the instantaneous deformation

$$\varepsilon_0 = \frac{\sigma}{M_2}$$

$$\therefore \quad \varepsilon = \frac{\sigma}{M_2} + \frac{\sigma}{M_1}[1 - \exp(-t/\tau_2)].$$

At $t = 0$

$$\varepsilon = \frac{\sigma}{M_2} = 0.002.$$

As t tends to infinity

$$0.006 = 0.002 + \frac{\sigma}{M_1},$$

$$\therefore \quad \sigma/M_1 = 0.004.$$

Therefore at $t = 1000$ s

$$0.004 = 0.002 + 0.004[1 - \exp(-t/\tau_2)]$$

$$\therefore \quad 0.002 = 0.004[1 - \exp(-t/\tau_2)],$$

$$1 - \exp(-t/\tau_2) = \tfrac{1}{2}, \qquad \therefore \quad \exp(-t/\tau_2) = \tfrac{1}{2},$$

$$\exp(t/\tau_2) = 2, \qquad t/\tau_2 = \ln_e 2 = 0.693.$$

$$\therefore \quad \tau_2 = \frac{1000}{0.693} = 1440 \text{ s.}$$

3. The maximum stress occurs at 1 per cent extension and will be $(2 \times 10^9 \div 100) = 20$ MPa. Because of the initial extension the amplitude of the vibrating stress (σ_0) is ± 10 MPa, and the amplitude of the vibrating strain (e_0) is $1/200 = 0.005$.
    The stored elastic energy will depend on the dimensions of the specimen.

Its maximum value is $W_{st} = \frac{1}{2}$ (area) (length) $\sigma_0 \varepsilon_0 \cos \delta$. $A = 10^{-5}$ m^2, $L = 0.2$ m

$$\tan \delta = 0.1 \sim \sin \delta \qquad \therefore \quad \cos \delta = (1 - 0.01)^{1/2} \sim 0.995.$$

$$\therefore \quad W_{st} = \frac{1}{2} \times 10^{-5} \times 0.2 \times 10^7 \times 5 \times 10^{-3} \times 0.995 \sim 5 \times 10^{-2} \text{ J.}$$

Energy dissipated per cycle is

$$\Delta W = \pi A L \sigma_0 \varepsilon_0 \sin \delta = \pi \times 10^{-5} \times 0.2 \times 10^7 \times 5 \times 10^{-1} \sim 3.1 \times 10^{-2} \text{ J.}$$

At 0.5 Hz, logarithmic decrement $\Lambda = 0.2 = \pi \tan \delta_{0.5}$.

$$\therefore \quad \tan \delta_{0.5} = \frac{0.2}{\pi} = 0.064.$$

Therefore phase lag $= 0.064$ radians.

The damping is less at a lower frequency. Lower frequency is equivalent to testing at a higher temperature. Hence at 5 Hz the damping is less at a temperature somewhat above 20 °C, which implies the presence of a relaxation maximum somewhat below 20 °C.

4. Linear viscoelasticity—hence at any time stress is linearly related to strain, e.g. if extension is x for load m, t minutes after loading, then extension for load am will be ax, at the same time after loading.

    For step loading: m at $t = 0$, plus am at t_0; the subsequent creep is that for m for time t *plus* am for time $(t - t_0)$.

    Unloading is equivalent to adding a negative load; hence subsequent effect of unloading at time t_0 is that of m for time t, *minus* m for time $(t - t_0)$.

    (i) Residual extension at $T = 240$.

        Extension due to 0.1 kg for 240 min $= 0.600$ per cent.

        Extension due to -0.1 kg for 200 min $= -0.585$ per cent.

        Extension due to 0.1 kg for 160 min $= 0.555$ per cent.

        Extension due to -0.1 kg for 120 min $= -0.514$ per cent.

        Therefore net residual extension $= 0.056$ per cent.

    (ii) (a) Extension due to 0.1 kg for 80 min $= 0.462$ per cent.

        Extension due to 0.2 kg for 40 min $= 2 \times 0.390$ per cent.

        Therefore total extension after 80 min $= 1.242$ per cent.

(b) Immediate recovery is the same at any time: 0.300 per cent due to 0.1 kg, plus 0.600 per cent due to 0.2 kg.

Therefore total immediate recovery = 0.900 per cent.

(c) Extension due to 0.1 kg for 240 min = 0.600 per cent.

Extension due to 0.2 kg for 200 min = 2 × 0.585 per cent.

Extension due to −0.3 kg for 40 min = −3 × 0.390 per cent.

Therefore residual extension at $T = 240$ is $0.600 + 1.170 - 1.170 = 0.600$ per cent.

5.

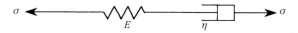

The relaxation time is the time taken for the stress to fall to a value $1/e$ times the initial value $\sigma = \sigma_0 \exp(-t/\tau)$, where

$$\tau = \frac{\eta}{E} = \frac{10^{10}}{10^8} = 100 \text{ s.}$$

At time $t = 0$ a strain of 1 per cent is applied

$$\varepsilon_v = 0 \quad \text{and} \quad \varepsilon_E = \varepsilon_{\text{total}} = 1\%$$

$$\therefore \quad \sigma_0 = 0.01 \times 10^8 \text{ Pa} = 10^6 \text{ Pa.}$$

At time $t = 25$ s the stress σ_0 has dropped to

$$\sigma_0 \exp(-25/100) = \sigma_0 e^{-1/4}.$$

On application of a further 2 per cent strain an additional stress of

$$0.02 \times E = 2 \times 10^6 \text{ Pa}$$

is developed in the element. Therefore the total stress is

$$2 \times 10^6 + 10^6 e^{-1/4}$$

which has decayed after a further 25 s to

$$(2 + e^{-1/4})10^6 \times \exp(-25/\tau) = (2 + e^{-1/4}) \times e^{-1/4} \times 10^6 \text{ Pa}$$

$$= 2.78 \times 0.78 \times 10^6$$

$$= 2.17 \text{ MPa.}$$

6. Strain due to stress σ_1 applied at $t = 0$ is given by

$$\varepsilon_1(t) = \sigma_1 J_0(1 - \exp[-t/\tau]).$$

Strain due to stress $(\sigma_2 - \sigma_1)$ applied at $t = t_1$ is given by

$$\varepsilon_2(t) = (\sigma_2 - \sigma_1)J_0(1 - \exp[-(t - t_1)/\tau]).$$

Therefore total strain at time $t > t_1$ is

$$\varepsilon(t) = \sigma_1 J_0(1 - \exp[-t/\tau]) + (\sigma_2 - \sigma_1)J_0(1 - \exp[-(t - t_1)/\tau])$$

$$= \sigma_2 J_0(1 - \exp[-(t - t_1)/\tau]) + \sigma_1 J_0(\exp[-(t - t_1)/\tau] - \exp[-t/\tau])$$

$$= \sigma_2 J_0(1 - \exp[-(t - t_1)/\tau]) + \sigma_1 J_0 \exp(-t/\tau)(\exp[t_1/\tau] - 1)$$

$$= \sigma_2 J_0 - \sigma_2 J_0 \exp(-t/\tau)\exp(t_1/\tau) + \sigma_1 J_0 \exp(-t/\tau)(\exp[t_1/\tau] - 1)$$

$$= \sigma_2 J_0 + [\sigma_1(\exp(t_1/\tau) - 1) - \sigma_2 \exp(t_1/\tau)]J_0 \exp(-t/\tau)$$

$$= \sigma_2 J_0[1 - K \exp(-t/\tau)]$$

where

$$K = \left[\exp(t_1/\tau) - \frac{\sigma_1}{\sigma_2}(\exp[t_1/\tau] - 1)\right]$$

7. For the Maxwell element

$$\tau = \frac{\eta}{E} = \frac{10^{11}}{10^9} = 100 \text{ s.}$$

In a Maxwell element the instantaneously applied strain appears across the spring. With time the dashpot strains, but since the applied strain is constant the strain developed across the spring drops.

At time $t = 0$, $\varepsilon_E = 1\%$ and hence $\sigma = \sigma_E = \sigma_v = 0.01 \times E$. At time $t = 30$ s this stress has dropped to

$$\sigma(30) = 0.01E \exp(-30/100) = 0.01Ee^{-0.3}.$$

At this stage an additional strain of 1 per cent is applied to the element. This strain appears across the spring giving an additional stress of

$$\sigma_a = 0.01 \times E.$$

The total stress at time $t = 30$ s is therefore

$$\sigma_t(30)0.01E(1 + e^{-0.3}).$$

The stress at a total time of 100 s is then

$$\sigma(70) = 0.01E(1 + e^{-0.3})\exp(-70/100)$$

$$= 0.01E(1 + e^{-0.3})e^{-0.7}$$

$$= (0.01E)(1.74)(0.50)$$

$$= 8.7 \text{ MPa.}$$

8. For a Maxwell element

$$\tau = \frac{\eta}{E}$$

For a rubber $E \propto T$

$$\therefore \quad E_{100\,°C} = \frac{373}{293} E_{room} = \frac{373}{293} \times 10^6 \text{ Pa.}$$

The viscosity at 100 °C is

$$\eta_{100\,°C} = \eta_{room} \exp\left(\frac{-4000}{8.3 \times 373}\right) \text{Pa s,}$$

$$\therefore \quad \tau_{100\,°C} = \frac{10^8}{10^6} \frac{293}{373} e^{-1.29}$$

$$= 21.6 \text{ s}$$

$$\text{cf } \tau_{room} = \frac{10^8}{10^6} = 100 \text{ s}$$

Chapters 7 and 8

1. (a) $E_3 = 10$ GPa, $E_1 = 1$ GPa(10^{10} N m^{-2}, 10^9 N m^{-2}); (b) $E_{45} = 1.59$ GPa (1.59×10^9 N m^{-2}).

2. $\varepsilon_{11} = S_{11}\sigma_{11} + S_{12}\sigma_{22}$; $\varepsilon_{22} = S_{12}\sigma_{11} + S_{22}\sigma_{22}$; $S_{22}\varepsilon_{11} - S_{12}\varepsilon_{22} = S_{11}S_{22}\sigma_{11} - S_{12}^2\sigma_{11}$.

$$\sigma_{11} = \frac{S_{22}}{S_{11}S_{22} - S_{12}^2}\varepsilon_{11} \quad \frac{-S_{12}}{S_{11}S_{22}S_{12}^2}\varepsilon_{22}.$$

For composites

$$\sigma_{11} = \frac{57}{2}\varepsilon_{11} + 7\varepsilon_{22}$$

and $S_{22} = S_{11}$. Equate terms

$$\left.\begin{array}{ll} S_{11}^2 - S_{12}^2 = \dfrac{2S_{11}}{57} & (1) \\[2mm] S_{11}^2 - S_{12}^2 = \dfrac{-S_{12}}{57} & (2) \end{array}\right\} \Rightarrow S_{12} = \frac{-2.7}{57}S_{11}.$$

In (1),

$$S_{11}^2\left(1 - \left(\frac{2.7}{57}\right)^2\right) = \frac{2S_{11}}{57}$$

$$\Rightarrow S_{11} = \frac{2}{57(1 - (2.7/57)^2)}$$

$$\Rightarrow E_1 = \frac{1}{S_{11}} = 26.8.$$

3. Take A and B in series. Call the combination X.

Here λ is common to A and B, therefore $\sigma_x = \sigma_a = \sigma_b$. For the same reason ϕ is the volume fraction of A and $(1 - \phi)$ the volume fraction of B.

$$\Delta l_x = \Delta l_a + \Delta l_b \qquad \left\{ \begin{array}{l} \varepsilon_x = \dfrac{\Delta l_x}{1} \\[2ex] \therefore \quad \varepsilon_x = \phi \varepsilon_a + (1 - \phi)\varepsilon_b \qquad \varepsilon_a = \dfrac{\Delta l_a}{\phi} \\[2ex] \text{Hooke's law: } E = \dfrac{\sigma}{\varepsilon} \qquad \varepsilon_b = \dfrac{\Delta l_b}{(1 - \phi)} \end{array} \right.$$

or $\varepsilon = \sigma/E$.

$$\therefore \quad \frac{\sigma}{E_x} = \frac{\sigma \phi}{E_a} + \frac{\sigma(1 - \phi)}{E_b} \quad \text{or} \quad E_x = \left(\frac{\phi}{E_a} + \frac{1 - \phi}{E_b} \right)^{-1}$$

Now take X and B in parallel. Length is the same for each; strain is the same for each; but force is shared

$$F_c = F_B + F_X$$

$$\sigma_c = F_{c/1}; \qquad \sigma_B = F_{B/(1-\lambda)}; \; \sigma_X = F_{X/\lambda}$$

Areas, and volume fractions of X and B are λ and $(1 - \lambda)$.

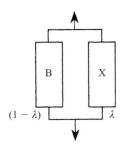

$$\frac{F_c}{1} = \sigma_c = \sigma_B(1 - \lambda) + \sigma_X \lambda.$$

Modulus $= \sigma/\varepsilon$, but ε is common.

$$\therefore \quad E_c = (1 - \lambda)E_b + \lambda E_X$$

$$= \lambda \left(\frac{\phi}{E_a} + \frac{1 - \phi}{E_b} \right)^{-1} + (1 - \lambda)E_b.$$

Calculation

(i) $\phi = 1, \lambda = 0.5, E_a = 10$ GPa; $E_b = 1$ GPa.

$$E_c = \frac{0.5}{(1/E_a + 0)} + 0.5E_b = 0.5 \times 10 + 0.5$$

$$E_c = 5.5 \text{ GPa}.$$

(ii)

$$E_c = \frac{0.5}{(0.8/10 + 0.2/1)} + 0.5 \times 1 = \frac{0.5}{(0.28)} + 0.5 = 2.29 \text{ GPa}.$$

A relatively small amount of the weaker component will reduce the modulus severely when it is in series with the stronger component.

Chapters 11 and 12

1. Keep compressive normal stress positive, take positive shear stress; $\sigma_N = \sigma \cos^2 \theta$, shear and shear stress on plane $\tau_N = \sigma \sin \theta \cos \theta$, yield stress $\tau_Y = \tau_0 + \mu \sigma \cos^2 \theta$.

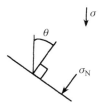

Yielding when $\tau_N - \tau_Y$ is a maximum,

$$\frac{\mathrm{d}}{\mathrm{d}\theta} \{ \sigma \sin \theta \cos \theta - \tau_0 - \mu \sigma \cos^2 \theta \} = 0,$$

$$\tfrac{1}{2} 2 \cos 2\theta + 2\mu \cos \theta \sin \theta = 0,$$

i.e.

$$\cos 2\theta + \mu \sin 2\theta = 0,$$

$$\tan 2\theta = -\frac{1}{\mu}$$

By assumption $\theta > 0$. Values of 2θ for $\tan 2\theta = -1/\mu$ are, with $\mu = 0.1$, $-84.29° = 180 - 84.29°$, etc. Here $2\theta = 180 - 84.29°$ gives $\theta = 47.86°$, i.e. $47° \, 51'$.

Yield occurs when $\tau_Y = \tau_N$, i.e. $\sigma \sin \theta \cos \theta = \tau_0 + \mu\sigma \cos^2 \theta$, i.e.

$$\sigma = -\frac{\tau_0}{\sin \theta \cos \theta - \mu \cos^2 \theta}$$

$\tau_0 = 10$ MPa gives $\sigma = 22.1$ MPa.

2. Put

$$\sigma_1 = \frac{pr}{t}, \qquad \sigma_2 = 0, \qquad \sigma_3 = \frac{pr}{2t}$$

The equation becomes

$$\left(\frac{pr}{t}\right)^2 + \left(\frac{pr}{2t}\right)^2 + \left(\frac{pr}{2t}\right)^2 = 6k^2 + \mu\frac{3pr}{2t}$$

$$6\left(\frac{pr}{2t}\right)^2 - 3\mu\left(\frac{pr}{2t}\right) - 6k^2 = 0$$

$$2\left(\frac{pr}{2t}\right)^2 - \mu\left(\frac{pr}{2t}\right) - 2k^2 = 0$$

$$\frac{pr}{2t} = \frac{\mu \pm 1(\mu^2 + 4.2.2k^2)}{4}$$

take positive root to maintain positive p:

$$p = \frac{t}{2r}(\mu + \sqrt{(\mu^2 + 16k^2)}).$$

3. Dependence of yield stress on hydrostatic pressure. Assume $\sigma_y = \sigma_{y0} - \alpha p$ where $\alpha > 0$ and $p > 0$ for tension. In tension,

$$35 = \sigma_{y0} - \tfrac{1}{3}\alpha \times 35.$$

In compression

$$40 = \sigma_{y0} + \tfrac{1}{3}\alpha \times 40.$$

Subtract

$$5 = \tfrac{1}{3}\alpha(40 + 35) \Rightarrow \alpha = \tfrac{1}{5}$$

$$\Rightarrow \sigma_{y0} = 37\tfrac{1}{3} \text{ MPa}$$

In biaxial tension, $P = \tfrac{2}{3}\sigma$

$$\sigma_y = \sigma_{y0} - \frac{\alpha 2\sigma_y}{3} \Rightarrow \sigma_y = \frac{\sigma_{y0}}{(1 + 2/15)} = 32.9 \text{ MPa.}$$

Uniaxial strain (ε_u, $-\tfrac{1}{2}\varepsilon_u$, $-\tfrac{1}{2}\varepsilon_u$) assuming incompressibility. Biaxial strain (ε_b, ε_b, $-2\varepsilon_b$). Equate maximum shear rates

$$\tfrac{3}{2}\dot{\varepsilon}_u = 3\dot{\varepsilon}_b \quad \text{i.e.} \quad \dot{\varepsilon}_b = \tfrac{1}{2}\dot{\varepsilon}_u$$

so use half the strain rate in biaxial tension

$$\dot{\varepsilon}_b = 0.5 \times 10^{-3} \text{ s}^{-1}.$$

4. Given that $V = kT\Delta(\ln \dot{\varepsilon})/\Delta\sigma$,

$$\sigma(300) = 120 \text{ MPa at } 10^{-1}\,\text{s}^{-1}$$

$$\sigma(300) = 100 \text{ MPa at } 10^{-3}\,\text{s}^{-1}$$

then

$$V = 1.38 \times 10^{-23} \times 300 \times \ln(100)/20 \times 10^6 = 9.5 \times 10^{-28} \text{ m}^3$$

$$\sigma = (\Delta U + kT \ln(\dot{\varepsilon}/\dot{\varepsilon}_0))/V$$

so

$$d\sigma/dT = k \ln(\dot{\varepsilon}/\dot{\varepsilon}_0)/V$$

Since

$$d\sigma/dT = (200 - 100) \times 10^6/(50 - 300) \text{ Pa K}^{-1}$$

$$k \ln(\dot{\varepsilon}/\dot{\varepsilon}_0)/V = -4 \times 10^5 \text{ Pa K}^{-1}$$

At $T = 50$ K,

$$\sigma = 200 \text{ MPa} = \Delta U/V + 50k \ln(\dot{\varepsilon}/\dot{\varepsilon}_0)/V$$

so

$$\Delta U/V = 2 \times 10^8 + 2 \times 10^7 = 2.2 \times 10^8 \text{ Pa.}$$

giving

$$\Delta U = 2.2 \times 10^8 \times 9.5 \times 10^{-28} = 2.09 \times 10^{19} \text{ J.}$$

$$\dot{\varepsilon} = \dot{\varepsilon}_0 \exp(-(\Delta U - \sigma V)/kT)$$

giving $\dot{\varepsilon}_0 = 9.1 \times 10^8 \text{ s}^{-1}$.

5. $$K_{IC} = \sigma_B (\pi c)^{1/2},$$

$$K_{IC} = 1 \text{ MN m}^{-3/2} \qquad c = 10^{-2} \text{ m,}$$

$$\sigma_B = 1 \text{ MN m}^{-3/2}/(\pi \times 10^{-2})^{1/2} = \frac{10}{1.77} = 5.64 \text{ MPa.}$$

6. $$G_c = \frac{P^2 \, dC}{2B \, dc}$$

$$K_{IC} = 4\sqrt{6} \frac{P_c}{Bb^{3/2}} \qquad P = \frac{K_{IC} Bb^{3/2}}{4\sqrt{6}c} = 0.433 \text{ N}$$

$$= \frac{2 \times 0.003 \times 0.02^{3/2}}{4\sqrt{6} \times 8 \times 10^{-2}}$$

$$= 21.65 \text{ N} = 2.21 \text{ kg.}$$

7. $$\sigma_c \delta_t = \frac{K_{IC}^2}{E^*} \qquad \sigma_c = \frac{10^3}{6} \text{ MPa}$$

$$= 166 \text{ MPa}$$

$$R = \frac{\pi}{8} \frac{K_{IC}^2}{\sigma_c^2} \qquad R = \frac{\pi}{8} \frac{36}{10^6} = 1.414 \times 10^{-5} \text{ m}$$

$$= 14 \ \mu\text{m.}$$

Index

An Introduction to the Mechanical Properties of Solid Polymers I. M. Ward and J. Sweeney
© 2004 John Wiley & Sons, Ltd   ISBN: 0471 49625 1 (HB); 0471 49626 X (PB)